THE 1870 AGRICULTURIST.

FOR THE

Farm, Garden, and Household.

"Agriculture is the most Healthful, the most Useful, the most Noble Employment of Man."—Washington.

Edited by Steven J. Rakeman & Donald Berg

ISBN 0-937214-02-7
Lib.of Cong. Card #80-67096

Antiquity Reprints Box 370 Rockville Centre, N.Y. 11571

Foreword

In 1870 THE AMERICAN AGRICULTURIST was a monthly tabloid, owned, edited and published by Orange Judd, the former agricultural editor of THE NEW YORK TIMES. Judd had increased the paper's circulation from less than 1,000 to over 100,000 in the eight years from the time he acquired it.

Judd was a pioneer of the science of farming. His journal dispensed know-how and ingenuity to farmers and their families. It served as a forum where readers could swap ideas or refute theories on subjects that ranged from how to compost manure to how to raise a child.

As an editor, Orange Judd kept the needs of the working farmer ever in mind. He maintained that "Farmers who have their work properly planned will find little time for reading." So his articles were practical and to the point.

Judd's contemporary, Solon Robinson, the agricultural editor of The New York Tribune, had this to say:

> "The American Agriculturist is a large, beautifully printed monthly...filled with matter designed to advance the science of agriculture and horticulture, and improvements of all that pertains to rural life. This paper... probably has the largest circulation of any agricultural paper in America, and is published in an office with rent of $3,000 a year. This indicates success in a purely agricultural journal."

Orange Judd retired from the paper in 1833. THE AMERICAN AGRICULTURIST, which merged with THE RURAL NEW YORKER, has been published continuously since then and is still one of the most popular farm journals in the country.

About This Book

All of the type, illustrations and ads included in this book were reproduced from 1870 issues of THE AMERICAN AGRICULTURIST. We selected the articles that we thought would be of most interest to modern readers and arranged them to give the visual and chronological feeling of the original monthly issues.

We hope you enjoy looking backwards over a century to when America was done fighting with herself and was busy getting back to the earth, to a time when the majority of Americans were farmers who lived and worked on small family farms, a time when petroleum was something that was used to paint fence posts, fertilizer was produced by the horsepower that pulled the plow, Chicago was in the Northwest and beyond that was the frontier.

THE 1870 AGRICULTURIST

Index

JANUARY, 1870.

AMERICAN AGRICULTURIST,

FOR THE FARM, GARDEN & HOUSEHOLD.

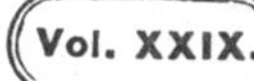

Vol. XXIX.

Number 1.

PUBLISHED BY
ORANGE JUDD & CO.,
245 BROADWAY,
NEW-YORK.

PLATFORM AND COUNTER

SCALES.

For Farmers, Merchants, Manufacturers, Railroad Companies, Machinists, Druggists, &c., &c., in great variety.

Every Farmer should have a **FAIRBANKS' SCALE.**

Send for Circular. FAIRBANKS & CO.,
252 Broadway, New York.
246 Baltimore St., Baltimore, Md.
53 Camp St., New Orleans.
FAIRBANKS & EWING, Masonic Hall, Philadelphia.
FAIRBANKS, BROWN & CO., 118 Milk St., Boston.

HORSFORD'S SELF-RAISING BREAD PREPARATION

Makes the most wholesome and best of BREAD, BISCUIT, CAKES, &c. Unlike some other yeast, it contains no POISON to create DYSPEPSIA, and the bread may therefore be eaten hot without detriment. Resolving itself into Phosphate of Lime and soda, it prevents RICKETS, CHOLERA, and decay of TEETH, and promotes the growth of Muscle and Bone. In "*raising*" the dough it does not, like other yeast, decompose the flour, but adds nutriment to the bread, and otherwise improves it in quality and quantity. Each package contains full directions for use. Send for pamphlet, supplied gratis. Ask your Grocer for "Horsford's Bread Preparation." WILSON, LOCKWOOD, EVERETT & CO., Wholesale Agents, 201 Fulton-st., New York.

"MAPLE SHADE FLOCK."

THOROUGH-BRED COTSWOLD SHEEP.

This justly-celebrated flock was selected from the flocks of the *most noted breeders in England*, by John D. Wing, Esq., of Washington Hollow, N. Y., who gave personal attention to its collection, with reference to the best wool-producing and mutton qualities. It is pronounced by competent judges to be the finest flock in America; and the present leader, "*Champion of England*" and some of the ewes, are believed to equal the best in any country.

The wool is long, fine, and lustrous, yielding from 8 to 20 pounds per head. They are full and square-bodied, very strong in the loins, and weigh from 200 to 300 pounds at maturity—sometimes exceeding even this weight. They are hardy and vigorous, and for breeding pure or crossing with other breeds, are believed to promise more profit than any other sheep. The wool is in good demand at remunerative prices, and the *thorough-bred* rams crossed with any other sheep produce a good combing wool, and lambs of such size as bring a large price early in the season in market.

Every sheep at present in the "Maple Shade Flock" was either *imported* or bred direct from *imported sire and dam*, and has a perfect pedigree.

This flock took the first prizes in the long wool classes at the New York State and Dutchess Co. Fairs, in 1867 and 1869.

Having purchased of Mr. Wing his **Entire Flock,** we offer for sale Choice **Ewes, Rams** and **Lambs.**

Address, LUCIUS A. CHASE, 245 Broadway, New York.

Cahoon's Broadcast Seed Sower,

For Sowing All Kinds of Grain and Grass Seed.

The unprecedented success of this Seeder the past year is without a parallel in the history of Agricultural Implements. It has probably received more First Premiums at State Fairs in the fall of 1869, than was ever awarded any other machine of any name or nature, in one year. Joseph Harris, author of Walks and Talks in *American Agriculturist*, says: "I like the Cahoon Seeder very much indeed."

There is one continuous voice of praise ringing in our ears from all parts of the country.

We warrant the Hand Machine to sow 50 acres of wheat in 10 hours, and the Horse Power Machine to sow 120 acres in the same time, and the work to be done with greater precision and accuracy than it can be done by any other means whatever. No farmer who has an acre of grain to sow yearly can afford to do without one of these Seeders.

Price of Hand Sowers, $10.00; Horse Power, $50.00. Send for Circulars, and name of Agent near your residence.

D. H. GOODELL & CO.,
Sole Manufacturers, Antrim, N. H.

BEST CABINET ORGANS

AT LOWEST PRICES.

That the MASON & HAMLIN CABINET and METROPOLITAN ORGANS are the BEST IN THE WORLD is proved by the almost unanimous opinion of professional musicians, by the award to them of SEVENTY-FIVE GOLD and SILVER MEDALS or other highest premiums, at principal industrial competitions within a few years, including the MEDAL at the PARIS EXPOSITION, and by a sale very much greater than that of any similar instruments. This Company manufacture ONLY FIRST-CLASS INSTRUMENTS, and will not make "cheap organs" at any price, or suffer an inferior instrument to bear their name. Having greatly increased their facilities for manufacture, by the introduction of new machinery and otherwise, they are now making BETTER ORGANS than EVER BEFORE, at increased economy in cost, which, in accordance with their fixed policy of selling always at least remunerative profit, they are now offering at PRICES of INFERIOR WORK. FOUR OCTAVE ORGANS, Plain Walnut Case, $50. FIVE OCTAVE ORGANS, Double Reed, Solid Walnut Case, carved and panneled, with FIVE STOPS (VIOLA, DIAPASON, MELODIA, FLUTE, TREMULANT), $125. Other styles in proportion.

Circulars, with full particulars, including accurate drawings of the different styles of organs, and much information which will be of service to every purchaser of an organ, will be sent free, and postage paid, to any one desiring them.

MASON & HAMLIN ORGAN CO.,

154 Tremont St., Boston; 596 Broadway, New York.

THE MAGIC WATCH-KEY.

Will fit any watch, and last a lifetime. Sent by mail for 50 cts. Address J. S. BIRCH, 14 Maiden Lane, New York.

$100 to $250 *per Month guaranteed. Sure pay.* Salaries paid weekly to Agents everywhere selling our *Patent Everlasting White Wire Clothes Lines.* Call at, or write for particulars to, the Girard Wire Mills, 261 North Third-st., Philadelphia, Pa.

CISTERNS PREVENTED from overflowing, and from bursting, by Muss' Patent Automatic Water Leader. Agents wanted. State and County Rights for sale. For particulars inquire of J. MUSS, Box 726, Quincy, Ill., or J. R. MITCHELL, Salem, Columbiana Co., Ohio.

American Vinegar Generator.

New plan, just patented. For full description, &c., send stamp to A. D. STRONG, Ashtabula, Ohio.

STENCIL PLATES and Checks of all sizes, very reasonable. Enclose stamp for samples and price circular. Address WM. POTTS, Cochranville, Chester Co., Pa.

Lilium Auratum.

The Japanese Queen of Lilies.

A large importation from Japan, just received by the subscribers, in fine, healthy condition. Flowering Bulbs mailed to any address at following prices:

No. 1, $1.00 each; $9.00 per doz. No. 2, 75 cents each, or $6.75 per doz. No. 3, 50 cts. each; $4.50 per doz. Prices by the hundred or thousand will be given to dealers upon application. B. K. BLISS & SON,
Nos. 41 Park Row, and 151 Nassau St., New York.

Marblehead Mammoth Cabbage!

This is the largest cabbage in the world, sometimes weighing over sixty pounds each, and averaging as high as thirty pounds by the acre! It is not only large, but cannot be surpassed for reliability for heading, tenderness, and sweetness. As some cultivators have an impression that this cabbage cannot be fully matured outside of Marblehead, I invite their attention to the following extracts from among the many commendatory letters which I have received.

"Your Marblehead Mammoth Cabbage cannot be excelled. There were heads weighing 50 lbs., and heads of Fottler's Improved Brunswick (from your seed) weighing 40 lbs.—John H. Howlett, Charlestown, Ill."

"I raised from your seed Marblehead Mammoth Cabbages that weighed 50 lbs.—A. H. Mace, Clintonville, N. Y., Nov. 14th, 1869."

"Your Marblehead Mammoth Cabbages were very fine, they all headed well, and weighed 27 to 40, and 47 lbs.—W. Llewellyn, Red Wing, Minn., March 12, 1869."

"Your Marblehead Mammoth Cabbages are wonderful; they grew to the size of an umbrella.—Thomas Flanigan, Palermo, Kansas."

"The Marblehead Mammoth Cabbages were a perfect success. They headed well, and were three times as large as any cabbage I ever raised before.—J. F. Butt, Kosciusco, Miss."

"I have raised your Marblehead Mammoth Cabbage for two years, and it has proved the tenderest and sweetest cabbage I ever saw.—S. S. Groves, Stones Prairie, Ill."

John Van Wormer, Springs' Mills, Mich., raised some weighing 30 lbs. John Humphreys, Titusville, N. Y., 33½ lbs. M. D. Clark, Elyria, Ohio, 37 lbs. H. A. Terry, Crescent City, Iowa, 40 lbs., measuring 56 inches around the solid head. Thos. A. Lambert, Becancour, C. W., exhibited three cabbages weighing respectively 40, 42½, and 44 lbs. John W. Dean, St. Michael's, Md., has grown them weighing 33 lbs. S. M. Shuck, Preston, Minn., 33 lbs., when trimmed. E. H. Ellis, Etna Green, Ind., over 30 lbs. A. E. Garrison, Des Moines, Iowa, 30 lbs. James S. Allen, Union Springs, N. Y., 30 lbs., when stripped of loose leaves. Wm. Lee, Jr., Denver, Colorado, has grown heads weighing 45 to 50 lbs., as a penalty for which the miners of the mountains call him the "Big Cabbage Man." Leonard Choat, Denver, Colorado, raised one which weighed 45 lbs. when trimmed of waste leaves. Collins Eaton, Ogdensburgh, N. Y., 50 lbs. P. Sweeney, Loretto, Pa., 43 lbs. Sam'l B. Ornsbee, Rolling Prairie, Wis., 53 lbs. Chas. W. Oden, Little Sioux, Iowa, produced quite a lot weighing from 50 to 60 lbs. A. C. Van Tassal, 39 lbs. trimmed. A. C. Goodwin, Kennedy, N. Y., 45 to 50 lbs. W. H. Spera, Euphrata, Pa., raised fifty heads that averaged over 31 lbs. each. Wm. D. Munson, Burlington, Vt., raised some weighing 46 lbs. Mary B. Sellman, Galesburg, Iowa, 28 to 43 lbs. stripped of loose leaves. Hundreds of others have written me that they have "taken all the prizes at the County Fairs." "Raised the largest cabbage ever seen in the country." "Astonished all their neighbors." "That in sweetness, crispness, and tenderness, they were unequalled," etc., etc. As the original introducer of the Mammoth Cabbage, I am prepared to supply seed grown from *extra large heads*, at following prices, by mail, post-paid: Per package, 25 cts.; per oz., $1; 4 ozs., $3.50; per lb., $12.

Full instructions for cultivation accompanying the seed.

Catalogues free. JAMES J. H. GREGORY,
Marblehead, Mass.

FERRE, BATCHELDER & CO.'S ILLUSTRATED CATALOGUE of Seeds and VEGETABLE AND FLOWER GARDEN

MANUAL for 1870, will be ready for distribution early in January. It will contain about 100 pages, with numerous illustrations—a complete list of Vegetable and Flower seeds, to which will be added a list of **Summer-Flowering Bulbs, Plants and Small Fruits.** Sent to all applicants **enclosing 10** cents. Our customers supplied without charge. Address

FERRE, BATCHELDER & CO.,
231 Main St., Springfield, Mass.

To Market Gardeners.

Your business involves an immense outlay over small areas, and it is of first importance that the seed you plant should be *fully reliable.* My enterprise is of special interest to you, as I am engaged in growing, directly under my own eye, a great variety of vegetable seeds, many of which seedsmen usually import or gather together from growers scattered through the country. Because I grow these seed myself I am enabled to warrant them *fresh and true to name*, and to bind myself to refill the same value gratis, should they not prove so. All seed warranted to reach each purchaser. I also import and procure from first growers, choice varieties that cannot be grown in my locality, giving me thus one of the largest if not *the* largest variety of vegetable seed sold by any dealer in the United States. Give my seed a trial. Catalogues gratis.

JAMES J. H. GREGORY, Marblehead, Mass.

Evergreen Broom-corn.

I have the genuine. The brush is worth from $5 to $20 per acre more than the old sort. I will send enough to plant an acre *post-paid*, for $1.00, one bushel as freight, $5.00. My Catalogue describes it, and 150 varieties of seed. Early Rose potatoes, Alton Nutmeg Melon, Liberian Cane, Choice Flowers, &c., &c. Send for it. Address

WALDO F. BROWN, Box 75, Oxford, Butler Co., Ohio.

300,000 HORSE-RADISH SETS.—The best in market, at $3.00 per 1,000.
EDWARD WHEELER, Box 724, Kalamazoo, Mich.

AMERICAN AGRICULTURIST

FOR THE

Farm, Garden, and Household.

"AGRICULTURE IS THE MOST HEALTHFUL, MOST USEFUL, AND MOST NOBLE EMPLOYMENT OF MAN."—WASHINGTON.

VOLUME XXIX.—No. 1. NEW YORK, JANUARY, 1870. NEW SERIES—No. 276.

"SEVERE WEATHER."—FROM A PAINTING BY R. ANSDELL.—*Engraved for the American Agriculturist.*

The shepherds of the mountainous regions are liable to have their flocks exposed, both on the coming on of winter and in the spring, to storms of great severity accompanied by rain, snow and cold, and often by driving wind which allows no escape from cold or wet. Even so hardy an animal as the Highland sheep often suffers greatly. Separated from the flock, weighed down by their ice-loaded fleeces, weakened by lack of food, and chilled through and through, it is no wonder that the most vigorous sometimes give up in their struggle with the elements, and die. When the storm breaks away or a lull for a few hours comes, then is exercised the sagacity of that most intelligent of brutes, the shepherd dog. Every old country in the agriculture of which sheep form a prominent feature, has its own race or breed of shepherd dogs. Those used in mountainous regions resemble each other a good deal, because the same arduous duties are required of them. They must be intelligent, kind, hardy and docile, of good size, fleet, well coated, and enduring. The most remarkable instances of intelligence have been manifested by females. We are most familiar with the English and Scotch shepherd dogs. The latter, called the *Colly*, is the breed of the dog in the engraving. These, like other shepherd dogs, are from their birth familiar with sheep. They have the constant companionship of the shepherds and manifest the strongest attachment for their master, generally being entirely indifferent to other men. Their care of the flock is not a cultivated natural impulse, nor second nature, even after centuries of breeding, but simply a business to which they are trained and in which they find unlimited opportunities for the display of their wonderful sagacity, aided by remarkable powers of vision and scent, and probably also of hearing. Untrained they make affectionate, companionable dogs, quick to learn tricks, and to understand language addressed to them.

NEW YORK, JANUARY, 1870.

It was wise in the Ancients to begin their social year in January. The Jewish year begins in September; the Mohammedan in May; the Ecclesiastical at "Advent," about the first of December; and our National or Civil year on the 4th of July. At no time have farmers, gardeners, professional or business men so much leisure to wind up the affairs of the closing year, and to begin aright the new as at this season, when traffic is, in a measure, suspended, when the ground is frozen, when the hours of light are few, and the evenings are long. We are receiving the income from the harvest in one way or another, or know what to calculate upon, in a measure, at least; weeks must pass before the activity of field work presses us again, and we have time to prepare for a successful year. Success does not come unsought, at least not to farmers following their ordinary avocations. It must be won by well-studied plans, and thorough preparations, judiciously carried out. To plan and prepare is, therefore, the most important work for January. Perhaps we ought to make an exception to this statement in favor of that mental culture for which the winter offers such opportunities. Nothing is so important to a farmer, as a man, as to be well informed,—as a farmer, he may be tolerably successful, without education except in the routine and labor of his profession, but every year makes it harder for those farmers who depend exclusively on native wit and innate shrewdness, and easier for those who study farming, and follow their profession with all the aids they can get, with the printed experience of a thousand neighbors, the best thoughts of men of science, and the best implements and methods they can afford to get and learn, and can carry out to use.

Hints About Work.

The Work-shop.—The winter affords time to work a good deal at repairing tools, making and mending articles, and if one has a room which may be warmed and lighted, and with a good work-bench, furnished with common carpenters' tools, a soldering iron, a little kit of saddlers' tools, some leather, rivets, etc., it will be found a very attractive place for the boys; and not only would there be a good deal of good work done, but good habits formed, and skill gained, worth a great many dollars more than an expensively fitted up work-shop would cost.

Frost and Snow.—Never delay path-making and road-breaking after every fall of snow. The work is much lighter, and it is better done. Never trust to mild nights, and leave the water standing in pumps or pipes. Sudden changes of weather, with accompanying damage, occur without warning.

Building may often be done by contract in winter cheaper than in warm weather, especially if the contracts be made in the autumn; but builders will give moderate figures for good work, if one can present well-made plans and specifications early in the new year, and give them time to do the finishing after the house is enclosed, and the roof on, when their other work does not press upon them.

Timber.—Cut and haul to the saw-mill. In selecting trees, take such as have nearly stopped growing, but are still sound and healthy. Such wood is quite as good as that which is growing fast, and stiffer. Young timber is elastic, old is stiff; that which has stopped growing, and has many dead limbs, is brash, though good for fuel.

Ice.—Those who have ice-houses should not neglect to fill them whenever sufficient thickness of ice forms to make it possible to handle it economically. At the lowest latitude where ice-houses are found, and where ice is usually gathered, it often happens that good ice can be obtained only for a very few days. Ice one inch thick may be very profitably handled, being dragged out upon a clean platform, running into the water, and from this shoveled into carts. It should be packed by pounding into as solid a mass as possible in the ice-house. An intermixture of a moderate quantity of pure, fresh-fallen snow is an advantage if it be well pounded. When thick ice can be obtained, chip off all the porous snow-ice, and pack only the clearest, filling the crevices with fine ice chipped from the top of each layer. Cut the cakes to fit.

Orchard and Nursery.

Animals must be kept out of young orchards. Tight fences and securely closed gates will do much to exclude domestic animals, including man.

Mice and Rabbits are the most troublesome among the wild animals. Tramp down the light snows around the trunks of the trees. Sprinkle blood near the base of each tree, to keep off rabbits.

Pruning.—Though winter pruning is objected to by good authorities, if we had an old orchard that needed treatment, we should go at it in winter when there is plenty of time. Painting over large wounds, or covering them with melted grafting wax will prevent injury from rotting. If trees are properly shaped when young, it will seldom be necessary to do much pruning.

Insects.—One of the worst enemies of the orchardist is the Tent-caterpillar, but it is, fortunately, one of the easiest to keep in check. The eggs may now be seen near the ends of the twigs, glued in a broad band-like cluster. Remove the eggs and there will be no caterpillars, as another crop will not be laid until next summer. A pole pruning implement of some kind is convenient for this.

Manure may be spread upon the orchard. We sometimes see trees manured by a heap placed directly around the trunk, where, if not a positive injury, it is of no use. Spread it evenly over the surface.

Fruit Garden.

Those who have only a limited space must plant their fruits and vegetables as they best can; but where circumstances will admit of it, by all means have a separate garden for small fruits and dwarf and other trees of small growth. It is impossible to grow vegetables properly in the close neighborhood of trees and large shrubs; besides this, the trees and shrubs are likely to suffer for the want of proper nutriment. Select a well-sheltered spot, with good deep soil; drain if necessary, and set it apart as a fruit garden. Strawberries, raspberries, currants, gooseberries, grapes, pears, etc., can be produced in abundant supply. Were small fruits in abundance, there would be fewer discussions of the question, "Why do boys leave the farm?"

A mild spell will allow of the pruning of such grape-vines and gooseberry and currant bushes as were left at the regular autumn pruning.

Kitchen Garden.

At the North we can do but little this month, except when vegetables are forced under glass. In Southern States the hardy vegetables may be sown whenever the soil is in good condition, and the temperature averages about 45°. The beet, carrot, parsnip, parsley, cauliflower, radish, turnip, onion, leek, cress, spinach, cabbage, etc., are hardy.

Seeds.—If the seed be poor in quality, or not true to its kind, it is worse than useless. Look over the stock on hand, and reject all about the identity which there is any doubt. The vitality is easily tried by planting a given number in a box of earth, and keeping in a warm room until they germinate. If three-fourths come up, the seed may be considered good. Decide early what seeds are wanted, and order. Stick to tested kinds for the main crop, and invest in novelties for experiment only.

Manure.—The heaps should be turned over when the steam issues copiously. Cart manure to those points where it will be handy for use.

Hot-beds are to be prepared for. Sashes may need repairs in the way of painting and glazing. In some parts of the South they may be put in operation. The safest rule for all latitudes is to start the hot-bed six weeks before the time at which the plants can be set out with safety.

Straw Mats will be needed to cover the glass of the hot-beds during cool nights. We have frequently given directions for making them.

Cold-Frames.—If the snow covers them while the plants are frozen it need not be removed, but if the weather is mild when the fall takes place, it should be swept off. Give air every mild day, and endeavor to keep the plants as dormant as possible.

Flower Garden and Lawn.

Many plans for the improvement of places will be formed. If one has abundant means, and but little knowledge of such matters, it will be safest for him to employ some landscape-gardener of acknowledged taste to lay out his grounds. The great mass of people are, however, obliged to both plan and execute themselves. To such our advice is, do not attempt too much in the way of adornment at once. Do not copy the plans for expensive places which will be difficult to carry out, and which, if laid out, are not likely to be kept in order. Make a plan of the place as it is, and see what can be done to improve it, and at the same time retain as many of its present features as are desirable. Houses are generally set too near the road, which much restricts the space in front, and this space is generally divided by a path directly from the road to the front door. It is often the case that a side entrance from the road can be so arranged as to avoid breaking up the space directly in front of the house, and thus a lawn of moderate dimensions can be had. A smooth, level lawn is the first thing to be provided for. Make only such paths as are necessary, and if pleasing curves can be given to them, it is better than to have them straight. Shrubs, trees, and flower beds are to be provided for, and for these no definite directions can be given. A common mistake is to so surround the house with trees that distant views are quite shut out. A little judicious cutting away will often open a beautiful landscape to view. All such plans should be made long in advance of the working season, and be well considered in all their relations before the work of laying out is commenced.

Plants in pits, cold-frames, and cellars, should be looked to, and aired when the weather will allow. Plants thus stored will seldom need water, but should they become very dry, give a small quantity.

Rustic Work will afford interesting employment for those who have some taste and mechanical skill. Seats, flower stands, and the like, can be made out of very unpromising material. Cedar, Laurel, the Wild Grape, and others, afford available stuff for the purpose.

Stakes, Labels, and all garden conveniences, should be prepared in advance in abundant supply.

Green-house and Window Plants.

The temperature of the green-house, where plants are expected to grow and flower, should reach 60 or 65° in the daytime, and may sink 10 or 15° lower at night. If plants are only stored for the winter away from frost, then the night temperature may go down to within a few degrees of freezing.

Ventilation is to be given whenever it can be done without injury by cold. Plants in rooms suffer greatly for the want of fresh air, and the success with plants in modern, close-built houses, is much more rare than it used to be in less carefully closed structures. In providing for a change of air, sudden chilling of the plants must be avoided.

Dust is one of the great drawbacks to the health of house plants. The table on which the plants stand may be so arranged that wires or rods, to support a cloth, may be attached to it, and thus provide a dust protector while sweeping is being done. Shower the plants as often as practicable.

Insects.—The Green-fly, or Aphis, and the Red-spider are the principal insects that infest house plants. The first is to be treated to abundant smoking with tobacco, and the other will soon be rid of if the foliage is thoroughly wetted every day.

Water only when the earth shows signs of dryness. More plants are injured by over watering than by too little water.

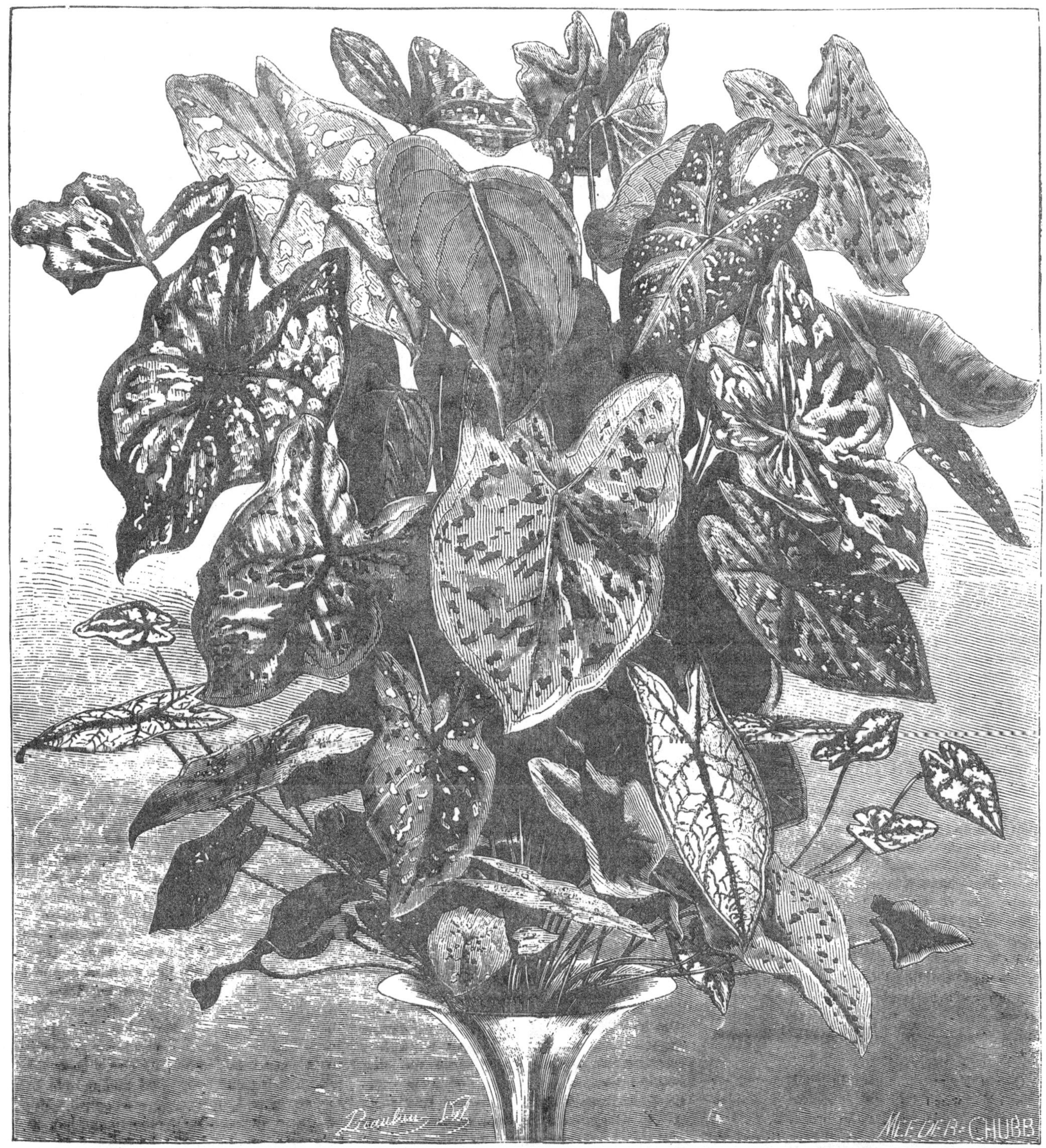

A GROUP OF CALADIUMS.—Drawn from Nature.—*Engraved for the American Agriculturist.*

Caladiums as Garden Plants.

The attempts to follow the styles of garden ornamentation which are so successfully carried out in England are with us, for the most part, if not downright failures, at least quite unsatisfactory. Our florists import bedding plants, and with them the glowing descriptions of English dealers; the plants are tried by a few seekers after novelties, and are generally heard of no more. A few, indeed, stand the test to which our hot summers subject them, and we shall in time find a sufficient number of plants suited to our climate, with which we can produce all desirable effects. This list will, however, be quite different from the plants which flourish so finely in the moist summers of England. One after another, the plants which we formerly supposed could be seen only in a hot-house, or at least a green-house, are found to answer admirably when planted in the open ground during summer; our weather seems to remind them of their native tropics, and they flourish accordingly. Those who read of the wonderful beauty of the sub-tropical gardening of Europe need not envy the cultivators across the water. Our own climate allows of the use of a larger number of sub-tropical plants for garden decorations than does theirs. It needs only a few good examples here and there to create a public taste for this style of gardening, and as soon as there is a demand for plants of a tropical habit our florists will not be slow in supplying it. The Caladiums are now attracting attention for decorative purposes out of doors. The great beauty of their leaves, in both form and color, have long made them prized ornaments of the hot-house, but they are now to be more widely known, and we hope to see them before long as popular as their relative, the well-known Calla. The *Caladium*, (or *Colocasia esculentum*,) a very large species with enormous leaves, the bulbs of which are the chief article of food of the natives of the Sandwich Islands, has been more or less cultivated for years. The leaves of this are of a soft, light green, but other species present us with foliage of most brilliant colors and exquisite markings. A group of the leaves is given in the engraving, where it is difficult to show much more than the form. The imagination must fill up the markings of some with rose, carmine and purple, give a pure white ground, traced with delicate green lines to others, and to a few it must give the metallic lustre of bronze. The florists offer some eighty varieties at prices ranging from 25c. to $4.00 each, according to their rarity and ease of propagation. The lower priced ones, which comprise most of those shown in the engraving, include some of the most desirable sorts. The Caladiums are mostly natives of the tropics of Asia and America; they flourish in good garden soil, but all the better if it is partially shaded. They must not be put out until about the first of June, and when the early frosts come the bulbs must be taken up and kept in a warm place. The derivation of the name Caladium does not appear to be settled, but if the plants become as popular as we think they may, it will present no more obstacles to those who have an aversion to botanical names than do Rhododendron, Magnolia and Geranium.

The Sylvester Apple.

At the State Fair held in Rochester in 1868, we saw an apple which was remarkable for its beauty, and when its originator, Dr. E. Ware Sylvester, of Lyons, N. Y., gave us a specimen to taste, we found that its quality kept the promise made by its exterior. Again, this year, we have been able to try other specimens of the variety, and considering it as deserving a wider popularity than it now enjoys, have had it engraved. The tree is said to be vigorous and an abundant bearer. The size and shape of the fruit are shown in the engraving. The skin is white and of a most delicate waxy appearance, which is hightened by the beautiful markings of crimson that are found upon the specimens, which have been well exposed to the sun. The flesh is white and very tender and juicy; indeed, upon cutting, the juice follows the knife as it does with a well-ripened pear; flavor, a pleasant subacid. Excellent for cooking Sept. and Oct. Dr. Sylvester should feel gratified at having his name attached to so good a fruit.

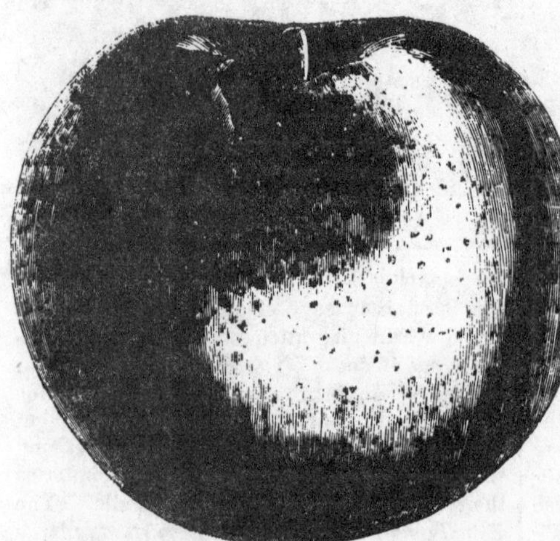

SYLVESTER APPLE.

The Climbing Fern. (*Lygodium palmatum.*)

The Climbing Fern is so unlike others of our native ferns, that one at first sight would hardly class it with them. Its peculiar form and climbing habit are shown in the reduced engraving, where a fragment is also given of the full size. The light brown stalk is very slender and wiry, and twines closely around small shrubs and other plants, climbing to the hight of two or three feet. Small branches are placed alternately on the main stem; these are forked, and bear at each division what passes for a single leaf, but which, in a botanical description, would be called a *frond* or *frondlet*. Each of these leaf-like bodies is several lobed, the upper ones being very much divided. The small upper ones are the fertile fronds, and upon the under side of them the spore-cases, or what correspond to seed-vessels, are borne. This is among the rare plants of the country, but it is found in several localities, from Massachusetts to as far south as Florida. One of the most abundant localities is at East Windsor Hill, Conn. The striking delicacy and beauty of this fern adapt it to decorative purposes, and it is used in its green state, or pressed and dried, when it is formed into graceful wreaths to surround pictures, attached to white curtains and the like. So popular had it become in Connecticut that there was danger that the locality at East Windsor Hill would become extinct, as large quantities were carried off yearly, until an act was passed by the Legislature forbidding its wanton destruction. The root-stalks are very slender, and the plant does not, as a general thing, bear removal well, though we have known it to be successfully transplanted and to establish itself thoroughly. *Lygodium* comes from the Greek word for flexible, and the specific name *palmatum* is in reference to the manner in which the fronds are lobed, like an outspread hand.

CLIMBING FERN—(*Lygodium palmatum.*)

Propagating the Larch from Seed.

BY D. C. SCOFIELD, ELGIN, ILL.

[Mr. Scofield, who is a warm advocate of tree-planting, considers the European Larch the most valuable tree for timber. He gives the following as his method of treating the seed.—EDS.]

"First. Two prominent difficulties are encountered in this country, which I believe are unknown in Europe; the hot rays of the sun having the double tendency to scald or heat the soil, so that it causes the plant to die at the collar, or as the phrase has it 'damp off,' as well as to scorch the tender plant as it emerges from the earth. These are overcome first, by selecting a light, sandy, dry, though rich soil, for the seed bed; and secondly, by a partial shading the first season, which may be best done by nailing strips of lath one inch apart, and placing them, one foot in hight, over the seed bed, so as to partially obstruct the rays of the sun. Any other material that will render the same amount of shade, will answer.

"Second. The soil where the seed is to be sown should be as clean from weed-seed as possible. The usual time of sowing onion seed, or from middle of April till first of May, is the time to sow Larch. It may be sown in drills, four or six inches apart, or broad-cast, and covered just enough to retain moisture until it germinates. It may be sown as onion or other seed without preparation. Clean culture is indispensable. The shading may be dispensed with after the first season. The seedlings should be transplanted either at one or two years old from the seed-bed. This should be done as soon as the weather and soil will permit, in the spring, in a similar soil to that in which they were grown, and better at one year old than two. Transplant in beds, in rows six

EARLY WYMAN CABBAGE —(*See next page.*)

inches apart and four inches in the row, and shade, as directed for the seed-bed. The roots should be kept from the atmosphere as much as possible in transplanting. With good seed, a satisfactory success will be realized. This method will apply with equal success in propagating any variety of hardy evergreens.

The Early Wyman Cabbage.

The market of each large city presents peculiar features not to be found elsewhere. In New York an article will sell if it looks well, no matter if its quality is inferior to that which is less showy in appearance. In Boston, people are more particular; they take the pains to learn the names and characters of the best varieties of fruit and vegetables, and are governed more by quality than by external appearance. The gardeners who supply that market have obtained, by careful selection, several varieties or sub-varieties which have a local popularity, and are worthy the attention of growers elsewhere. Boston Market Cauliflower, Boston Market Celery and Boston Market Tomato are all among the best, if not the very best, of their kinds. The popular early cabbage in the Boston market is the Early Wyman—which, for a wonder, is not called "Boston Market." The variety originated with Mr. John Wyman, of Arlington, Mass., and the seed was introduced by Washburn & Co., of Boston. The engraving on page 8 gives the shape of the head. Specimens furnished us last spring by Messrs. B. K. Bliss & Son were of large size, solid for an early variety, and very crisp and tender. It is said to bring a higher price in market than any other variety.

Use of the Plow in Digging Ditches.

No ditch-digging machine has yet been introduced. There are some for which great claims are put forth, but we must wait before they become common articles upon the farm. Meanwhile ditch digging must go on by spade and mattock, pick and scoop. We may, however, greatly facilitate the operation by employing the labor of horses or oxen with plows. There are several difficulties to be obviated. It is hard to plow a furrow on a sufficiently straight line; this may be accomplished by stretching a cord and turning over a narrow line of sods with a spade, exactly where the ditch should be. The cattle will follow this line of themselves, when they could hardly be driven exactly along a line of stakes. To use the plow economically, one needs a pretty strong force of diggers and pickers, and to have them well scattered throughout the line plowed. Two pairs of oxen make the best team, probably. Two furrows are turned out, 12 inches wide, and 9 inches deep. Then if the sods are cleared away, the plow may lift another furrow-slice out on each side, but probably it will only loosen the dirt and make it easy shoveling. This will stir the soil down some 4 to 6 inches, and when it is cleared out the ditch will be 13 to 15 inches deep, with ridges of earth and stones along both sides. The oxen will not easily be made to walk longer in the furrow, and without very long yokes they will not go one on each side of it. These are sometimes used, but we are informed that another plan is practised in some places with great success. The two yokes of oxen are attached abreast to the pole of a cart, the body being removed. They draw by chains made fast to the axle, and the tongue is supported by a light ash or hickory pole, lashed firmly to the yokes of each pair of oxen. The earth is thrown out from the ditch on each side, and the length of this pole is such that the inside ox of each pair walks on the inner side of the ridge. The plow is attached to the axle-tree. The chain may be fastened to the axle-tree itself, but it is far better to use an oak knee, as shown, which is lashed forward to the tongue and hangs below. In this consists the chief merit of the whole arrangement, for by it the plow can be drawn at any depth below the surface, provided it be wooded so that the handles do not interfere with the sides of the trench. Any wheelwright will alter the wooding of a plow, or new wood it so that it shall have but one handle, and that directly above the furrow. The plow is fastened to the end of the knee in which a pin is set, by a rather short chain at first, which may be lengthened afterwards. The hight at which the draft for the plow should be is regulated by raising or lowering the knee, which is not only chained by one end to the tongue, but also to the axle, and is made higher or lower by blocks laid across under it, resting upon the spreading fork made by the tongue, where it joins the axletree,

KNEE FOR DITCHING PLOW.

and shoved forward or drawn back, according as one wishes the draft chain higher or lower.

It will be necessary to change plows as the depth increases. There are several plows so arranged that very narrow mould-boards may be attached, and after these, the subsoil plow may be used until little besides "finishing" remains to be done. In stony land men with crowbars and pickaxes must attend and take out stones as fast as they are touched by the plow.

Farmers Should Take Enough Sleep.

Said one of the oldest and most successful farmers in this State, "I do not care to have my men get up before five or half-past five in the morning, and if they go to bed early and can sleep soundly, they will do more work than if they got up at four or half-past four." We do not believe in the eight-hour law, but, nevertheless, are inclined to think that, as a general rule, we work too many hours on the farm. The best man we ever had to dig ditches seldom worked, when digging by the rod, more than nine hours a day. And it is so in chopping wood by the cord; the men who accomplish the most, work the fewest hours. They bring all their brain and muscle into exercise, and make every blow tell. A slow, plodding Dutchman may turn a grindstone or a fanning-mill better than an energetic Yankee, but this kind of work is now mostly done by horse-power, and the farmer needs, above all else, a clear head, with all his faculties of mind and muscle light and active, and under complete control. Much, of course, depends on temperament, but, as a rule, such men need sound sleep and plenty of it. When a boy on the farm, we were told that Napoleon needed only four hours sleep, and the old nonsense of "five hours for a man, six for a woman, and seven for a fool," was often quoted. But the truth is, that Napoleon was enabled, in a great measure, to accomplish what he did from the faculty of sleeping soundly—of sleeping when he slept and working when he worked. We have sat in one of his favorite traveling-carriages, and it was so arranged that he could lie down at full length, and when dashing through the country as fast as eight horses, frequently changed, could carry him, he slept soundly, and when he arrived at his destination was as fresh as if he had risen from a bed of down. Let farmers, and especially farmers' boys, have plenty to eat, nothing to "drink," and all the sleep they can take.

The Cord-Grass or "Spartina."

The materials used in making paper are numerous. We have linen and cotton in the shape of paper rags, but the supply from these sources was long ago found to be inadequate. There have been numerous inventions for using the fibre of basswood and other soft woods, as well as for the cane of the Southern States. Fibres from these sources have more or less helped to supply the great demand for paper stock. Lately, the "Esparto," a curious grass growing on the shores of the Mediterranean, has been largely used in England, and to some extent in this country, as a paper making material. As yet we know but little of its cultivation, but we believe experiments are being made with it in the Southern States. Meanwhile, one of our native grasses has come into notice, the Cord-grass, or "*Spartina cynosuroides*. This, which is shown in the engraving, is abundant along our fresh-water rivers and lakes, especially at the North, and actual experiment has shown it to be a valuable paper stock. In a report to the Commissioner of Agriculture, Mr. Jas. Woodruff, of Quincy, Ill., says: "It is much superior to straw, yielding, when properly treated, a much stronger, longer, and softer fibre, and a much larger percentage of stock. Its cost, delivered at my mill, during the past two years, has been about $5 per ton." Mr. W. says that experts who have worked the two consider the Cord-grass a better material for paper than the "Esparto." There are doubtless many hundreds of acres of otherwise unavailable land that might be devoted to this grass, and the matter is worthy the attention of both paper makers and those who have land adapted to its culture.

CORD-GRASS.

For other Household Items, see "Basket"

An Efficient Rat Trap.

BY M. QUINBY.

The engraving represents the best rat trap I ever saw. It is not uncommon for two or three of the vermin to be caught in it at once. I have caught five. The trap is open on every side so that even an old rat, cunning as he is, suspects no danger. When he is fast there is no sight or sound to tell the tale; so that after getting one you can keep right on trapping more in the same place.

The trap consists of a platform of 1½ inch plank, 2 feet square, with a low curb around it; two posts, 14 inches high and 2 inches square, are set inside the curb, midway of opposite sides of the platform; there is a second platform, a trifle smaller than the first, made to drop easily inside the curb; it has several inch holes bored in it so that in falling the enclosed air may easily escape. Notches must also be cut midway of two of its opposite sides to fit the posts which act as guides when the trap springs. The upper platform is raised by a 1×1½ inch standard of hard-wood, mortised into its centre, and passing through a hole cut for it in the cross-piece above. A hook or staple of stiff hoop iron must be screwed to this standard 7 inches from the bottom, to receive the short end of the lever *A*, when the trap is set. Just behind *A* is a rod passing up from the "pan," which is a piece

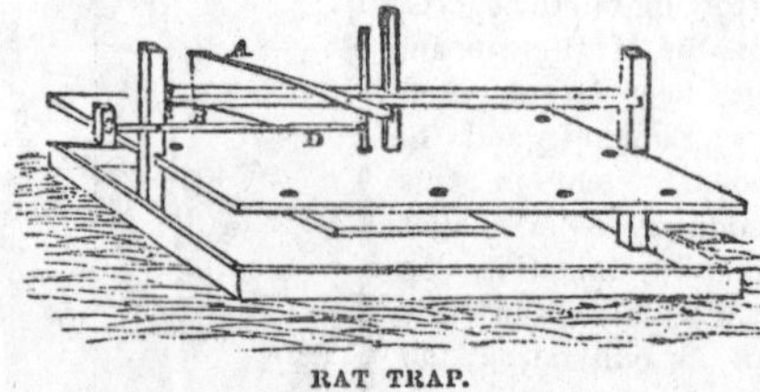

RAT TRAP.

of thin, light wood, 6 inches square, and fitting very loosely in a shallow box cut for it in the lower platform. There should be a quarter-inch hole in two corners of this "pan" through which pass short, upright wires, which are fixed in the bottom but do not reach above the level of the floor. These wires will keep the pan steady when the trap is set. *A* is a lever turning easily and connecting at its extremity by a cord with the lighter lever *B*. The nearer the fulcrums of these levers can be brought to the weights, the easier will the trap spring. Just below *B* is another lever connecting at one end with the upright rod from the "pan" below, playing freely on a wire pivot in the post, and bearing at *C* a weight so adjusted as to balance the weight of the "pan" and rod at the other end.

In the cut the trap is represented as set; the upper platform raised, supported by the short end of the lever *A*, the other being connected with *B*, which in turn is so slightly caught in the notches at *D* that the least disturbance of the "pan" below will detach it and spring the trap. The distance from lower platform to upper when raised is 5 inches

Winter Clothing for Little Ones.

BY FAITH ROCHESTER.

A young child in winter needs garments warm enough to allow of its sitting comfortably on the floor. To be sure it ought not to sit there long at a time. I know it would be an uncomfortable seat for me. I should be cold and should soon get weary in back and limbs. No, baby shall sit mostly in her crib, or in a high chair beside the table. But she will want to creep before warm weather, and I shall wish to have her do so. Creeping soils the clothes, but it strengthens the back, and is a good preparation for walking. Her new garments should reach only to her toes, that they may not be in the way when dear old nurse Nature (who knows better than any one else when these advances should be made) begins to pull the little one upon her feet. She shall have long woolen stockings and home made cloth shoes. These are softer and better than the shoes usually bought for babies who have not learned to walk. It is easy to cut a pattern from a little morocco shoe. Very nice ones can be made of thick broad cloth or cloaking, lined with drilling or strong thin flannel, with the seams laid open and felled down. Her active little legs should not depend too much upon her skirts for warmth. She shall have "knee breeches" such as were invented for her little brother before it was convenient to put drawers on him. I made them of flannel cut in the shape shown in the diagram, with four little gores in the bottom (instead of gathers) and a strong tape fastened to the top, on the end of which was a button to fasten through a hole in the petticoat waist. Being outside the diaper, this strap helps to keep that garment in place. A soft, sleeved, flannel shirt to envelop the arms, chest back and bowels; a warm woolen skirt, with a loose waist buttoned behind, and suspended from the shoulders by easy straps; a lined flannel dress, cut in the pretty style called Gabriella, and a sleeved apron, complete a winter suit for our half-year-old baby.

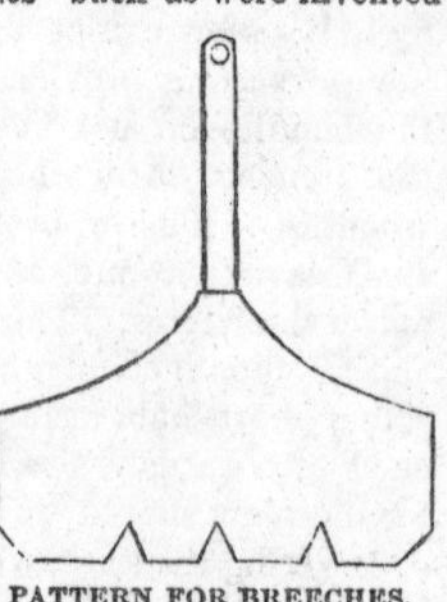

PATTERN FOR BREECHES.

There are mothers who will think this quite too much clothing for a child of that age, especially upon the arms. It is not unusual to see little girls of eight or ten years, who have never in any season worn long-sleeved dresses, a sleeved apron or sack being deemed sufficient protection from cold. Who would think of dressing a boy in that way? What sense is there in such a distinction of dress? Of late, fashion seems to be coming to the rescue, but we cannot put our trust in fashion. We must be guided by common sense. When I see the bare blue arms, and chapped knees, and pale faces of the little ones, I think there is need of mothers who are "strong-minded", in the best sense of the word.

A Codfish Dinner.

An American resident in Rome, a consul we believe, became celebrated for his codfish dinners. He carried one of the staple articles of New England food to the Eternal City, and made famous a dish which some affect to despise. A dinner of codfish cooked by an average Biddy, and one prepared by a good New England housekeeper, present just the difference that is always to be found between things well done and badly done. So simple a thing as codfish can be spoiled. Biddy puts the fish into a pot of boiling water and lets it "wollop" until she thinks it is done. It comes out generally too salt and very tough, is served with potatoes and some greasy paste which she calls "d'hrawn butter". No wonder that such a dinner is not relished. The codfish should first be soaked in cold water, changing the water every few hours until it is fresh enough. It should not be entirely freshened, but left just salt enough to be palatable. It is then put into a kettle, covered with water and brought up to the boiling point, but not *boiled*. Let it simmer gently for 15 or 20 minutes and it is ready to serve. As accompaniment to the fish there are of course potatoes and many add plain boiled beets and carrots, and hard-boiled eggs. We have sometimes seen small bits of salt pork fried crisp and served with some of the pork fat. The proper sauce is butter, which is sometimes simply melted, but more usually made into a sauce with flour and water. This may be made into an egg sauce by adding chopped hard-boiled eggs. The fish, vegetables, etc., are served separately, and each one, being helped to his liking, mixes them upon his plate according to his fancy, the sauce being served last. Mustard, pepper and horse-radish are the condiments most frequently used. In many New England families the regular Saturday dinner is of codfish. What is left is converted into fish balls or minced fish for Sunday morning's breakfast.

Cooking of Game Birds.

A correspondent in Saratoga Co., N. Y., requests us to give some recipes for cooking game birds. This is a difficult matter, as there is so much difference in opinion as to the way in which it should be done. The old directions for a wild duck are for the cook "to run through a warm kitchen with it and serve,"—a way of expressing the idea that the duck should be very rare. Some prefer duck as rare as underdone beef, while at most tables it is served well done. In whatever way it is cooked the duck should be basted frequently in order to preserve its flesh in a juicy condition. Our correspondent asks particularly about woodcock and partridge. The manner in which woodcocks are cooked by epicures is one not likely to be adopted by persons at all squeamish. The birds are plucked without being drawn, and are hung up before the fire to roast, a piece of toast being put beneath them to catch the drippings. They require about 20 minutes, and are basted with butter and served very hot upon the toast. It is considered by "the authorities" as quite the improper thing to remove the entrails of a woodcock, but persons of other tastes cook them differently. Split and broiled quickly, well buttered and served on toast, they are so good that we can forego the delicious morsel of the "trail," as the intestines are called. Partridges, or grouse, and quail, are dry birds, and are cooked in the various ways in which chickens are prepared. If roasted or broiled, the cooking should be done very quickly, and a plenty of butter used for basting. In broiling they are split in the same manner as a spring chicken. Some dip them in melted butter and cover with bread crumbs before broiling, as this, in a measure, prevents them from becoming dry. Either of these birds makes an excellent pie, or they may be stewed with such seasoning as is preferred. A very good way to cook them as well as ducks is to prepare them as for roasting, using stuffing if desired, and place them in a stew-pan with a few slices of pork and a little water, and cook slowly. When the gravy is simmered down and browned, add more water. The birds should be turned occasionally. In this way the birds are kept enveloped in steam, and are much juicier than when cooked in the drying heat of an oven.

Wedding Entertainments.

A correspondent writes: "I would like to inquire with regard to wedding entertainments. What kind of refreshments should be served, if in the forenoon or afternoon. If not taken at the table should the guests stand or sit? Should coffee be served when wine is excluded?"—Among the few sensible decrees of fashion is that which allows the entertainments at weddings to be as simple or as elaborate as the givers are inclined to make them. A glass of wine and a piece of cake are all that are offered in many instances; in others, a table with sandwiches, salads, cakes, etc., as for an evening party, is provided, and this, especially if wines are included, is often a costly affair. A large and increasing number serve no wine, the drinks being lemonade and coffee. The character of the entertainment should depend in a great measure upon the company. If the guests come from a long distance—as is frequently the case in the country, something substantial should be provided in addition to delicacies. Where the wedding takes place early in the day the old English custom of giving a "Wedding breakfast" is a pleasant one, especially if the company is not so large that all cannot be seated at table. At a feast of this kind, cold fowls, tongue, oysters, and other substantials, are given, besides which there are salads, jellies, ice-cream, and the like, with cakes, confectionery and fruit. Coffee, excellent in quality and abundant in supply, is provided. Where it is not convenient to seat the guests at table, they stand or sit as there may be room. At all events there should be chairs for such elderly people as may be present. Where the majority stand while taking refreshments there is likely to be less of that stiffness and solemnity which frequently attends social gatherings. We do not know that we have given our correspondent a very definite answer. Let her consider what will be the most to the comfort of her guests, and what will comport best with her means, and she cannot go very far wrong.

Improvement in Farmers' Homes.

There has been a very great change for the better in the homes of all classes of our people within the last thirty years. We think quite a large share of this improvement is found in the strictly rural districts, and is fairly the result of the opportunities and privileges of agricultural life. The average rural parish is the equal of the city parish in intelligence, in good morals, and piety. Social life has not so much show and brilliancy, but quite as much solid happiness. The children coming up in the country have a much better chance of sound health, of a good education, and of a useful career in life. The division of labor, brought about by the introduction of manufactures, has blest the farmer almost as much as any other class. He no longer provides his own clothing or makes his own furniture. He can buy cheaper. The thrifty farmer in the older States has an architect to build his house, and there is taste displayed in the building and in its surroundings. He knows a good deal about fruits and flowers, and what he does not know his wife or daughter does. The flower border is quite up to the town standard. The upholstery may not be quite so attractive, but the floors are carpeted, and the windows have blinds and curtains quite enough for cheerfulness and health. The table is neatly spread, and the chinaware and other appointments come from the same manufactories that furnish city homes. The cookery, especially that part of it furnished by the mistress of the mansion, is above the average in cities. Pianos and melodeons are very common, and the same songs are heard there as in the town, and they are sung about as well. Professional singers do not go to the country for their audiences, but the country comes to them and furnishes a fair share of their appreciative listeners and admirers. There is leisure in the country, time for reading and reflection, plenty of newspapers and magazines, and the village library has its numerous patrons in farmers' homes. Farm life in this age of railroads and steamers is quite different from the life led by our fathers. It moves in the right direction.

Hints on Cooking, Etc.

Pumpkin Pies.—Mrs. S. Hannahs, Portage Co., O., says: "Pare the pumpkin, then grate it, and add sugar and ginger to taste, and milk enough to make it of the proper consistency; then line your pie-tins with crust, put in your pumpkin, and bake in the ordinary way. After trying this once, no one will, I think, wish to go back to the old way of making pies of stewed pumpkin."

Cream Pudding, by Miss M. M. F., Westchester, Mass.: 6 tablespoonfuls of flour, 1 quart milk, 3 eggs, 1 teacupful sugar, salt; take a little of the milk and stir with the flour, to make a batter, and boil the remainder. When the milk boils, add the batter and when sufficiently cooked, take it off and stir in the eggs, beaten. Sift a part of the sugar in the pudding dish, then pour in the pudding, and put the rest of the sugar on top. Flavor to taste, and cover tightly until cold.

Mangles.—Miss E. A. C., Flatrock, Pa. The English mangles take up too much room and require too much power to find favor with our housekeepers. Some small ones have been invented, but their operation so far as we know, has not been such that they have met with any considerable sale. Here is a chance for inventors,

containing a great variety of Items, including many good Hints and Suggestions which we throw into smaller type and condensed form, for want of space elsewhere.

Wisconsin Horticultural Society.—The Annual Meeting will be held at Madison, on the first Tuesday in February, and continue for two days. The Secretary, O. S. Willey, sets a good example to other secretaries of such societies. We received his announcement of the meeting in December, and extract from it the following: "First, we don't want any one to say he did not have a timely notice. Second, we want every body to know that there is a Wisconsin State Horticultural Society, and that its members, and all interested in fruit growing, either as a luxury or profession, are invited to be present and participate in its discussions, or at least to add their mite by way of encouragement by their presence. Third, we want any one who has raised any fruit, and still has it on hand, to have this timely notice, that they may have some of their fruit still farther kept, and send or bring such samples as will add a share of interest to the occasion," etc.

Hedging.—An article upon Hedges for the West, by our Iowa contributor is unavoidably put over to another month. It is by a practical hedge-grower and will answer several who have inquiried about hedges.

European Larch.—Most nurseries keep these in moderate supply. Those who have seedlings in quantity should advertise.

The Kittatinny.—This Blackberry seems to do splendidly in Iowa, to judge from a photograph of a cluster sent by B. Larned of Eddysville.

Opium.—G. D. Cramer and others. We have no evidence that opium has yet been made in this country. An extract of the poppy plant, a very different thing, was exhibited as opium. This was subjected to analysis, and found variable, and of poor quality. The opium excitement started in Vermont, and its apparent object was the sale of directions for poppy culture. If any one can furnish us reliable information upon the subject we shall be glad to receive it. Opium is not, as many suppose, to be made by pressing the juice from the poppy and evaporating it. It is the dried milky juice which exudes when the unripe seed-vessel of the poppy is carefully scarified. In India each head only yields about two grains, and we doubt if at the present prices of labor and our uncertain climate it can be made to pay. It is, however, a good field for experiments, and we would like to hear of something more reliable than the newspaper accounts heretofore published.

Fine Grapes.—Wm. Hamilton, Gardener for Wm. Hoyt, Stamford, Conn., has been very successful with his vinery. The cluster he sent us attracted much attention and—we are sorry to say it—proved too much for some fingers.

Peaches.—"W. L. S." Currituck Co., N. C. It is customary to plow the orchard after the fruit is gathered and cultivate it as long as weeds grow. We do not see how it can affect the starting in the spring. A mulch put on after the ground is frozen, might retard the buds somewhat, but not much. The buds being exposed to the influence of spring will be apt to start without much reference to the condition of the soil.

Vines in an Orchard.—J. B. asks if we would advise planting grape vines or raspberries in a peach orchard where the trees are 15 feet apart each way. —Decidedly not. Annual hoed crops may be grown and the breadth given to them diminished until the third year, after which they should be stopped and all the ground given up to the trees. Grape vines will be well established about the time the orchard comes into full bearing, and each will be in the way of the other.

Rabbits, Mice and Trees.—Dr. M'Cannell, Adams Co. O., finds that a mixture of asafœtida and soap, painted upon the bark of trees, will preserve them from the attacks of rabbits and mice. Will the Doctor kindly give us the prescription, as he has forgotten to say what proportions he has found best, or the form in which the asafœtida is used.

Swindling the Nurserymen.—A well-known dealer writes: "Last spring, a man in Ohio issued circulars stating he wished to purchase a certain amount of small fruit stock, as he was going into the berry business. He sent these circulars through the country, asking nurserymen to bid for filling the proposals, reserving to himself the right to accept all or any part of the bid. In time an answer came, stating, 'Your proposal has been accepted for a certain amount of stock, to be forwarded to S——, and to draw on the First National Bank of that place for the amount of bill.' All that any one got was a protest from the bank, with costs. I understand he is doing the same this fall, only in other places." We hope that our friend did not purchase his experience at too high a price. Hereafter, he will ask for references of those who order and are unknown to him.

Smoking Plants.—"F. C. S.," Boston. The length of time a plant should be smoked will depend upon the density of the smoke. In half an hour the lice, if not dead, will generally be so stupefied that they will loose their hold, and may be shaken off.

Dwarf June Berry.—Dr. M'C. This variety came from the West, and we look for information concerning it from that quarter. We, as well as others, have the plant on trial, but it has not yet fruited.

Watering and Manure Water.—"R. J. H." writes: "How is a person who has never had experience to teach him to know what amount a moderate quantity of water is, or what is a moderate quantity of manure water?"—The want of "experience" is just our correspondent's trouble, and which he can, with a little patience, overcome. Dust and mud are the two extreme conditions to which the earth in flower pots can be brought, and both, with rare exceptions, must be avoided. A rapidly growing plant will require an amount of water that would ruin a slow growing or a dormant one. Now, no written rules will instruct one just how much water to give a particular plant. The best way is to experiment. More plants are killed by overwatering than by drying up, and it is better for the health of the plant that the soil should get dryish occasionally. As to manure water, use liquid cow manure, diluted so that the water is slightly colored; apply this once a week, and if the plants seem to do well under it, use it twice a week, making it a little stronger if the plants appear to demand it....Fuchsias, except a few winter-flowering ones, are better stored in the cellar for the winter, as they are generally poor parlor plants.

An Erratic Grass.—Daniel Noble, Shairano Co., Wis., sends specimens of Timothy in which the *palets*, or envelopes of the flower, are developed as small leaves about half an inch long. We have seen the same condition in this grass several times before.

The New York Fruit-Growers' Club.—We were informed by an officer of this Club that it had become mortified at its inability to pay its premiums, and had committed *hari-kari*, and made an announcement to that effect. It seems that it was only cut in two, and like a polyp, its separated parts are struggling into individual existence. Nine men and one woman met at the rooms of the American Institute and made an attempt to organize. As all could not be chairmen at once, this portion—which represents the head, as it contains whatever of brains there was in the concern—adjourned to a more favorable season. The tail end has since shown signs of vitality and has called itself—of all things—the "Horticultural Society of New York." Well, there is something in a *name*. We do not observe that any one ever suspected of horticultural knowledge is identified with the movement—but the name looks well in print.

The "Mexican Everbearing" Strawberry.—The Michigan Farmer quoted us triumphantly as endorsing the claims of this variety to be new. We gave its statement the positive denial it deserved, whereupon it follows with a column, the purport of which is, that our opinion is not worth anything either way. The Farmer mistakes greatly if it thinks to draw us into a controversy with it. Abuse will not help the strawberry nor will it hurt us; and the editor of that paper may feel just as badly as he chooses. The question which interests the public is this—Is the "Mexican Everbearing" Strawberry worth growing? To which we answer, if the old Red Alpine is worth growing, that is, for we are unable to see the difference. The inability to see the difference between the "Mexican" and the Red Alpine may be stupidity or it may be an Eastern prejudice, for both these charges are made against those who think them as alike as two white beans. For ourselves, we consider the Alpine and the "Mexican," be they alike or different, as not worth growing. The fruit is small, soft, pasty, and without the flavor we are accustomed to associate with the strawberry, and have no doubt that nine out of ten who should plant either would be sadly disappointed. There are a few who like a fruit of this kind and they will find it in the so-called "Mexican." The names of some persons whom we highly esteem are quoted as stating that the "Mexican" is a distinct variety. Neither of these have ever grown the two sorts together. We know them well enough to be sure, that should they find upon trial that the "Mexican" is not a new variety, they will frankly say so, and should our experience with the two side by side, show the plants in any respect different, we shall admit that our present opinion is not well founded. Mr. B. Hathaway of Little Prairie Ronde, Mich., publishes in the Michigan Farmer and elsewhere the most positive testimony as to the identity of the two sorts. Mr. H. grew the "Mexican" and an Alpine, which he had cultivated for many years, side by side. He says, "And carefully comparing in size, form, flavor or fruits, and in habit of productiveness, there has been no appreciable difference. Had I not set them myself I could not tell them apart." The testimony of so well-known a fruit grower as Mr. Hathaway is worthy the consideration of the Michigan Farmer. The Farmer says: "Why did not the editors of the *Agriculturist* and the *Rural New Yorker* in the convention respond to the remarks of Mr. Meehan? or why not call the attention of Mr. Elliott or Dr. Warder to the identity of the two varieties? Simply because they dare not!" If that conundrum was not answered by its propounder we should say that the only reason we did not do these things—we can't answer for our friend Bragdon—was the perhaps insufficient one, that at the time we were some 30 miles away....Since the foregoing was in type, we notice that Mr. B. Hathaway has written to the Country Gentleman an article giving his experience with the "Mexican." He states that he has the "Mexican" and the Alpine planted together, and that he will give any one $500 who will at any time of the year pick out the plants of the "Mexican." Here is a chance for the Michigan Farmer.

Norway Spruce Hedge.—"G. H. F.," Montgomery Co., Pa. The Norway Spruce can be kept at any desired size by clipping. Cut a young hedge in October, to secure a strong growth, and when the hedge is once well established, prune the young growth in June, and trim again in October to bring it to the desired shape.

Bark Louse.—A. P. Lark, Millersburgh, Pa., sends a twig covered more thickly than we ever before saw with Harris' Bark Louse. Mr. L. says that the whole tree is covered with them, from root to top, and that they even were on the fruit. On this point we have a letter from D. A. Norris, Greenville, Conn.: "In the spring of 1868 I purchased a house and lot; there were about a dozen pear trees on the lot, and one of them (the Duchesse) was badly infested with scale lice. The former owner of the place had been trying for a number of years to get rid of the pests, but could not succeed, and I decided to cut the tree down. I had some painters at work on my house about the first of June, and I thought I would try an experiment; so I took a paint keg and brush and painted that tree from stem to stern (as a sailor would say), covering the leaf buds and everything else with a thick coat of white paint, lead and oil, such as they were painting the outside of my house with. All of my neighbors (and some of them knew very much more about trees than I did) said I had killed the tree. I told them, 'better dead than lousy.' The tree leaved out very well, though some buds couldn't break the crust of paint. It made a good growth of wood, some of the shoots growing from twenty-four to thirty inches. What is better than that, I could find no lice on them. I thought I would say nothing to you about it until I had tried it another season. I have watched it closely this season, and can find nothing on it that has any resemblance to a louse, and I am well satisfied that I have exterminated them. Some of my neighbors are troubled with them, and they say they 'shall try the paint next season.' The tree which I painted was seven or eight years old, and has never fruited until this year, when it bore one pear. I am in hopes to get a good crop from it next year."—A tree in the condition of Mr. Lark's is fit for no other use than some such heroic experiment as here detailed. If it is killed under the treatment, never mind, as it should be cut down if it cannot be cured. It is likely that linseed oil, without the paint, would do as well. We have no doubt that soap, made thin enough to work, would have answered as well as the paint, if applied early in June, just before the insects hatch. The experiment is interesting as showing how much abuse a tree will stand.

New Vegetables are described by J. J. H. Gregory and others in the Horticultural Annual for 1870.

Seed Peas.—"J. M. C., Ohio. Our seed dealers have their peas raised in Canada, and in parts of New England, where the "bug" is not troublesome. It is stated that peas sown in June will be so late that they will escape the attacks of the beetle.

AMERICAN AGRICULTURIST

FOR THE

Farm, Garden, and Household.

"AGRICULTURE IS THE MOST HEALTHFUL, MOST USEFUL, AND MOST NOBLE EMPLOYMENT OF MAN."—WASHINGTON.

VOLUME XXIX.—No. 2. NEW YORK, FEBRUARY, 1870. NEW SERIES—No. 277.

GROUP OF ORNAMENTAL PHEASANTS.—FROM LIFE, BY EDWIN FORBES.—*Engraved for the American Agriculturist.*

We have often wondered that this strikingly beautiful class of poultry is not more frequently bred by our amateur poultry fanciers. The pheasants breed readily in confinement, and are reported quite hardy. The common pheasant of Europe is said to have been originally (hundreds of years ago) brought from Asia. It is a very beautiful bird, though not so brilliant, from contrasts of color, as the golden and silver species which come from China. The Common Pheasant is known in several variations, one of which, the Ring-necked Pheasant, is seen with his mate at the right of the engraving. The cock is a pugnacious fellow, armed with sharp spurs, and weighs about 3¼ pounds. His plumage glows with all the colors of the rainbow. Above this pair is a pair of Albinos of the same species. The Silver Pheasant, seen in the foreground, with his hen, of more sober colors, behind him, is gay in his variagated plumage of black and white. This bird is a good deal larger than the common one; the markings are well shown in the engraving, and the head of the cock is adorned by remarkable crimson, velvety carunculations. The Golden Pheasant, seen above the last, is much smaller than the others. The general color of the plumage of the cock is crimson; the crest, which is erectile, is golden yellow, contrasting with the orange colored cape, barred with black. Other colors in the plumage are chestnut brown, black, blue, and green, with glossy iridescences. Pheasants do not sit well in confinement, but drop their eggs about, which are usually set under bantams or Silky fowls. The young are delicate, and require careful attention and a variety of food. No doubt some of the varieties of pheasants might be introduced, and if set at liberty in retired spots, below latitude 41°, become acclimated.

AGRICULTURAL IMPLEMENTS

Garden and Field Rollers,

of all sizes, from $9 to $500.

R. H. ALLEN & CO.,
P. O. Box 376, New York.
189 & 191 Water St.

Norway Oat Premiums.

In accordance with arrangements now completed, in connection with the sale of this seed another year, we are enabled to announce the following grand Premiums for the best crop next year. For the best acre will be awarded a cash premium of

Five Hundred Dollars.

For the best 10 acres, a cash premium of

One Thousand Dollars.

For the best 50 acres, a cash premium of

Twenty-five Hundred Dollars.

Those competing must be prepared to give full particulars of mode of culture, &c., with affidavits, if required. These premiums will be awarded by a committee of impartial and widely-known gentlemen. All who buy seed of us this year, can compete, and some farmer or farmer's son, will certainly get them. Order at once. Price: peck, $2.50; half bushel, $4; per bushel, $7.50, by the standard of 32 lbs. Remit by Post-office Order or Draft.

D. W. RAMSDELL & CO.,
218 Pearl St., New York.
Or, 171 Lake St., Chicago, Ill.

Send for our Large Illustrated Paper. FREE.

SEED OATS.

Ramsdell's "Norway."—Seed obtained from D. W. Ramsdell, Vt., in the spring of 1868. Warranted genuine. Four quarts, by mail, post-paid, $1; one bushel, by Express, $4; Ten bushels, or more, $3 per bushel. "Surprise," and "White Swedish," each, four quarts by mail, $1; one bushel, $3. *New Brunswick*, one bushel, $2.50; ten bushels or more, $2 per bushel. Circulars free. Address

S. B. FANNING, Jamesport, Long Island, N. Y.

CROSBY SWEET CORN.

Sample of illustrations in **our new Seed Catalogue for 1870**, which will be mailed **free** to all applicants on receipt of stamp.

Our Wholesale List

is now ready for the

Trade Only.

Address

R. H. ALLEN & CO.,
P. O. Box 376,
New York.

SUGAR TROUGH GOURD, 25 cts. per package. They grow to average from one to two bucketfuls, and I have one that I exhibited at the Agriculturist office that holds over eleven (11) galls. Send for Catalogue. Address WALDO F. BROWN, Box 75, Oxford, Butler Co., Ohio.

New Tomatoes.

ALGER.—This new variety has the same potato-like foliage as Keyes', but the fruit is larger, of good market size, early and very productive, 15 cts. per package.

GEN. GRANT.—Remarkably solid, round, flat in shape, handsome, and of excellent quality. Received the first premium for the two past years at the Annual Exhibition of the Mass. Hort. Society, 15 cents per package.

CRIMSON CLUSTER.—Early, grows in large clusters bearing handsome fruit, oftentimes elegantly spotted with gold, 15 cents per package.

MAMMOTH CLUSTER.—Very large, round, crimson, 15 cents per package.

BOSTON MARKET.—The result of most careful selection by the Boston market-men for a series of years; large, flat, round, solid; enormously productive, 15 cts. per package.

EARLY ORANGEFIELD.—An English sort, yields its fruit in large clusters. This Tomato is of a peculiarly rich and sweet flavor, and excellent as fruit for dessert, 15 cts. per package.

Also the following excellent varieties at 10 cts. per package. Early York, Dwarf Scotch, Keyes' Early Prolific, Yellow Fig, Maupay's Superior, Cherry, Large Yellow, Feejee, Cook's Favorite, Lester's Perfected, Large Smooth Red, Tomato De Laye, Tilden, New Mexican, Strawberry, or Ground Cherry.

All of the above are of my own raising, each grown isolated, scattered over three square miles of territory. Catalogues free.

JAMES J. H. GREGORY, Marblehead, Mass.

---

300 BARRELS of Early Rose Potatoes for sale; all *must be sold* before the middle of March. Price—large size, $4, and second size, $3 per bbl. CHAS. COLLINS, Moorestown, N. J.

THE STORY OF THE BIG CABBAGE that a sheep eat into and did not get out until spring, I doubt, but I have a Gourd that holds over 11 gallons. Read advertisement above.

SECURE A HOME.—The best Fruit and Garden lands for sale, in a mild and healthful climate, 30 miles from Philadelphia. Good Markets and Society. Pure soft water. Price, 30 dollars per acre, payable ½ cash, balance in 10 years. A rare opportunity. Thousands are settling. Address R. J. BYRNES, Hammonton, N. J.

"Cabbages, and How to Grow Them."

This is the title of a very thorough treatise that I have written, giving all the minute details so invaluable to the beginner. It treats on selecting and preparing the soil, preparing and applying the manure, the best varieties to raise, (illustrated by fine engravings), the hoeing, cultivating, and the protection of the plants from their insect enemies, how to market, and how to keep over winter, and on other subjects.—Also, my treatise on "ONIONS, WHAT KINDS TO RAISE, AND HOW TO RAISE THEM," and "SQUASHES; HOW TO GROW THEM." Either of these will be sent postpaid, to any address, for 30 cts. I intend that each of these Treatises shall be the most thorough of their kind published, and I believe the public will find them to be decidedly so.

Catalogues of my Garden and Flower Seed sent gratis to any address.

JAMES J. H. GREGORY, Marblehead, Mass.

AMATEUR CULTIVATOR'S GUIDE TO THE Flower & Kitchen Garden

Twenty-fourth Edition of this popular work, which has met with so much favor in the past, is now ready. It has been re-written and improved, printed with new type, and on fine paper; illustrated with a beautiful Lithograph, and many other fine engravings from nature. It contains full description and the culture of over 1,500 leading varieties of Flowers and Vegetables; also descriptive list of the novelties of the present season; to which is added a collection of 200 choice French Hybrid Gladiolus. This work, we feel confident, will compare favorably with any similar one.

From LEVI BARTLETT, Warner, N. H.

"I have received a copy of your superbly gotten-up Amateur Cultivator's Guide. I think it far ahead of anything of the kind ever before issued from the American press."

Sent to any address upon receipt of 25 cents for paper cover, and 50 cents for tastefully bound in cloth.

WASHBURN & CO., Boston, Mass.

---

200 Bbls. Early Rose Potatoes, Clinton, Boyden's No. 30, and other varieties of Strawberries; Clarke, and Philadelphia Raspberries. For price list address JOHN CRANE, Union, Union Co., N. J.

SEEDS! SEEDS! SEEDS!

GENERAL GRANT TOMATO.—This new variety is, I think, the *best* Tomato I have ever grown. Fruit, remarkably solid, large, smooth, handsome, and of excellent quality; 15 cents per packet.

LYMAN'S MAMMOTH CLUSTER.—Grows in large clusters like *Orangefield*, but twice the size. The color is a beautiful pinkish-red; packet 15 cents.

CRIMSON CLUSTER.—Early; grows in large clusters; crimson, elegantly striped and spotted with gold; quality excellent; 15 cents per packet. The three varieties will be sent by mail, post-paid, for 30 cents, and my NEW ILLUSTRATED CATALOGUE for 1870, which contains descriptions of 14 of the *best* Tomatoes, and all other vegetables for the garden. Also a select list of 150 choice ANNUALS for the Flower Garden. Be sure and send for a copy. Address

J. F. MENDENHALL, Carmel, Ind.

For the Flower and Vegetable Garden.

Grass Seeds, Field Seeds, Plants, Roses, Dahlias, Verbenas, Gladiolus, Grape-vines, Small Fruits, Asparagus Roots, Early Potatoes, Onion Sets, Books, Implements, &c.

Dreer's Garden Calendar for 1870; containing full descriptive lists of the above (144 pages), *beautifully illustrated with engravings*, will be mailed on receipt of a postage stamp. Address HENRY A. DREER,

714 Chestnut Street, Philadelphia, Pa.

$5.00 Select List, No. 49.

3 varieties of Strawberries, 2 of Raspberries, 1 of Blackberries, 2 of Grape-vines, 2 of Currants, 1 of Gooseberries, all for $5. Send for select lists of Small Fruits, with stamp, to LOUIS RITZ, Plainville, Hamilton Co., Ohio.

The Sanford Corn a Success.

The experience of the past season fully confirms all that has been said in its favor. Testimonials from nearly every State endorse it as *the best field corn*. In many instances being planted in the same field, and having in all respects the same chance, it has ripened from two to three weeks earlier, and yielded double the amount of other varieties. It has taken highest premium in this (Suffolk) Co., for four successive years. Every Farmer should send for a descriptive circular giving History and Testimonials from reliable Farmers throughout the country. Prices—One Quart, by mail, post-paid, 75 cts.; Two Quarts, $1.25; One Peck, by Express, $2; One Bushel, $5. Address

S. B. FANNING, Jamesport, Long Island, N. Y.

BEAUTIFUL FLOWER SEEDS.

Send for our **ANNUAL DESCRIPTIVE CATALOGUE OF FLOWER SEEDS** for 1870.

J. M. THORBURN & CO.,
15 John St., New York.

EARLY ROSE POTATOES.—A large stock of *genuine*. Also, Bresee's Prolific and other seed potatoes. 2d size very low. Southern planters should order at once, so as to have them shipped the first good weather. I deliver them on board Railroads or Steamboats in Philadelphia, free of extra charge, thus saving a large freight bill from other places. For points East or North, I will deliver at R. R. in Moorestown. Send for prices by pound, barrel, or 100 barrels. THOS. C. ANDREWS, Moorestown, N. J.

Buist's Large Late Flat Dutch Cabbage.

No variety will produce more weight of solid heads to the acre, and as a reliable and profitable variety for Market Gardeners, it is unequaled. WE WARRANT every Seed to grow under favorable circumstances, every plant to produce a solid head. Price $5.00 per pound, and 40 cents per ounce. Address ROBERT BUIST, JR.,

SEED GROWER, PHILADELPHIA, PA.

The Charter Oak Prize Tomato,

The Earliest, Smoothest, Most Solid, Prolific and Best

FLAVORED TOMATO for Private or Market Gardens ever before introduced. Send for Descriptive illustrated circular.

WETHERSFIELD (CONN.)
LARGE RED ONION SEED,
1,000 POUNDS
RED, YELLOW AND WHITE,

All of the growth of 1869, and warranted in every respect. Our Descriptive Catalogue of Fresh and Genuine Wethersfield (Conn.) Garden and Agricultural Seeds is just published, and will be mailed to all applicants free of charge. This Edition contains a valuable treatise on Raising Onions as practiced by one of our most successful growers in Wethersfield. Orders from Truck Growers will receive our careful attention, and satisfaction guaranteed. Address for Catalogue,

R. D. HAWLEY,

Seed and Implement Warehouse,

Established in 1842. **Hartford, Conn.**

AGRICULTURAL IMPLEMENTS

GEDDES HARROW,

Of 6 sizes, from $12 to $30.

R. H. ALLEN & CO.,
P. O. Box 376, New York.
189 & 191 Water St.

CURTIS & COBB'S

New Illustrated Seed Catalogue, and Flower & Kitchen Garden Directory.

The SEVENTEENTH EDITION of our popular and comprehensive Catalogue is now ready, and will be mailed to all applicants enclosing us Twenty-five cents. Regular customers supplied without charge.

Also, now ready our Division Catalogues, of Flower Seeds, Vegetable Seeds, Small Fruits, and Gladiolus Bulbs separate—either of these last named, free of charge, on application. No pains or expense have been spared in preparing these Catalogues, and we invite our friends and the public generally to make application for the same.

Address **CURTIS & COBB,**
348 Washington St., Boston, Mass.

American Sage Seed.

True Broad-leaved Sage Seed, grown by the famous Danvers growers who raise Sage by the acre. *Warranted growth of 1869.* Price—10 cts. per package; 35 cts. per oz.; $1 per ¼ lb.; $3.50 per lb. *Sent by mail, post-paid, with full directions for cultivation. Catalogues free.*

JAMES J. H. GREGORY, Marblehead, Mass.

GARDEN SEED.

Fine Gladiolus and Spring Bulbs. "Much the largest packets for the price I have seen for some years," wrote Mr. Geo. Faulkner, of Flemingsburg, Ky. "Good as the best, and cheaper than the cheapest." (T. B. Shepherd, Mannsville, N. Y.) Catalogues, "*illustrated*"—*with low prices—free.* Address WARDWELL & CO.,

West Dresden, Yates Co., N. Y.

NORWAY OATS.—We are now prepared to place seed with a limited number of responsible farmers, on contract for the crops, next fall, to supply our European trade. Address D. W. RAMSDELL & CO., 218 Pearl St., New York, or 171 Lake St., Chicago, Ill.

SEEDS and FERTILIZERS.—Best Garden Seeds sent by mail or express. Pure Fertilizers delivered in New York. Send for Circulars, address EASTERN SEED & FERTILIZER CO., Hartford, Conn.

AMERICAN AGRICULTURIST.

NEW YORK, FEBRUARY, 1870.

We approach the time, with the passage of this month, when farmers must have their plans made, and know pretty well just what they purpose, and how they will carry it out. The lengthening days make farmers impatient for field work. It is high time that good farm hands were engaged for the summer. The best make the earliest engagements. As there has been quite a dearth of employment for laboring men, both in town and country, we anticipate engagements at lower rates than ruled last year. Money is "tight," in commercial phrase, and farmers have not been so well paid for their products as in the past few years. This should not influence us to decrease our operations; to extend them with discretion, and to employ still more labor, would be better policy. The prosperity of the country depends directly upon large, good crops, of our staple productions. Labor well employed, and manure well applied, will surely pay in the long run. The price of produce is influenced greatly by the European markets, and, of course, by the harvests of the rest of the world. We are at peace, and the irregularities consequent upon a state of war have nearly passed away. Our population is rapidly increasing, land is growing in value, and the prosperity of our country and of the farming interest was never more certain.

Hints About Work.

Orchard and Nursery.

Planting.—The time for setting trees will be governed by the locality. In the Southern States, planting will be done this month, but at the North nothing is gained by planting too early, even if the ground happens to be open for awhile. The cold, drying winds are very injurious to trees that have not yet recovered the use of their roots.

Varieties.—In planting for family use, the selection should comprise varieties from the earliest to the latest. In orchards, for marketing, there should be but few varieties, and those of popular market kinds, known to succeed in the neighborhood. In making a selection, local experience is the only safe guide. Do not buy from the extravagantly colored pictures shown by traveling agents. If unfamiliar with the sorts found to do best, make it a business to go about among those who grow fruit, and learn.

Young Trees, that are vigorous and healthy, are to be preferred to larger ones that have become checked in their growth by being crowded in nursery rows. Some planters prefer trees only one year old from the bud or graft. If trees are frozen in transportation, let them thaw very gradually.

Old Trees, that have become established, may be treated, during a damp, foggy time, to a wash of strong soft-soap, thinned with water enough to work, or a lye of potash. This destroys moss, loosens old scales, and leaves the bark smooth.

Injured Trees, such as have been, broken by storms or otherwise, should have the ragged wound pared smooth. Those slightly injured by mice and rabbits will recover if earth be drawn up to cover the wound. If the bark is completely gone, the only way to save the tree is to connect the bark below and above the wound by cions, inserted in the bark so as to span over the injured part.

Grafting should be done only when the swelling of the buds shows that vegetation is starting. Cions may be cut and preserved in moss or sawdust.

Pruning is to be done before growth begins. In pruning neglected trees, the object should be to get an open and well-balanced head. Take care that a bad wound is not made by the falling of the limb when partly sawed off. Pare wounds smooth, and cover them with melted grafting wax or paint, which may be tinted, to be less conspicuous.

Insects.—Those which need particular attention at this time are the Tent-caterpillar and the Canker-worm. The first named is still to be attacked in the eggs, which will be found attached in bands to the twigs, near their ends. The Canker worm issues from the ground in spring, and often in warm days this month. The females are wingless, and can only ascend the trees to deposit their eggs by climbing. Some obstacle must be presented to their ascent. The simplest is a band of stout paper tied around the tree, to which tar is applied. This must be looked to every few days, and be renewed if the surface has become hard. There are a great many contrivances for surrounding trees with a gutter or barrier of oil or other liquid, impassable to insects, some of which are given in back volumes. The success of all these depends upon frequent inspection and care. See back volumes for details.

Manure may be spread upon the surface of the orchard. It should never be put in a heap around the trunks, where it does no good, but harm.

Fruit Garden.

Trees, and there should be only dwarf ones in the fruit garden proper, will need pruning, washing with soap or lye, protection against insects, etc., and such other care as has already been suggested for trees in the orchard.

Grape-Vines may be pruned when not frozen. It frequently happens that, in the pressure of fall work, the vines are left until now. If the coldest of the winter is over, go over those trimmed last fall, and remove the extra buds that were left as a precaution against the severity of the winter.

Blackberries and Raspberries should be set as early as the condition of the soil will allow. The underground shoots, which will form the canes of next season, start very early, and are likely to be injured if the setting is left until late.

Strawberries may be planted in those localities where the frost is out of the ground.

Kitchen Garden.

Manure is the main question, and it will be needed in large quantities, not only to apply to the soil, but for hot-beds. The heaps should be so large that the generated heat will not allow them to freeze. When the heaps become heated, which is shown by the issuing of steam, or may be ascertained by thrusting a stake into them, they should be re-built; water if the interior is dry.

Cold-Frames.—The plants will now bear full exposure during sunny days, but they must be covered in the afternoon, even if the nights are mild, for fear of a sudden change and snow storms.

Hot-beds should be started six weeks in advance of the time for planting in the open air; hence they are now needed only in the warmer States, where Tomatoes, Egg Plants, etc., may be sown. Preparations should be made. The common size of sash is 3 x 6 feet, glazed with 8 x 10 glass.

Straw Mats will also be needed to cover the sash, to protect plants from frost or too much sun. The mats should be 7 feet long, and 4½ feet wide, so that two will cover three sashes. We have, in former volumes, given directions for making them. One of the simplest is to stretch five strands of strong twine or "marlin," of the proper length, to form the mat, then lay on straw, with the but ends towards the edges of the mats, and about an inch in thickness; then put five other strings over the straw, and directly above the first ones, and take a large needle and twine and sew through the straw, taking care that the loop of the stitch catches both the upper and lower strings. The sewing should be done at each pair of strings.

Brush, and Poles for peas and beans. Cut while there is leisure, and before the leaves start.

Potatoes.—A few for very early planting may be kept in a warm room to start the sprouts, and then be planted in a warm place in the garden. A few days may be gained, as they may with

Peas, by a little coaxing. Plant a row or two in a sheltered place, laying a board over the rows at night, and on very cold days. When the peas are up, raise the boards by means of bricks or something else that will keep them clear of the plants. Two boards, nailed together like an eaves-trough, are sometimes used for a cover at night. They may be placed, in the day, near the plants, to break off the wind and reflect the sun's heat.

Parsnips and Salsify.—Dig as soon as the ground is thawed, and before the plants start.

Rhubarb.—Roots may be forced by placing them in earth at the bottom of a barrel in a warm room, or, where there are cold-frames or green-houses, they may be forwarded easily.

Seeds.—Test their vitality as directed last month. Trust to none that are of doubtful identity.

Flower Garden and Lawn.

Plans for new improvements should be completed before the working season begins. We give, on page 21, some suggestions about laying out flower beds, and shall probably have something to say on the subject next month. Whether the place be large or small, a considerable extent of unbroken turf should be secured, unless one's taste for flowers is so strong as to require that all the available land be appropriated to them.

Shrubs should be taken into account in the plan. Many of them are beautiful in foliage all the season, and some of them produce exquisite flowers. Prune established ones if they have become overcrowded and out of shape. Let the trimming conform to the natural habit of the plant, and do not try to make one with naturally curving branches grow upright. Those which flower only on the new wood, like the Rose of Sharon, need to be cut back, to induce a strong new growth, while shrubs upon which the buds for next year's flowering, are ready formed, as the Lilac, need only to be thinned.

Ornamental Trees, if they need pruning, should be treated with the same care as fruit trees. Sometimes it will be necessary to remove lower limbs which are in the way, but, as a general thing, it is better to leave the tree to take its natural form.

Half Hardy Plants, which have been stored for winter in pits or in cellars, will need looking to, to guard against their being started into growth by the warmth of the sun. Give air, and keep as cool as possible, without severe freezing. Plants in cellars, if too dry, will need a little water.

Dahlias, Cannas, and other roots, stored for the winter, should be examined occasionally. They are more apt to suffer from dampness than dryness, and should be removed to a drier place if there are any signs of decay.

Wood-work, such as trellises, garden seats, rustic ornaments, etc., will need painting or oiling.

At the South, where the climate permits it, trees and shrubs may be planted, perennials lifted and divided, lawns made, and other spring work executed.

Glass-Covered Run for Early Chickens.

It is a great object with breeders of choice fowls to have a few broods, at least, of very early chickens. If these do well, they will probably be the prize-winners at the fall shows, or they will be ready for use, if for market, as broilers, in May and June, when prices are the highest, or the pullets will be laying from September, or earlier, to Christmas, when eggs are scarce and high. To raise early broods with success, the chickens must have no pullbacks, but a steady healthy growth from first to last. It is easy enough to get the eggs hatched, and to rear the chickens until they are a week old, but then commences a series of trials which few early broods live through in common hands. In June, the old hen, left uncooped, will take care of her brood almost without care from us, and with little feed, and the chicks grow and thrive, but in February and March the case is different. At hatching, the greater part of the yolk of the egg fills the stomach of the chicken, and is gradually absorbed as food, so that the first day it needs no food, the next day but little, and it is only on the third or fourth day that the little things experience real hunger when deprived of food. This period comes sooner in cold than in warm weather, and quicker, too, if they are not properly brooded. It is absolutely necessary to keep chickens warm and dry. They should have clean coops, and a clean, dry, sunny run. These requisites are very conveniently provided by using hot-bed sashes in the manner shown in the accompanying engraving. Three large

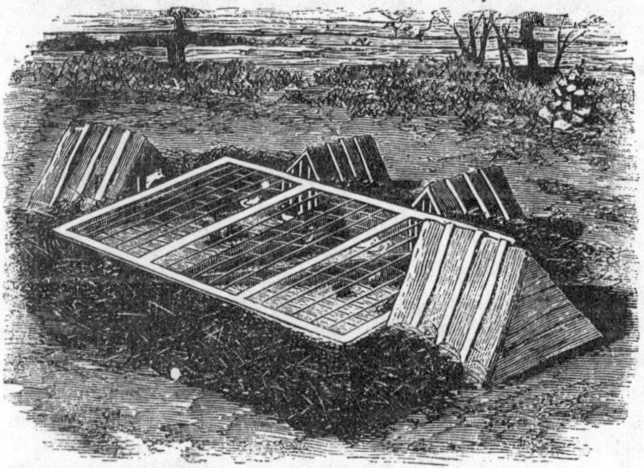

Fig. 1.—CHICKEN-RUN AND COOP.

hot-bed sashes cover a space of about 72 square feet, abundantly sufficient for four clutches. The frame on which the sashes are laid is made of 2×3-inch stuff, and supported by stakes driven into the ground at the corners and such other points as may be needed to give firmness, and the frame is nailed through to these posts. The slant given to the sashes need not be more than one foot in the six feet, the rear posts being therefore 18 or 20 inches high, and the front ones 6 or 8. Boards are placed between the coops and at the ends, and earth or litter is banked up and packed firmly against them. If the coops open at the back, it will be found convenient for cleaning them out, and removing the hen if necessary. Ventilation takes place through the openings in the peaks of the coops, which, however, should not be so large that chickens can fly up and get out. The chickens are fed and watered by slipping the sashes up or down, and it might also be convenient to have a pane of glass arranged so as to be removed at each end. A mat should be provided, to give shade in case the place gets too hot, and the sashes may be lifted a little at the upper end, to give freer ventilation. Thus the temperature may be perfectly and easily controlled. We indicate in the engraving two partitions; one crosses the middle of the space, dividing it

Fig. 2.—DIAGRAM OF SHELTER.

into two equal, square parts; the other divides one of the squares in two equal, triangular parts. This gives the chickens of each coop 18 square feet of space. A warm, shady shelter is easily made by laying a board against the front posts, as shown in the accompanying diagram (fig. 2). This, if covered with warm manure, packed above it, and covered with a layer of earth, will warm through and give a very pleasant place for the chickens to run under in cool weather. When the weather becomes warmer the manure will have lost its heat.

Salt-Muck.—Will it Pay to Dig it?

For shore farmers, who have creeks penetrating their meadows, we have no doubt that salt-muck is one of the cheapest sources of manure. In an analysis made some ten years since by Prof. Johnson of Yale College, it was shown to contain 5.41 per cent. of potential ammonia, standing at the head of all the 33 specimens of peat and muck analyzed for the State Agricultural Society. This sample was taken from a ditch in Stonington, where the tide-water flowed daily, and was probably a fair specimen of what is now found in inexhaustible quantities in the creeks all along our coast. The analysis was made on account of the very noticeable results, which followed its use, both upon grass and corn, in that town. This deposit is made up very largely of decayed marine plants, and the silt brought into the sea from fresh-water streams. River deposits and marine plants are well-known fertilizers. If the kelp and rock weed are so valuable in the fresh state they ought to be still more so when they have been rotted down in the water, and their bulk so much reduced. J. J. Day used it freely upon his meadow at Stonington, and the result was not only remarked upon the corn crop, to which it was applied, but upon the subsequent grain and grass crops. The mud was raised by hand into a scow, and from thence carted upon the land. J. D. Fish, of the same town, has improved upon the method of raising it. He uses a steam derrick rigged like a common mud digger. The mud is dropped from the bucket, immediately into the cart, and is driven off and dumped upon the meadow. After lying a few weeks to dry, the heaps are spread with a shovel. We see no reason why the common mud digger, used to deepen channels, could not be made available for this purpose. It could readily be taken into many of our creeks and the mud be dropped directly into carts, backed a little into the water. It would require of course a good many extra teams, and extra expense for help, but if it is worth only half as much as yard manure, it will pay abundantly. The whole surface dressed with this mud upon Mr. Fish's farm this autumn was a few weeks after its application distinctly marked by the greenness of the aftermath. It is too early of course to tell what the next crop will be, but there is every indication that it will be abundant, and that the muck will pay for digging.

MANURE in a bulky form is less liable to damage; it goes further; is better incorporated with the soil, and it produces more effect than that which is more concentrated.

Clod Crushers.

As a rule, harrows and rollers, used one after the other, are the best clod crushers, but there are many soils which no reasonable amount of rolling and harrowing will bring into good tilth, which a good clod crusher would put in good shape with once going over. Clod crushers for such work, however, like the best rollers, are not home-made articles. The common clod crushers, such as almost any one can make, are, nevertheless, fair substitutes for the roller in subduing land before it is sown with grain, and on some lands may even be used after grain is sown and harrowed. A clod crusher of very simple construction and which can be made in a few minutes is shown in fig. 1, which represents 4 logs, about 10 inches in diameter at the largest end, and eight feet long, fastened together by a chain. This, having quite a large "bight" to hitch to, is run through holes bored about a foot from the ends of each log. The logs are fastened apart by keys put into the links, and should be separated about half their diameters. The keys may be rings or pieces of large wire bent so as not to slip out. Convenient forms are shown in figure 2. They must be made of wire strong enough to stand a heavy strain, so that if the implement runs against a fence, tree or stump, it will not break. The engraving shows sufficient length of chain behind to attach another log if desired. This makes a very good smoothing implement. It may be drawn even,

Fig. 1.—CLOD CRUSHER OF CHAINED LOGS.

or the clevis for attaching the horses may be at one side, in which case the earth will be more or less shoved to one side to fill dead furrows, etc. In figure 3, a form of clod crusher is shown which is superior for smoothing

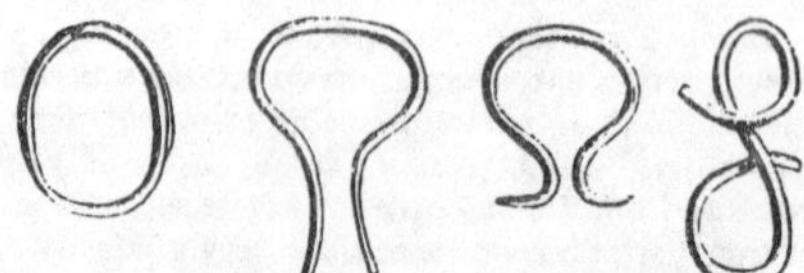

Fig. 2.—WIRE LINKS FOR CLOD CRUSHERS.

rough, irregular land. This also is made of logs about 10 inches in diameter, which are set in a frame, being morticed in, as shown in figure 4. In passing over little knolls and hummocks nearly the whole weight of the affair bears upon

Fig. 3.—CLOD CRUSHER OF LOGS IN A FRAME.

the highest points, and levels them down considerably. This clod crusher may be made either with or without teeth. If employed, they should be inserted at an angle of 45° to the mortice, and may be longer or shorter, according to the work to be done. They should, however, never be long or numerous enough to prevent the logs resting upon the ground. The operation of the teeth is to crowd strawy manure out of sight if the soil be moderately soft. This implement is drawn by a chain fastened to a clevis on the end of each side beam of the frame. The tool may be reversed by turning it over and hitching to the other end. It is not necessary to sharpen the teeth much; they will wear rounding very soon. These may be omitted altogether from the rearmost stick, if smooth work not marked by furrows be desired; but in this case the implement could not be so well inverted and used the other side up, for then the smooth log would be at the forward end. The same result is attained by attaching a light log or brush to the back end when in use.

Fig. 4.—SHOWING CONSTRUCTION OF FIG. 3.

Wet Hog-Yards a Nuisance

Where hogs are kept on a large scale, say where thirty or forty 300 or 400-pound hogs are fattened every year by one farmer, they are seldom kept in yards, but have the range of an acre or more, with a brook or water-trough, and their manure is made little account of. Farmers who feed fewer swine, and reckon the manure a very important, if not the chief source of profit, confine their hogs, and keep them working over vegetable matter of all kinds. The yards, if not continually supplied with large quantities of fresh litter or weeds, are fitly described by a correspondent of Somerset Co., New Jersey, whose letter we quote as follows:

"Hog-yards, in connection with hog-pens, as most farmers have them, are a nuisance. The hogs root up the bottom of the yard until they get a great hole, and every rain fills it with a slush, into which no one wishes to go to clean out the manure, and it is too offensive to be near the house. My neighbors have tried paving with stone and brick, but the hogs rooted up the pavements. About a year ago, I laid mine

ROOFED HOG-PEN YARD.

with oak plank, on a level with the surface of the ground, so that the water could run off when it rained. The ends of the planks run under the lower boards composing the sides of the yard, so that the hogs cannot root them up; they are not laid on timbers; it is unnecessary. We can now go into our yard and clean out the pen with no more inconvenience than we clean the plank floor of a horse stable. I have no doubt that the reason why farmers make so little manure in their hog-yards is because they are such filthy places to go into, and have no doubt the planks will last in such a situation a lifetime, they being always wet."

This is very well so far as it goes, and it is just about as far as most farmers will go who have to get the manure out of their own hog-pens. We propose a plan for those whose pig-stys are already built, and have the usual inconvenient open yards. It would be best to have the hog-yard sheds open to the south, and accessible to carts at this side. A simple roof, however, something like the hay-barrack roofs of our Dutch neighbors in "Jersey," will answer. A yard 12 or 14 feet square is large enough for half a dozen hogs, and better than if larger. Such an one may be roofed as represented in the engraving. The four corner posts of the yard are 12 feet long, or a little less, and set nearly 3 feet in the ground. There are four rafters mortised upon these posts, meeting in the middle, at a hight to give the desired pitch, and extending 3 feet over the posts, to give wide eaves. Before putting up the rafters, 3 x 4-inch plate-pieces are nailed or pinned near the tops of the posts, all around, and tied by similar pieces mortised into the middle of each side, crossing at right angles. The roof may be of boards, thatch, or shingles, and an old tin roof, taken off from some other building, may be made to turn rain for several years by an occasional coat of some roofing—pitch or paint. A yard so covered will receive rain enough usually to keep it moist; it will not become wet unless water flows into it. Manure will make better and faster than in an open yard; none will be washed away, and little or none will be lost by evaporation, as during the warmest part of the day the yard will be in shade. In dry weather it might need wetting.

Tree Labels

A tree label that will not require too much trouble to make it, and that will remain legible for a series of years, has long been a desideratum among fruit-growers. It may be that this is supplied by the simple zinc label written with a common black-lead pencil. Several gentlemen inform us that they have had labels of this kind remain legible for ten or more years, and that though the writing makes but little show when recently done, in time it becomes more distinct. We suppose that the surface of the zinc just under the writing is protected by the black-lead or plumbago of the pencil, and that while the rest of the surface is oxidized by the action of the weather this remains intact; or it may be that the carbon—the best black-lead is nearly pure carbon—unites in some way with the zinc. The only objection we see to these labels is the ease with which they may be effaced when the writing is fresh, but a few weeks' exposure fixes it. The

ZINC LABELS.

Gardeners' Monthly gives a convenient form for the label, which is shown in the engraving. The zinc is cut in the form of an elongated triangle, the point of which, when wrapped around a twig, will hold the label, and at the same time expand as the tree increases

Maple Sugar Making.

BY W. T. CHAMBERLAIN, HUDSON, O.

During the last twenty-five years many improvements have been made in the apparatus for, and methods of, making maple sugar. I give those now in use by the best sugar-makers of Northern Ohio.

THE BUCKETS are made of the best "IX" tin. They are straight, (not flaring) and are of three sizes—three buckets fitting together and forming a "nest." The nest occupies only the space of the largest bucket, thus securing convenience in handling, and economy in storing. The three sizes, too, permit the size of the bucket to be adapted to the flowing capacity of the tree. The dimensions are—Largest, circumference 34 inches; depth, 9 in.; capacity, 15 quarts. Medium, circumference $32\frac{1}{2}$ in.; depth, 9 in.; capacity, $13\frac{1}{2}$ quarts. Smallest, circumference 30 in.; depth, 9 inches, capacity, 11 quarts.

Three buckets require eight sheets of 10 x 14 inch, and two sheets of 10 x 10-inch tin. The sides of each bucket require two 10 x 14 sheets and a *piece.* The piece in the largest is $7\frac{1}{2}$ inches wide, and in the medium 6 inches. Both these pieces are made of the seventh sheet of 10 x 14 tin. The bottoms of these buckets are made of the two 10 x 10 sheets. The *piece* in the smallest size is $3\frac{1}{2}$ inches wide, and this, with the bottom, is made of the eight sheet of 10 x 14 tin. Figure 1 shows the smallest of the three buckets; in the other two the piece is much wider. Just below the wire rim an inch hole is punched, so as to hang the bucket on the "spile." The hole is in the middle of the piece above mentioned, as in fig. 1, as the seams strengthen the bucket, and prevent its bending when it hangs full of sap. Such buckets cost now about $40 per hundred. The buckets should be painted outside with yellow ochre, or other durable paint, to protect them from rust. The sap does not rust the inside. The buckets will last thirty years or more. Tin is better than wood, as it is more easily kept clean, does not sour the sap so much, and does not shrink, get leaky, and require hoop-driving every spring, when one is in haste to be tapping. It is also more easily handled, and stored. It is better than earthen-ware, which is heavy to handle, and cracks when the sap freezes.

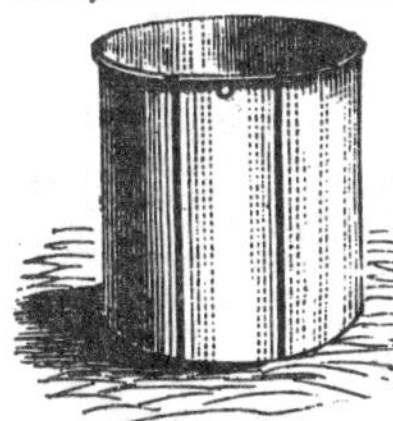

Fig. 1.—SAP BUCKET.

Fig. 2.—SPILE.

Hanging the bucket on the tree is preferable to setting it on the ground. It saves hunting for a block or stone; the bucket is more conveniently emptied, as will be seen hereafter; the wind cannot blow the sap away as it drops, nor blow the bucket away; and, what is of most importance, *the bucket can be covered.*

THE SPILES, fig. 2, are made of beech or maple, turned and bored by machinery. They are better than elder with the pith punched out, because they are not so liable to sour, and are stronger and more durable. They should be $\frac{7}{8}$ of an inch in diameter, 3 or 4 inches long, tapering for an inch of the length, and only $\frac{3}{8}$ of an inch where they enter the tree. Each should have three notches turned around it, about an inch apart, to keep the bucket from sliding off. There should be three notches, so that the bucket may be hung in one or the other of them, and be level, whatever way the tree may lean. The spiles cost about $1.25 per hundred.

The Covers.—The buckets should *always be covered.* This is the greatest single improvement yet made. It keeps out rain, snow, dirt, insects, and prevents the effects of heat and cold. The sap is not so liable to sour during the

Fig. 3.—INTERIOR OF SUGAR HOUSE, WITH ARCH AND BOILERS.—(*Scale 8 feet to 1 inch.*)
A, Arch; b, b, Boilers or Pans; c, c, Car, hooked to front pan, ready to lift it; d, d, Track; e, Store-trough; f, f, Conductors, with Self-feeding Attachment; g, g, Conductor from outer Store-trough; h, Outside Store-trough; i, i, Ventilator; k, k, Iron Supports for the wood, best made of rail-road iron, set three feet apart.

warm days, nor to freeze in cold nights. Keeping out the rain, however, is the chief thing. Sometimes in a sugar season, four or five inches of water fall, as snow or rain. This, in a "camp" of five hundred trees, would make about twenty-five barrels of water to be boiled away. Nor is the useless labor and expense of boiling this water all. The rain trickles down the trees, carrying with it coloring matter and dirt. Syrup or sugar of the first quality can never be made from sap and rain-water. The covers are made of $^3/_4$-inch lumber, 1 foot wide, and planed on one side.

The Arch and the Boilers (fig. 3).—Select a dry, level spot, near the center of the sugar orchard or "camp," and, if possible, just at the foot of a small hill, that slopes at least three feet in twenty. Dig below the frost, and lay a good foundation of stone. On this build an "arch" of hard burnt brick, laid in lime-mortar. The wall should be 12 or 16 inches thick as far back

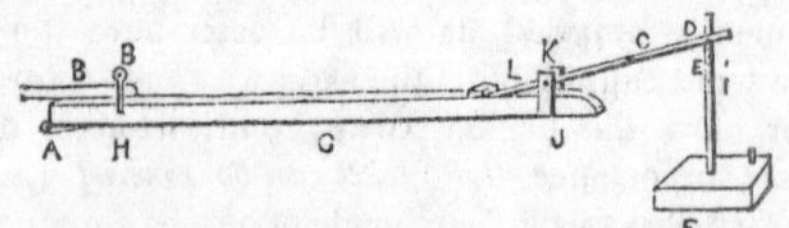

Fig. 4.—SELF-FEEDING ATTACHMENT.

as the wood reaches, beyond this then 4 inches less will do. An arch of good brick and mortar, on a good stone foundation, with walls 16 inches thick, will last fifteen years with occasional repairs about the mouth. But if the bricks are laid in mud for mortar, or if the arch is not built upon the rock, it must be rebuilt each year. The wall should be 2 feet high, and, for five hundred trees, 15 feet long. Beyond the point where the wood reaches (five feet from the mouth), the arch should be filled in with dry [illegible] five inches of the top. This throws the flame and heat all close to the back pan, as it passes it, and makes this boil about as fast as the front one. The arch has a sheet-iron door (not shown in the engraving, fig. 3), which is closed except when wood is put in.

The Pans (boilers) are made of heavy Juniata sheet-iron, and are 7 inches deep, $3^1/_2$ feet wide, and usually from 6 to 8 feet long. Two sheets are riveted together lengthwise, and the corners are cut, lapped, and riveted. The edge is strengthened by a thick band of strap-iron, and four strong wire handles are attached near the corners (see *b*, fig. 3). Directly above the pans, and parallel with them, runs the track (*d*, in fig. 3), and on this is the car (*c*, fig. 3), arranged with crank, windlass, ropes and pulleys, to lift the pan from the arch a few inches, and roll it towards the front, away from over the fire, when you wish to take off syrup. Those who would avoid the expense of the car, divide the large front pan into two small ones, the front one $3^1/_2 \times 3$ feet. Then they dip out, with a flat-edged dipper, all the syrup but a pailful, when the pan is easily lifted off by two men, and the syrup poured from one corner. These broad, shallow pans evaporate the sap fully twice as fast as the old kettles used to do, even when they were set in an arch. Kettles belong to the days of wooden plows. Some maple sugar makers use the patent sorghum evaporators instead of pans; but the ordinary pan here described answers perfectly well, and only costs one-third as much, or about $10 for an 8-foot pan.

The Floats and Faucets.—Quite a convenience is the "self-feeder" *f f*, fig. 3, which is shown upon a larger scale in fig. 4. It consists of an ordinary wooden conducting trough (*G*, fig. 4), attached to the store-trough at *A* by a hinge. The sap enters, through the faucet *B*, at the point *H*; *F* is a tin float, resting in the sap or syrup in the pan; *C* is a lever, having the fulcrum at *L*, the weight at *H*, and the power at *D*. This lever and the conductor together form a compound lever, so proportioned that when *D* rises or falls an inch, *H* rises or falls *half an inch*. *I* is a perpendicular (vertical) bar from the float *F*; at *E* are holes for the horizontal pin at *D*, joining it to the lever *C*. When the pin is in the upper hole, there cannot be more, nor much less, than an inch of sap in the boiler. At the second hole there will be about two inches, and so on. The flow of sap is regulated thus: When the pan is empty, the float rests on the bottom of the pan; the bottom of the faucet, near *H*, is half an inch from the conductor, at *H*, and the sap flows freely. As the sap in the pan rises, the float rises with it, and gradually lifts the conductor until the point *H*, presses against the bottom of the faucet and stops the flow of sap. As the sap in the pan boils away, the float sinks, and the sap flows again. The chief advantage of the self-feeder is that you can build a large fire and leave it with safety as long as necessary. The sap or syrup can never burn, and the sap can never overflow the pans. Those who have not the self-feeder, when leaving a large fire, must either fill the pans, or leave the faucets turned so that the sap will flow into each pan about as fast as it will boil away. In the former case, if they are gone too long, their sap will boil to candy, or burn. In the latter case, the sap will overflow. A self-feeder costs about $2.

The Store-troughs (*e*, in fig. 3), are made of long, wide, clear, well-seasoned 2-inch plank—pine or white-wood (poplar), rabbeted and spiked together, and bolted horizontally at top and bottom with six bolts, (two at top, and four at bottom) having nuts to loosen or tighten, as the bottom and ends swell or shrink. Three coats of paint are needed. At least five barrels of storage are required for each hundred trees. The larger trough or troughs should be outside, as it keeps the bulk of the sap coolest. From the outer trough or troughs the sap flows through a wooden or tin conductor to the inner trough. The sap is gathered in barrels, rolled up nearly horizontal skids from the stone-boat sled, and emptied through the bung-hole into the outer trough. The arrangement of skids,

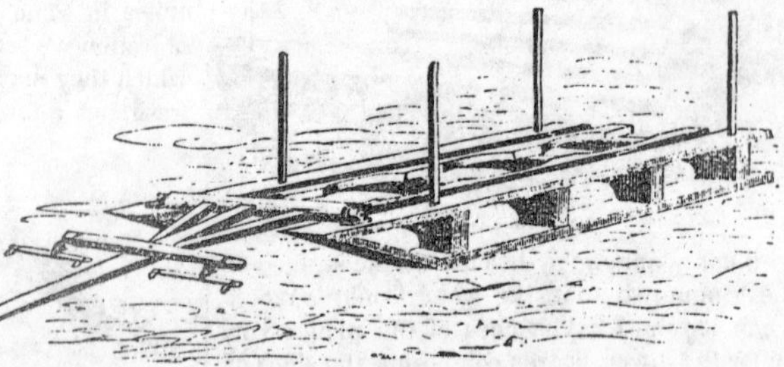

Fig. 5.—SLED FOR HAULING SAP.

troughs, conductors, etc., is best seen in the picture engraving of the camp on the next page.

The Wood-shed and Wood.—It saves half the time of boiling to have the wood prepared and housed during the dry weather of the previous September or October. With green or wet wood you may succeed, with a good 15-foot arch, in evaporating a barrel to the hour. But with good dry wood, and far less labor, you can evaporate more than two barrels to the

A MAPLE SUGAR CAMP.—*Drawn and Engraved for the American Agriculturist.*

hour. It is always best, and sometimes very important, to have the sap boiled rapidly. Once my seven hundred trees yielded seventy-five barrels in twenty-four hours. My wood was not the best; Sunday stopped our boiling, and Monday I had twenty-five barrels of sour sap, which would only make a second-class syrup.

It is best to have three forty-gallon casks for gathering sap, and two twenty-gallon casks for syrup. It is also best to have a stone-boat for drawing the casks in gathering sap. It is shown in fig. 5, and is made much like an ordinary wood-shod farm sled, only the runners and shoes are eight inches wide. It has two boards running parallel with the raves, fitted for holding the casks. There will also be needed a four-gallon funnel, large at the top, and made to fit the barrel, and not shake about; also two or three cone-shaped strainers, made of rather loose white muslin, and held open at the mouth by a hoop or strong wire, one foot in diameter; also two yards of loose muslin to tack tightly across the vats. Other small articles will be needed, such as skimmer (perforated tin, made square, so as to fit the corners of the pans), a dipper with a flat edge, large tongs and shovel, pails for gathering, a lantern, etc. When these things are prepared, you are ready for

TAPPING.—When a decided sugar-day comes, which is indicated by the appearance of a general thaw, late in February, or any time in March, it is time to be busy. Let one man load up two hundred buckets and two hundred covers at a time, in a large wagon or sled-box, and start out for distributing. He should be careful to leave the large buckets at the thriftiest trees. This is one advantage of having buckets of three sizes. Take your half-inch bit, set in a good bit-stock, select a sound side of the tree, where there are no old holes yet unhealed, and bore about two feet from the ground,

Fig. 6.—HUNG RIGHT. Fig. 7.—HUNG WRONG.

or as low as you can without having the curve of the roots interfere with the hanging of the bucket. Bore about two inches deep, drive your spile firmly, hang your bucket so that the top of the outer edge shall be on a level with the bottom of the spile hole. This requires some care, but if the spiles are notched as directed, there is no difficulty. If the tree stands plumb, use the middle notch (see fig. 2); if it leans towards you, use the inside notch; if from you, use the outside one. Figure 6 shows both buckets hung right, and fig. 7 shows both buckets hung in the wrong notch. If you have more buckets than trees, you can put two buckets to each large, thrifty tree.

See that the spile is driven firmly, put the cover on, and go to the next tree. Tapping requires judgment and care. A heedless hand will bore holes in a poor place, and hang half the buckets so that they will waste sap. A good hand will tap thirty trees to the hour where the trees stand thick. Two men, one boring, and the other driving, hanging and covering, will do more than twice as much as one. You must be wide-awake now, for if it is a good day, the trees first tapped will need gathering by the time you have tapped five hundred trees.

The Mexican Cobæa.—Variegation.

The Mexican Cobæa is a rapidly growing climber, soon covering a trellis or lattice-work with handsome foliage, and later in the season producing large bell-shaped flowers. It is interesting and attractive, and, though a quite old plant, is not very common in gardens, probably from the difficulty of starting it from the seed. It is properly a green-house perennial, but if it is started early enough, it will bloom the first season. Those who sow the seeds in the open ground, at the North, at least, will generally fail, or, if they do succeed in raising plants, they will be too late to bloom, unless in an unusually prolonged season. Those who wish to start the Cobæa from seeds, can do so in a hot-bed, or in a sunny window. The seeds, which are large and flat, should be thrust into the soil edgewise; the plants are to be put out in a rich place when frost is no longer to be feared. It being a perennial, the florists usually keep plants grown from cuttings. The readiest way to obtain it is to procure well-established plants from a green-house. The flowers, which are two inches or more long, appear late in summer, are at first green, and gradually change to a deep violet or purple, and last for a number of days. The large leafy calyx, and the long stamens, bent to one side of the flower, give it a striking air. In these days of variegated leaves, we are prepared to see any old floral friend wearing a "motley coat," and we were not altogether surprised when Mr. Peter Henderson pointed out to us a Cobæa which had taken a fashionable freak, and appeared in dappled leaves. Should the Variegated Cobæa appear as well out of doors as it does in the green-house, it will prove a very acceptable novelty, as its leaf markings are very well defined. The species generally in cultivation is *Cobæa scandens*, a native of Mexico. The genus was named in honor of one Cobo, a Spanish priest.

THE VARIEGATED MEXICAN COBÆA.

The Gigantic Water-bug.

The large Water-bug must have been unusually abundant last summer, as several have been

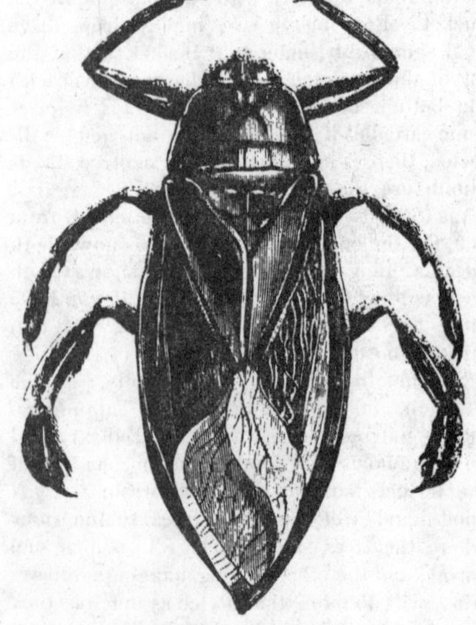

GIGANTIC WATER BUG.

sent us from different parts of the country. Mr. T. C. Grooms, Green Castle, Ind., from whose specimen the engraving was made, thus describes their appearance, in large numbers at his place: "About the middle of August, during a storm, after a rain ceased, large quantities of these bugs were found on the ground. My attention was called to them by the chickens and turkeys making a noise. A neighbor of mine who has a tan-yard, found thousands of them in the yard. The water in the vats was covered with them. By noon the next day after they came, they were all dead."——This Water-bug (*Belostoma grandis*) belongs to the sub-order Hemiptera, which includes the true bugs, plant-lice, fleas, locusts, and other disagreeable insects. There are several allied genera, which, like this, inhabit the water. Most of them swim on their backs, and their legs are admirably adapted for this kind of locomotion. They have a sharp sucker or proboscis, through which they take their food, and which, when not in use, is folded up under the chest. This insect lives on other insects, and is said to be very destructive to young fish; a friend informs us that he has been several times sharply wounded by this water-bug while wading.

The Egyptian Beet.

In beets, deep color and sweetness seem to go together, but it is difficult to associate these two qualities with extreme earliness. The Bassano is a very early variety, but it is not of as fine quality as the Blood Turnip, which is a week or ten days later. Besides earliness and good qualities for the table, a handsome shape, small top, and slender tap-root are desirable. Some think that perfection has been reached in Dewing's Early; others consider Hatch's the best of the early beets, and this year the Egyptian puts in a claim for superiority. The seed was introduced last year by a German seedsman who states that the variety came from Egypt.

It is said that the leaves are distinct in character and are very ornamental, as is the plant when the roots are set out the second year for seed. It must be recollected that beets are used in Europe in ornamental planting for the sake of the effects of color produced by their foliage. We have not seen the leaves, but the roots as raised by B. K. Bliss & Son, were of fine shape, and were within of an intensely red color. We give an engraving showing the form.

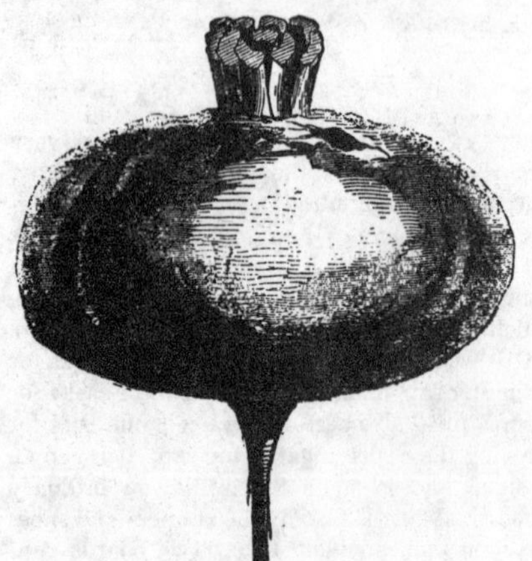

EGYPTIAN BEET

Sleeping and Eating as related to Health.

BY FAITH ROCHESTER.

How many healthy men and women can you count on your fingers?—grown up persons who have no dyspepsia, rheumatism, bowel diseases, headaches, nervousness, nor any other of the long list of ills that flesh is heir to. These "visitations of Providence" are penalties for violating the laws of health; or, they are the harvest that results from the sowing of bad seed in the way of wrong habits. For the old sinners there is not much help nor hope; but how shall we save the children from the diseased and enfeebled condition of their parents? It is time for American women to take the matter in hand. It would be ludicrous if it were less sad,—the idea of such a set of feeble and nervous creatures as American women generally are, setting themselves up as the guardians of health! But all must agree that this properly falls within the natural "sphere" of woman. It is a household matter. For, you see, we women almost have it in our power to make or ruin the health and usefulness of the best and wisest citizens by our management or mismanagement of our homes. Mental and moral power depend largely upon the physical condition; and this depends upon the food we eat, the rest and exercise we take, the air we breathe, and the cheerfulness of our homes. Good sleep is essential to health. Both brain and body need plenty of rest to keep them in good working condition for a series of years. Nothing makes young women grow old-looking so fast as keeping late hours. Nine o'clock is the old-fashioned, country bed-time; and it is the best bed-time in a majority of cases. Of course the little children should go earlier. If they go to sleep in a happy frame of mind it will help much toward refreshing slumber. A cheerful "good night," and an affectionate kiss (if there is sufficient spontaneity about it to make it worth anything) are decidedly *healthful* for the little ones.

The skin is full of pores, through which impurities are thrown off from the system. This exhalation goes on more rapidly during sleep than at any other time. For this reason, if for no other, the body should be washed after a day of sweating work or play in the dust. Attention to this will do much in the way of preventing bowel diseases and fevers. On account of the exhalation of impurities through the skin during sleep, the bedding should be well aired, each day, before making up.

I have nothing to say about feather-beds. None of our family like them; but I would willingly provide one for any elderly person to whom habit had made it seem a necessity. The spring mattrass is generally satisfactory, but it needs a warm mattrass over it in winter. For my own use I should not ask anything better than a bed of well-cured corn-husks, or, a good straw bed covered with a rather thin cotton mattrass or a thick comforter. Such a mattrass can be easily washed. Corn-husks wear longer than straw, make less dust, and are more elastic. A *clean* hair mattrass is an excellent bed; but it seems desirable to use bedding that can be cleansed without much trouble. The hair of a mattrass can be washed in cool water, dried in the sun, and put back in a clean tick. With a board frame the size of your mattrass, fastened up like a quilting frame, and with an upholsterer's needle, you can make, or remake, your own mattrass. We made one so; but I always dread to take hold of the thick heavy mattrasses in taking care of sleeping rooms. Feather pillows remain in favor after the pretty general "going out" of feather-beds—which is unreasonable, since the head should always be kept cool. An evenly-tacked, hair pillow suits me.

How to get Work Done on a Farm.—Every farmer is at times oppressed with a sense of the overwhelming number of things that requires to be done. We have seen a nervous man in such circumstances commence one job and before he had got fairly started, abandon it for something that seemed more pressing, merely to leave this in the same unfinished state; and when night came he had accomplished little or nothing, and passed hours tossing about on a sleepless bed thinking what he should do on the morrow. A young farmer could not adopt a better rule than to repeat to himself every morning, "Whatsoever thy hand findeth to do, do it with thy might," and act upon it. When a job is commenced, finish before beginning another; but, at the same time, attend to the little things. But if you happen to get behind, strike a lively gait, do one thing at a time, and when it is done, take a little rest by immediately commencing another, and in this way you will pull through in good time.

Laying Out a Flower Garden.

A number of letters have come to us this winter asking us to give directions and designs for laying out a flower garden, and it is gratifying to note that our friends are maturing their plans before the working season comes on. It is easy enough to give designs, but the trouble would be that no particular one would suit a half dozen of our readers, as their means and tastes vary as widely as do the shapes and sizes of the portions of land at their disposal. If one is to begin upon a new place without stint as to land or purse, he had better employ a first-class landscape gardener to both furnish the design and lay out the work. But the majority of our readers either do their gardening within the

Fig. 1.—Garden at the terminus of a walk.

boundaries of a village lot or in the usually too restricted space of the "front yards" to farm-houses, or at best have a moderate-sized country place with an acre or less of ornamental grounds. In laying out a flower garden one has first to consider the territory at command, and then what he wishes to cultivate flowers for. If he wishes to produce effects of color, which can only be done by masses of flowers and bright leaves, where all individuality of the plants themselves is lost, he will pursue a different course from those who wish to grow flowers as flowers—objects to be loved and cared for, and developed into things of individual beauty. Such an one cares not if a plant be new or old, fashionable or unfashionable, if it please him with its own beauty, or through some cherished association. Before describing the mixed flower border, let us consider the case in which flowers are grown for their effects in masses. Plants can be employed for this purpose in two distinct methods: in beds each containing but one kind, or in beds where plants affording distinct contrasts in the colors either of their flowers or of their leaves are grown in successive bands or belts. The first is properly called bedding, and the second, belt or ribbon gardening. A large class of plants adapted to both are known as bedding plants. Where the beds are planted with distinct colors, they must be of such form that each one will hold a proper relation to the other, and the whole group of beds form a pleasing and symmetrical figure. Here are opportunities for a display of taste in designing the forms of the group, and the beds of which it is composed, as well as in the proper choice of plants with which to fill them, in order that a pleasing effect of colors may be obtained. Squares, triangles, and other figures with straight lines, are less tasteful than those with curved outlines. In fig. 1, we give a design by the late Mr. E. A. Baumann, as an example of simple work of this kind. In this design a flower garden is represented at the terminus of a walk. The circular space is graveled; within it are three pear-shaped figures in grass, within which are cut the flower beds. Beds of this kind are set in the grass, as in the above example, or they are placed with very narrow graveled walks between them, and their outlines marked with an edging of box or other material. It will be seen that the fancy may suggest a great number of forms for the beds, or the figure which they compose. Mr. I. Pilat, gardener at Central Park, N. Y., has introduced with good effect, figures suggested by leaves and the parts of flowers. In fig. 2 is given a portion of the beds in the flower garden of the Park. A single oval or other shaped bed planted with one kind of flowers only, or filled with some plant of showy foliage, such as Coleus, is often made in a lawn with excellent effect. From beds of this kind it is but a step to the ribbon style of planting, in which several colors are used. To be effective, ribbon-planted beds should be of considerable size, and the plants selected with reference to similarity of hight and correspondence in time of blooming. Planting of this kind is also used in such groups of beds as those to which we have alluded, but requires nice management to prevent confusion. Ribbon beds may stand alone by themselves, or they may be arranged with a symmetrical relation to one another, as shown in figure 3, another of Mr. Baumann's designs, in which a walk passes quite around a central oval bed, and other beds of various shapes are placed in the lawn at a little distance from the walk. Another way of planting in the ribbon style is to run a narrow bed along each side of a walk and plant it with two or three colors. It will be seen that these styles of flower gardening may be carried out in a single bed in the small lawn of a front yard, or may be extended to ornament the largest grounds. All planting of this kind has its beauty much enhanced when framed by the green of a well kept turf, and is seen to the best advantage when it can be looked upon from a higher level. An enumeration of some of the annual and other plants best suited for use in beds of the kinds we have here described, as well as notes on mixed planting, must be deferred to another month.

Fig. 2.—Design for beds in Central Park, by I. Pilat.

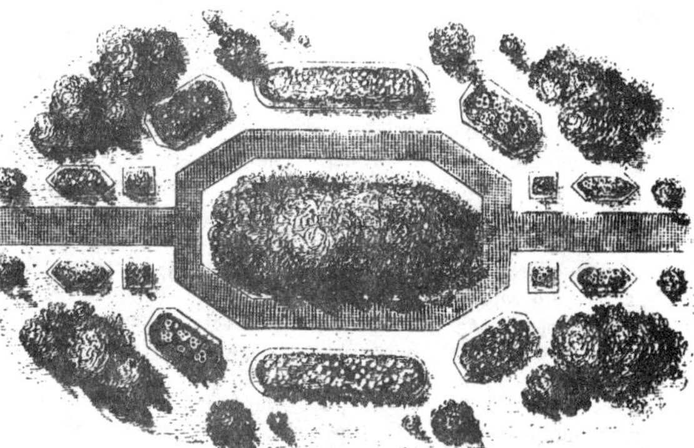

Fig. 3.—Design for flower beds.

☞ *For other Household Items, see "Basket"*

A Shoe-blacking Stand.

Well blacked shoes are a necessity, but the operation of blacking them is irksome, and the apparatus used is a nuisance in the eyes of the house-

SHOE-BLACKING STAND.

keeper. Hence the blacking and brushes are banished to some out of the way place, to which the one who would use them must follow them. The house-furnishing stores keep neat blacking stands, made like the one shown in the engraving. They are made of black walnut, and when closed no one would suspect their use. Upon lifting the lid we find a place for the brushes, one for the blacking, and a stand upon which to rest the foot while performing the polishing. Probably the majority of our readers do not find it necessary to black the boots in the house, but a stand of this kind, even roughly made, would be found a great convenience in the shed or other place, devoted to this part of the toilet. It would keep the brushes and blacking together, and free from dust, and prove a comfort in affording a foot rest of the proper hight. A person trying to black his boots with his foot in an inconveniently elevated position, shows himself in an attitude, the awkwardness of which is as amusing to others as it is uncomfortable to himself.

Bitter Butter.

Several have written in regard to bitter butter in winter, the communications being called out by an item which appeared in the "Basket" for December. The suggestions are essentially the same in all; we give one from Miss P. E. G., Lancaster Co., Pa., in which she addresses "G. W. S.," the correspondent who complained of bitter butter.

"You keep the milk in the cellar. The *Agriculturist* says, 'keep it at 60°.' Your cellar is probably quite as low as 35° during a part of the winter. Don't the potatoes freeze a little? ours do, without much care. The *Agriculturist* says: 'Keep both milk and cream where they will not absorb kitchen or other odors, especially smoke of wood fires, or of burning grease.' Dear me! What shall we do? Some of us, perhaps, can only afford one fire, and we make a dreadful smoke when we kindle it in the morning, and we fry sausage and mush for breakfast, and we boil pork and cabbage for dinner. What, then, shall we do with our milk? Answer—Make it *sour* before it can get bitter. In our region, the milk is sometimes kept in a cupboard, in the living room, or perhaps upon a table, nicely covered with a white cloth, or on the mantelpiece. This will make much better butter for you than your bitter milk. I have sold butter, in a city, at 65 cents, the milk for which stood in the broad kitchen window (where there was not a great deal of cooking), and was lifted occasionally to the mantel to sour. Don't let one pot be bitter. An experienced farmer's wife told me that one vessel of such milk would give churning a taste. In the case just mentioned of my own churning, the quantity of milk was so small, that I was not able always to churn in a week, if I recollect right. I will suggest another way.—Try it.—Have standing a vessel of sour milk, or buttermilk, not *bitter* sour milk. When you strain your milk, add to each pot or pan, intended for butter, a little of this sour milk—a skimmer full, perhaps. Your milk will be sour, and see what beautiful cream will rise. I have seen milk managed in this way kept in the winter in a spring-house with unglazed windows. Perhaps you will not have quite so much cream as if you keep your milk in a warm room, but try the experiment. I have further heard of setting milk pots upon the stove, and bringing the milk to a scald before setting it away. This extra heat may cause the milk to sour, and prevent that awful bitterness, of which I speak feelingly. But make your milk sour before it is bitter, unless indeed you can make all the cream rise before it is either sour or bitter. The evening before churning, if you have a coal stove, bring up your cream pot, or pots, and set them near the stove. Do not try to churn cold cream unless your time hangs heavily on your hands, or your name is Job. If you have a thermometer, you can vary your cream to 60°, or perhaps 65°, by stirring in warm water. Be careful of your thermometer, and do not plunge it into water so hot as to break it. After you have done these things, dear Illinois butter-maker, will you not give us in the *Agriculturist* the result of your effort?"

Soothing Syrup—Poisoning Made Easy.

There are mothers who use "Soothing Syrup" in perfect ignorance of its dangerous character. If it were labeled "Syrup of Morphia—Poison," as it should be, but very little of it would be sold. It ought to be very "soothing" indeed, if, as is stated in the California Medical Gazette, it contains very nearly a grain of Morphia to the ounce of syrup, and that the dose for a child three months old is equal to 10 drops of laudanum. It is ascertained that about 100,000 bottles of this stuff are sold annually in San Francisco, and it is also stated that one-third of the babies there die before they reach the age of two years. It seems to us most strange that a mother should give a child a medicine of any kind of the composition of which she was ignorant, unless she received it from the hands of a trusted physician. Years ago when certain worm lozenges were so popular that "children would cry for them," we made an analysis of them and found a good dose of calomel in each. Let secret remedies alone.

Washing Fluids Again.

Since last month's paper was made up, replies to our request for recipes for washing fluids have continued to pour in. It is, indeed, very gratifying to know that so many housekeepers are ready to assist their—what in view of washing-day we may well call—fellow-laborers. Some manufacturers of washing fluids and labor-saving soaps, have sent us samples of their goods, and some correspondents have written to recommend this or that soap or liquid. Those who have such preparations for sale can set forth their merits in the advertising columns, our object being to get some cheap and useful preparation that every one can make. Three-fourths or more of the recipes that have been sent, were the soda solution given last month, and the writers agree in assuring us that it can be used without injury, as the clothes require much less rubbing than when washed without it.... Here is another which several have sent, though the proportions of the ingredients vary. Sal Soda and Borax, ¼ lb. each; Gum Camphor, 1 oz.; Alcohol, ½ pint. Dissolve the soda and borax in one gallon of boiling rain-water, pour in two gallons of cold rain-water, add the camphor first dissolved in the alcohol, stir well and bottle for use. Four tablespoonfuls of the preparation are to be mixed with a pint of soft soap, and the clothes boiled in a suds made of this. It is all the better if the clothes are soaked over night, before putting them into the suds. We do not quite see what use the camphor can be in this preparation, though a solution of camphor and alcohol will dissolve some resinous substances that alcohol alone will not dissolve.... One lady adds a tablespoonful of Saleratus to the boiler of suds, which is no improvement over the generally used sal soda.... Another uses a mixture of Turpentine and Camphene, 1 pint each, and Ammonia, 4 oz. Three tablespoonfuls to a pint of soft soap, used in the first suds. Camphene, fortunately nearly out of use for burning, is only a very pure kind of spirits of turpentine, and the mixture is really only turpentine and ammonia.

Scouring Knives.—Miss H. M. S. says: Place a quantity of brick-dust on a board, and having the knife perfectly dry, press it down hard and rub it back and forth *crosswise of the blade*, when bright, turn over and scour the other side. Then wipe off with chamois leather. Knives thus treated will retain their brightness much longer, and have a *new* look after years of usage.

Household Talks.

BY AUNT HATTIE.

The fact of it is, I have not had time. Three months ago I sent Peggy away, and have not been suited with any servant since. I said something about Peggy in a previous "talk." She had lived in Ireland with her mistress for thirty years. About two years since Mr. Jackson died, and the farm had to be leased and their effects sold. The family, finding themselves quite reduced in circumstances, sought a home in this land of refuge. They tried to induce their old servant to stay in Ireland, but in vain; she determined to follow them wherever they might go. They went first to Canada, and Peggy had to work in another family for the first time in her life. The family afterwards moved to the States and left her behind. She was unhappy, and finally followed them here. Mrs. Jackson recommended her, for being faithful, excellent with young children, and possessing the rare virtue of never wanting to go out except once on Sunday. She was old, but neat in her appearance, and from her conversation, gave me the impression that she would be quite willing to learn anything, and that she lived only at my service. "Hattie," said the Doctor to me, when I had enumerated her qualifications to him, "you are fixed now, if you are wise; an old servant is what every woman with a family should have, and I hope you will not send her off for any trivial offence. Mark my word, these young girls are never to be trusted with children." I took the Doctor's advice kindly, especially as I felt confident that there was no difficulty in carrying it out to the letter. But, alas for human anticipations. Peggy had not been with me a month before I was tired of her. She could neither cook, bake, nor set a table. She could not, and I could not teach her to, do up Edward's shirts and collars, or any of the children's or my fine clothes, and I had to direct her in all she did. I tried to teach her to make bread, but she simply said she never had baked a loaf in her life, and she thought she was too old to learn now, so I had to bake the bread, pies, and cake, and help to set the table always. I had to cook beefsteaks, chops, and joints; in fact, I considered her incapable of completing anything she undertook. If she minded the baby, she did it faithfully and well—feeding him, tossing him, walking around with him—anything to amuse and keep him quiet; but the rest of the children might have been in Van Dieman's Land for any knowledge she would have been capable of imparting as to their whereabouts. In short, I found her to be the most one-thing-at-a-time person I ever saw. "Peggy," I said to her one day, "when you lived with the mistress in Ireland, what kind of work did you do?"—"Well, ma'am, Mistress Jackson had a large farm, and an illegant stone house, and it was not the likes of me that would be after sweeping her fine carpets and bedrooms, so I jest attended to feeding the chickens and calves, and washing the dairy pans." Peggy lived with me for two months, when I heard of a situation that I thought might suit her. She went, but only stayed one week; she afterwards lived with a sister of mine, who found her the same faithful, simple-minded, quiet person, but as her work consisted of up-stairs nursery, dining-room, and kitchen, the same difficulty was experienced as with me, and she left there also. For some weeks past Peggy has been living in a large seminary for young ladies, where I fancy she finds work suited to her exactly—washing vegetables and dishes on a large scale, requiring little ingenuity, and hardly any responsibility. The Doctor speaks from experience when he says an old woman for a servant is a good thing. He has living with him now a person who, within my recollection, has always been associated in my mind as old Jane; why so, I do not know, for she does not appear like an old woman, even now; perhaps the title is given to her as significant of her stability, uprightness, and maturity of thought and judgment. She would be a treasure in any family. I might challenge the town for a better bread and pie maker, or a better cook or housekeeper. Jane would be dreadfully mortified if the Doctor should come home to dinner and find it not ready. His candle, which he needs in the evening for attending to the furnace, is always ready. The water pitcher is always full, the kitchen is always clean, the steps are always clean, and her work appears always done, so that she is ever ready and willing to do any thing extra required. Give me such a servant as the Doctor's wife has, and I should have time to do my own sewing, to visit, and to write punctually my Household Talks.

"Any soap-grease to-day, ma'am?" "No, sir; I use my own soap-grease." "Make your own soap?" "No, sir." "How do you use your own soap-grease, then?" "I make it into eggs," I said, smiling. The man looked so astonished and half frightened that I thought it time to explain. "I feed the grease and fat which is unfit for cooking to the chickens."

The baby is cutting its teeth, and is at times so cross that it is difficult to amuse him, and sometimes on this account I am half tempted to regret having sent Peggy away. Although the weather has been quite cold, I take him out every day, sometimes in his little carriage; and when snow is on the ground, I take Willie's sled and fasten a box on the back part, spread over a small Afghan, and give him a delightful little sleigh ride.

This afternoon, while out with baby, I met a German or Holland woman. A good chance to try to get a girl, I thought. "Do you live around here?"—"Yaw; yust in Brighton." "Do you know of a girl to work?"—I shouted, (you always shout when speaking to a foreigner). "Yaw, yaw, a good one; you want one good one?"—"Yes," I said, "a large girl, to do anything—take care of the baby, wash, and iron." "Oh, yaw, she good girl; she wash, she iron, she mind the baby. Oh, she good girl, yaw; she fat." "Is she German?"—"Oh, no, she no German, she Hollands. Oh, she good, she fat." "When can I see her?" I said. "Yaw; I brings her round de day after yesterday." "To-day?"—"No, no to-day; yesterday I goes to wash for Mr. Calver. I can no come, but I prings her the next morning day." "Day after to-morrow?" I shouted. "Yaw, yaw, that is it; the day after to-morrow. I prings her; she clean all up for you; she no Irish; she no German; she Hollands. Oh, she be good; she fat."

When I came into the house I found the sitting-room fire nearly out, a very few coals only alive. To have put a large quantity of coal upon so small a fire would have extinguished it at once, so with the tongs I put on gently six or seven small lumps of coal; in about ten minutes I put on a shovelful, as it was now burning nicely, and I have just put on the usual allowance of coal for a good fire. How often I have had the fire put out by the girls pouring in a large quantity of coal when it was low.

Economical Cooking—Something out of Nothing.

BY FAITH ROCHESTER.

It is a favorite axiom with my father that "You can't get something out of nothing." I have sometimes felt that my experience in household matters almost proved the contrary. Perhaps few readers of the *Agriculturist* know what it is to be *too poor to economize.* It is such poverty as compels Nellie Kay to wear an old silk dress to school this winter. The mothers of her playmates think a good calico frock would be far more suitable, considering her father's circumstances, or suggest to each other that a warm flannel dress would be better and cheaper than anything else. Alas! At Nellie's house there is not a penny to be spared for buying any sort of a new dress, and every possible old dress of her mother's has been made over for her, until there is nothing left but the old silks of better days. They are learning the sad meaning of debt and interest, and are too poor to economize.

Taking health and contentment into account, as well as money, it is the best economy, under ordinary circumstances, to use a liberal diet, consisting largely of fruit and the finer grains. I think people are more likely to eat to excess habitually, after a few scanty and unsatisfactory meals, than when there is little variety in their fare from day to day. The stomach seems to be constantly unsatisfied and takes more than a proper amount in the effort to get what it craves but cannot find. There are lean times in the cupboards of so many homes, that I purpose to reveal a few secrets once taught me by a severe old schoolmaster named Experience.

When the cow has gone dry, and the stock of butter is getting low, and the hens don't lay, and the fresh apples are nearly or quite gone, and (worse than all, perhaps,) the family purse is very, very lean—then what shall a body do? Receipt books are very unsatisfactory at such times.

Most people use lard and even pork-fat for shortening. Some of us do not. Any kind of shortening needs to be carefully used or it will become an abomination. It is not so essential an article in cooking as many imagine. Of all kinds I give the preference to sweet cream. Pie-crust made with lard is whiter, but made with cream it is more wholesome. I have heard of shortening pie-crust with beans, but have never tried it. People *can* keep alive without pies, especially if they can get

DUMPLINGS.—Good crust for various kinds of dumplings may be made of a part of the dough when baking bread. It may be rolled out thin and wrapped around the apples, previously pared, quartered and cored, or cored without quartering, one in each crust. Leave the crust very thin under the apple (or it may be heavy), and when made let them rise a little before going into the oven. Bake slowly. Or, pare, quarter and core fresh apples, and put in the bottom of a dish with a little water (dried apple sauce is better than nothing, if fresh apples are wanting,) and cover with a crust of bread dough. It should rise a little before going into the oven, and then bake slowly. Graham mush is not a bad crust for such a pudding, and boiled rice makes a pretty cover when browned by the baking....Or bake a thin loaf of bread and split it open when done, and spread each half on the split side with canned or stewed fruit, or fresh berries in summer, placing one above the other. It is better, but not necessary, to spread the halves first with butter.

APPLE JONATHAN differs from the other dumplings chiefly from being made in a pot or kettle. I first made its acquaintance under the name of "Pot apple pie." Invert a plate in the bottom of the pot. Put in sliced or quartered fresh apples. Pour over them maple molasses if you have it, if not, add sugar enough to sweeten well. Add a teacup of water and nutmeg or allspice. Over this around the ridge of the pot put strips of crust, cover tightly and boil over a moderate fire. [This is what is called "Pan-dowdy," in some parts of New England.—ED.]

STEAMED BREAD AND BUTTER PUDDING.—Place slices of bread and butter in the bottom of a pudding dish, a layer of sliced apples with sugar and nutmeg, another layer of bread and butter, then one of apples, sugar and spice, until the dish is full, having bread and butter at the top, buttered side down. Cook thoroughly in a steamer.

PUDDING SAUCE.—Cream sweetened and flavored is a favorite dressing, and maple syrup is beyond praise; but when cream is scarce and maple syrup lacking, good sauce may be made after the receipts given in the last volume of the *Agriculturist*, only plainer as one's necessities require....For bread hastily made nothing can surpass

GEMS.—They are cheap, easily made, wholesome and palatable. Graham flour and water are stirred together to the consistency of a thick pancake batter, and baked in the iron or tin gem-pans. Everybody should have these bread-pans. Gems, whether of Graham meal, fine flour or corn meal, should be put into a *hot* oven. Success depends on this. Fine flour and sweet milk (skimmed milk is good enough), well beaten together, rather thicker than the Graham batter, makes a very sweet and good kind of warm bread. Corn bread of the best kind can be made without eggs or shortening, or sweetening. Simply scald the meal with boiling water, add a little salt, stir well and bake quickly in the gem-pans. We thought the Graham and white gems must have salt, until we found that its absence was not observed, and then we discarded it, as it seems an unreasonable amount of salt is eaten under the plea of a *little* salt being necessary.

CAKE.—This is not one of the necessaries of life, but good cake is seldom refused. There are a few ways of making it quite cheaply and satisfactorily.

BREAD CAKE of every grade is good if carefully made. The regular receipt as given me by a dear old playmate, reads: "One and a half cup of dough, one cup of sugar, one-half cup of butter, two eggs, one-half teaspoonful of soda. Raisins and spice to suit the taste. Mix with the hands until the dough seems thoroughly worked in, adding a little more flour if the dough is thin. Let it rise half an hour. It rises slowly and but little before going into the oven." This is very nice. We have eaten it with a relish when minus eggs and raisins, and with only a tablespoonful of butter, and a little clove, cinnamon or nutmeg for flavor.

BACHELOR'S CAKE is plain and good. "One and a half cup of sugar, one cup of milk, two tablespoonfuls of butter, two eggs, one teaspoonful cream of tartar, one teaspoonful of soda, three cups of flour." This makes two loaves....A queen among women, once treated me with cake made as follows: One cup of sweet milk, one cup of sugar, two cups of flour prepared with Horsford's powder. The loaf was split and the halves spread with canned strawberries, and one half laid above the other.

PLAIN RICE PUDDING.—Half a pint of rice, one quart of milk, half a pint (or less) of sugar, nutmeg or cinnamon. Bake it slowly two hours. Tapioca may be cooked in the same way, after soaking in warm milk for an hour or two; and Sago, after thoroughly washing and soaking over night, is good in the same fashion. It is possible to dilute the milk one-half and yet have the pudding good, if care is exercised in soaking and cooking.

SOAP.—It is convenient sometimes to know how to make soap quickly, when there is little grease and no leach set up. Make a white ley by boiling wood-ashes with water and pouring off the liquid after it has settled. When this ley is boiling, add all the grease (previously tried out—lard or tallow, or drippings) it will "take." Boil it together, trying frequently a little in a saucer, until you find it thickens as you stir it, and it cools. If you cannot make it "come," add a little water to a small quantity in your saucer. If that thickens it, do the same with that in the kettle. If not, try adding ley to a portion, and then to the whole, if that makes the small portion tried "come."

All the above is written for the benefit of those who know what it is to be "in a pinch." I don't envy those who have never had that experience. It quickens the wits and deepens the sympathies, but it is not a good way to live—long at a time

containing a great variety of Items, including many good Hints and Suggestions which we throw into smaller type and condensed form, for want of space elsewhere.

To Get Rid of White Birches without plowing or grubbing, cut in winter or spring, and feed close with sheep; or, cut about mid-summer (June 20th) close to the ground and subsequently, as new shoots start once in three or four weeks.

Ice-House. — A "Subscriber" writes to know how it will do to enclose an ice-house with three thicknesses of boards, leaving two air spaces, instead of the usual way here of having *one* space, and that filled with sawdust or tan-bark. He thinks it would be more durable, as there would be no moisture in contact with the frame, and it would be less pervious to heat. This plan would not work well, because there would be a constant circulation of air in the air spaces. The aim in making ice-houses and in filling them is, to prevent any circulation of air, and this is best effected by a non-conducting filling, like sawdust, in the air spaces, and straw or wheat chaff under and around the ice.

Hay and Straw Cutter.—"J. W. B.," Dutchess Co., N. Y., asks, "What hand machine, for cutting hay and straw, do you consider the best, and least liable to get out of order?"—We regard the Copper Strip Feed Cutter as decidedly the best as a hand machine, and know of some large stables in New York City where power cutters are set aside, and two men, in an hour's time, daily, cut all the hay for 100 horses, with a large sized Copper Strip cutter. They are extensively advertised.

Tan-bark Ashes. — "L. H. C.," New Madison, O. "Are ashes made by burning spent tan-bark at the tan-yard of any value as manure on stiff, clayey land?"—Yes; of high value.—"Will it pay to haul it two miles?"—Yes; ten.—"What crops are most benefited?"—Grass, potatoes, tobacco. It must be applied according to the crop—broadcast, in the hill, or upon the hill, just as the plants break the ground.

Rotary Harrows.—A "Subscriber," of Salem, N. C., asks for instructions how to make a Rotary Harrow. The only implements of this kind with which we are acquainted which work well are patented, and upon the patented device their usefulness depends. They possess several advantages over other harrows, in that they drag evenly on side hills, cover grain very well, smooth down the furrows in sod land, make a mellow seed-bed without cross-harrowing, and tear up no sods.

Nutritive Value of Beans. — "X." asks what quantity of small, white, old beans equals one bushel of good corn?—According to Wolff and Kopp's tables, given in Johnson's "How Crops Grow." Field beans contain 25¼ per cent of albuminoids, or nutritive substances, and maize contains but 10. On the other hand, maize contains 7 per cent of fat, or oil, and beans but 2 per cent. It is safe to say beans are worth something more than twice as much as corn if judiciously fed.

Boiled Beans, or Bean-Meal.—We know no reason why boiled beans should not be quite as healthy as bean-meal for any kind of stock except sheep.

Good Crops of Potatoes and of Corn.—John Kiernan has charge of the farm and garden of the Sisters of Charity, at Mt. St. Vincent, on the Hudson. He sends us a statement by a civil engineer in regard to his crop of Harison potatoes, which certifies to a yield varying, in different parts of the field, from 400 bushels to 640 bushels per acre. The total yield is not stated. They were planted early in May, in hills 4 feet apart each way, on clean land, in fine condition, cultivated twice, and kept clean of weeds until they stopped growing. The corn crop was planted 4 feet each way, dressed with top manure; four stalks were left to the hill, and 70 bushels of shelled corn were harvested. The land was plowed in the fall and again in the spring.

The Cow "Fancy."—In compliance with the request of a correspondent, we publish the following measurements of this animal (pictured in our December number). Length from base of horn to point of rump, 6 feet, 7 inches; length from centre of hip bone to point of rump, 1 foot, 6 inches; height at hip, 3 feet, 10¼ inches; height at shoulder, 4 feet; height at belly, from ground, 1 foot, 8 inches; girth around the chest, 5 feet, 2¼ inches; girth around the belly, 7 feet, 2 inches; circumference of fore leg below the knee, 5¾ inches.

AMERICAN AGRICULTURIST

FOR THE

Farm, Garden, and Household.

"AGRICULTURE IS THE MOST HEALTHFUL, MOST USEFUL, AND MOST NOBLE EMPLOYMENT OF MAN."—WASHINGTON.

VOLUME XXIX.—No. 3. NEW YORK, MARCH, 1870. NEW SERIES—No. 278.

THE AYRAULT FAT OXEN.—DRAWN FROM LIFE, BY W. M. CARY.—*Engraved for the American Agriculturist.*

These cattle, said to weigh more than 3300 pounds each, were over six years old, raised and fed by George Ayrault, of Poughkeepsie, N. Y., and slaughtered in February by Wm. Lalor, of Centre Market, N. Y. City.

METALIC BIRD HOUSE, enameled white, 13 inches high, 14½x10½ base. Manufactured by the

MILLER IRON CO.,

Providence, R. I. Manufacturers of Ornamental Iron Work for Gardens, Lawns, Parks, Cemeteries, etc. Illustrated Catalogue sent free.

COOPER'S STEAM ENGINE WORKS.

To keep pace with the growing demand for our Machinery, we are adding **$40,000** worth of new and improved Tools and Buildings to our present large Factories, and will continue to supply the following articles, after the best designs, on terms which will be found to be specially advantageous:

STATIONARY STEAM ENGINES,

For MILLS, FACTORIES, SHOPS, FURNACES, MINING, &c., of every required size, divided into three classes.

1. WITH SINGLE SLIDE VALVE, *Cutting off at two-thirds of Stroke by lap.*

2. WITH CUT-OFF VALVE, arranged so as to close at any part of Stroke and adjustable by hand-lever while engine is in motion.

3. WITH BABCOCK & WILCOX PATENT *Variable Cut-off, automatically adjusted by Governor.*

PORTABLE STEAM ENGINES,

Of 8, 10, 15, 20 and 25 Horse Power, combining all the improvements of the Slide Valve Engine. This is the only portable Engine and Boiler that has a *Combined Heater and Lime Catcher.*

Babcock & Wilcox Patent

Non-Explosive Tubulous Steam Boilers.

Grist Mill Machinery and Mills

Of any required size, with correct working drawings.

$1,500 WILL PURCHASE

A FIRST-CLASS COMPLETED TWO RUN GRIST-MILL, with erecting plans, &c., and guaranteed to give satisfaction and to be unsurpassed.

CIRCULAR SAW-MILLS

improved in construction and combining all modern improvements.

☞ MACHINERY DELIVERED at New York, Philadelphia, Baltimore, Chicago, Saint Louis, or New Orleans.
☞ Full particulars and circulars on application. Address in full, **JOHN COOPER & CO.,**
Mount Vernon, Ohio.

GREAT LABOR-SAVING Machine for the Million. For the Cotton Planters and all others. Lyman's New Patent Adjustable Weed Annihilator, a year of experience proves it to be by actual test the most practical, useful, cheap and reliable Weed Destroyer known, combining *strength, durability,* great ease in working, doing the work of two or more tools of the ordinary old styles now in use, with much greater ease and rapidity, cutting from 3 to 9 inches in width, with best cutlery steel tempered Knives, with cutting edges each side, and easily sharpened.

A. E. LYMAN & CO., Northampton, Mass.
Send for Circular. A. E. LYMAN, Patentee.

Galvanized Wire Cable Fencing.

Cheaper than wood; more easily erected; not liable to rust; will sustain a man's weight in clambering over it; made of 6 strands of galvanized wire; will not "set" like solid wire, in receiving a blow; will not snap in cold weather, even if strained tight, as the twist of the cable equalizes the contraction and expansion caused by cold and heat; exterior diameter nearly ¼ inch. One hundred pounds of this Cable will make about two-thirds more Fencing, than the same weight of solid wire of same diameter.

A 5-Bar or *5-Cable Fence,* with Straining Brackets, Screws, Nuts, and Staples, for Wood Posts, costs $30 per 100 yards, without the cost of the wooden posts.

A 5-*Bar* or *5-Cable Fence,* with Iron Posts and Standards, all complete, will cost $63 per 100 yards.

Send for Circular. Liberal discount for $1,000 and over.
Shops, 17 Coates St., Phila. PHILIP S. JUSTICE,
Offices: 42 Cliff St., New York; 14 N. 5th St., Philadelphia.

FARMERS—ATTENTION! Wire for Fencing and for Grapes, cheap, galvanized or ungalvanized. We are prepared to furnish merchants and farmers, everywhere, with a very superior quality of Annealed Fencing Wire at very low prices. Save money by sending to head-quarters. Price List free. Address

R. T. BUSH & CO., Manufacturers,
73 William St., New York.

IMPROVED STOCK. — THOS. B. SMITH & CO. are the most extensive and reliable breeders in the country.
See advertisement on page 116.

Lyman's Patent Transplanting & Weeding Machine.—(Just Out.)—This new Patent Implement, unlike anything heretofore used, is just what has been so long needed by Gardeners and Horticulturists; will include a sufficient quantity of earth and deposit it with the Plant in any desired position securely, with great rapidity and ease, giving double surety to the life and growth of the Plant. Steel Blades, adjustable.

30-inch or 2½ ft. Handle. **A. E. LYMAN & CO.,**
Northampton, Mass.

A. E. LYMAN, Patentee.

COMSTOCK'S
New Horticultural Implements
COMBINED.

(Patented June 1st, 1869.)

The Best in the World.

As a Cultivator.

Comstock's Hand Cultivator and Onion Weeder will do the work of SIX MEN with hoes. It is the only implement that *pulls the weeds* and thoroughly pulverizes the soil. As much superior to the hoe for all small drill crops, as the mowers and reapers are to the scythe and cradle. PRICE, $9.00. Boxing, 25 cts.

As a Weeder.

Comstock's Seed Sower.—The neatest and most perfect small seed sower yet invented—sowing Beet, Parsnip, and other difficult seeds with the greatest regularity. Combined with the Cultivator and Weeder, and can be separated in 5 minutes. PRICE, $15. The Seed Sower alone, $10. Boxing, 35 cts.

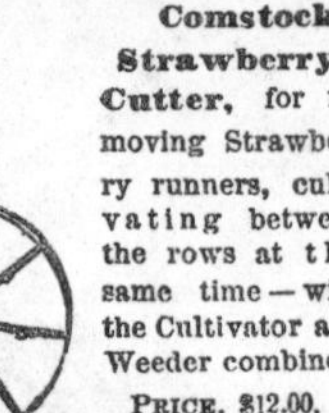

As a Seed Sower Combined.

Comstock's Strawberry Cutter, for removing Strawberry runners, cultivating between the rows at the same time—with the Cultivator and Weeder combined.

PRICE, $12.00.
Boxing, 25 cts.

As a Strawberry Cutter.

Comstock's Weeding Hook.—A little steel instrument with a hook at each end, one pointed and the other cleft, for hand weeding, and is just the thing for ladies in the Flower Garden.

PRICE, 50 cts.

The Weeding Hook will be sent by mail upon receipt of 65 cts.

The Set Complete, $13.50. Boxing, 35 cts.

The cuts will give a correct idea of the implements, with the exception of the handles, a portion of which it was necessary to cut off for want of space. Descriptive Circulars sent to all applicants. Address

B. K. BLISS & SON,
41 Park Row & 151 Nassau St.,
AGENTS for the Manufacturers. **NEW YORK.**

HAY SCALES!
FOUR TON!
$75.00.

Send for Price List No. 196. Make no mistake in my address, or number of List.
EDWARD F. JONES, Binghamton, N. Y.

For Family use—simple, cheap, reliable. Knits everything. AGENTS WANTED. Circular and sample stocking FREE. Address HINKLEY KNITTING MACHINE CO., Bath, Me., or 176 Broadway, New-York.

THE BEST
Family Sewing Machine.

"The Florence Sewing Machine is decided to be the best "on exhibition. It must also be stated incidentally that **"THIS IS BETTFR THAN ANY OF ITS CLASS "KNOWN TO THE JUDGES."**
Report of Judges, American Institute Fair, New York.

PRINCIPAL AMERICAN AGENCIES:

New York—505 Broadway;
Baltimore—140 Baltimore Street;
Boston—141 Washington Street;
Brooklyn—431 Fulton Street;
Charleston—382 King Street;
Chicago—43 Madison Street;
Cincinnati—28 West Fourth Street;
Cleveland—43 Public Square;
Detroit—158 Jefferson Avenue;
Galveston—234 P. O. Street;
Hartford—382 Main Street;
Indianapolis—27 N. Pennsylvania Street;
Louisville—110 Fourth Street;
Milwaukee—410 Milwaukee Street;
Nashville—32 North Cherry Street;
New Orleans—6 Chartres Street;
Philadelphia—1123 Chestnut Street;
San Francisco—111 Montgomery Street
St. Louis—612 North Fourth Street.

AGENTS WANTED.

We wish to get energetic agents in every section of the United States and Canada, where we are not now represented, to sell

The most simple and durable double thread Sewing Machine ever offered to the public.

This Machine is first-class in every respect, and at the same time furnished

At a price within the reach of all.

Licensed by Wheeler & Wilson, Grover & Baker, and Singer & Co. We will give parties with sufficient capital the exclusive agency of an entire State. To persons seeking a profitable business, we offer unparalleled inducements. The attention of Sewing Machine Agents is especially solicited. For terms address, GOLD MEDAL SEWING MACHINE CO., 334 Washington St., Boston, Mass.

SELF-ACTING GATES.

The American Gate Co. of Cleveland, O., are the sole manufacturers of Nicholson's *Improved* Self-acting Gates. These gates are beyond competition—the best ever made. They are recommended by the "Scientific American," the "American Agriculturist," Gen. Horace Capron, Commissioner of Agriculture; and many others. Send for Circular.

Address M. G. BROWNE,
Secretary Am. Ga. Co., Cleveland, O.

FOR EVERY HOUSEHOLDER.

Page's Portable Pump and Graduating Sprinkler, for $5.00.

"It has proved the best thing of the kind I have seen."—Chas. Downing, Newburgh, N. Y. It is unequalled or applying liquids to destroy insects on plants, vines, and fruit trees.

As a Fire Extinguisher, it is very effective.
Liberal discount to Clubs, and to Agents, for 1870.
NEW ENGLAND PORTABLE PUMP CO., Danvers, Mass.

For Drain Tile, Water Pipe, Garden Vases and Pedestals, Chimney Tops, Fire Brick, Fire Clay, Furnace Boiler and Grate Tile, Stove Linings, and all kinds of Clay goods made to order. Also, dealer in Roofing Slate, Hydraulic Cement, Calcined and Land Plaster, and Building Materials generally.

Address D. R. ECKER, Pittsburgh, Pa.

The Copper Strip Feed Cutter

For cutting Hay, Straw, and Corn Stalks. Warranted to please, or money refunded. Send for Illustrated Circular to

PEEKSKILL PLOW WORKS,
Peekskill, N. Y., or Cleveland, O.

AMERICAN AGRICULTURIST.

NEW YORK, MARCH, 1870.

We believe there never was a season which somebody did not regard as very remarkable, if not altogether without precedent. No one, however, on the seaboard, claims this winter as one of unusual severity. With us, it has been almost like summer for weeks together. We have seen a good deal of plowing done in both January and February, and the ground, up to the time of writing, has been bare of snow, except for a day or two at a time, since before Christmas. This comes, notwithstanding all the "signs." There hardly ever was such a crop of nuts—this betokened a severe winter; the corn husks were thick and abundant—this also indicated a long season of cold. Wild geese went south early; a good many bears, and other wild animals, were killed further south than usual, and there were a score of other "signs" of a hard winter. If it comes, it will be after we go to press. And now, almost everybody prophesies a cold March, and late spring. The fact is, nobody can tell, and he has been wise who has improved the winter to make his fences, clear off the stones, clean up fence rows, dig drains, and lay the tiles, so that whenever spring comes, he can go on with his work without interruption. March is, however, often best adapted for this kind of work, and we have yet to see the farm in America whereon there is not enough of it to do. Bright, sunny days, and a drying surface, are a temptation to begin plowing too early. Wait until the ground is settled, and the water is out of it, on all except sandy or gravelly soils which do not pack when wet.

The month is likely to be one of great changes in temperature; high winds will prevail, more or less, and the farmer's first thoughts should be to protect his stock and stores from exposure. The frost may derange foundations, loose weather-boards may be found, rain finding its way through the roof may make musty hay. All these things need a little closer looking to than was necessary during the winter. At the same time, the roads will probably be very bad, and the provident farmer will get all real necessities, including food and feed, lumber, nails for repairs, etc., beforehand.

Hints About Work.

Setting Fences.—Before the land is in condition to plow, and while yet it is mellow and loose, re-set old fences that have been blown or shoved out of place. The cheapest and best wooden fence a farmer can put up is one of ¾-inch oak strips, nailed to posts, set 6 feet apart, capped and battened.

Plowing.—Plow no water-soaked land. Many a field of heavy soil is damaged for the whole season by being plowed before the water is out of it, and the labor of subsequent tillage is greatly increased.

Poultry Houses.—We think it well to dig out for henneries, and set them a few (say 3) feet below the ground level, provided perfect drainage can be secured. Cement the bottom; set the studs upon the cemented floor; board up upon the inside above the ground level and fill in behind the boards with cement concrete. Outside nail hemlock boards against the studs as high as you have earth enough, or think best to raise an earth bank. We have given many plans for henneries in our back numbers. Light, warmth, cleanliness, and good feeding, will secure a supply of eggs all winter.

Orchard and Nursery.

Shriveling may take place with trees long on the way, and not well packed. Open a trench in sandy soil, lay in the trees, and cover them, tops and all. In a few days, or a week, they will be found to have regained their former plumpness.

Girdled Trees are to be attended to as soon as discovered, as recommended in previous months.

Grafting.—Cherry and plum trees should be grafted earlier than the apple and pear, which are best left until the buds commence to swell.

Pruning may still be done where vegetation is dormant. Cover the wounds with grafting wax.

Fruit Garden.

Grape-Vines.—Plant if the soil is in proper condition. Use no manure. Cut back the vines to three buds, but one of which is to grow into a shoot. Plow old vineyards, and use the hoe near the vines. Layers may be made from last year's wood. Set posts for trellises.

Blackberries.—Set new vines early; leave no old cane; the growth should be all from the buds near the root. Six feet apart, each way, is a good distance, if they are kept within bounds by pinching.

Raspberries.—Set from four to six feet apart, according to the size of the variety. Let no old canes remain on plants now set out. Uncover tender varieties when severe frosts are over.

Currants and *Gooseberries.*—Set early. Prune.

Cuttings of grape, currant, and gooseberry may be put out; pack the earth well against them.

Strawberries.—Set as soon as the frost is well out of the ground, and plants can be obtained. Put the plants in rows, two feet apart, and set them eighteen inches distant in the rows. Remove all decaying leaves at planting, and shorten the roots about one-third. Where pistillate sorts are grown, plant a perfect variety near by to fertilize them.

Kitchen Garden.

Every spring we are obliged, in reply to letters, to say something about plants under glass. By a little pains in this direction, many vegetables may be had several weeks earlier than from seeds sown in the open ground. Those who follow market gardening, make extensive use of glass to forward their crops, and it may be done in the family garden to great advantage, provided one will take the necessary care. If one will not take the trouble, then he had better go on in the old way, and let glass alone. The easiest plan, and one attended with fewest risks, is to start seeds in

Window Boxes, which may be about four inches deep, and of any convenient length and width. Soap, and other boxes may be obtained at the grocer's, which may be cut in two, and answer well. Fill the boxes with light, rich soil, and sow seeds of tomatoes, cabbages, etc., and place in a sunny window—all the better if in the kitchen, where the air is usually moist. Have similar boxes of earth ready to receive the young plants when they are large enough to transplant. Water when needed. The next step is by the use of the

Cold-Frame.—The regular sash is six feet long, and three feet wide, but any other size will answer. Even old window sashes may be made to serve, if the cross-bars have channels cut, to allow the water to run off. The frame to hold the sash is made by setting down stakes, and nailing on boards. It should be eighteen inches high in the rear, and twelve inches in front. The slope should be towards the south, and the bed placed where it is sheltered from cold winds. The soil within the bed must be light, fine, and rich. Expose the glass to full sun during the day, and in the afternoon cover it with mats or board shutters. Work over the soil every few days, and when it has become well warmed, sow the seeds. When the plants are up, the sash must be raised at one end, to air the bed during the day; water must be given as needed.

Hot-beds are like the cold-frames, but with a mass of fermenting manure at the bottom, to give more heat than that supplied by the sun. The old way was to make a heap of manure, three feet high, and place the frame upon it. It is much more economical of manure to place it below the surface. A pit is dug about two feet deep, and of a size corresponding to the number of sashes to be used. A frame is made within the pit by boarding up, and the rear of it is to be eighteen inches, and the front twelve inches above the surface of the ground. Bank up around the outside of the frame. Fill the pit with fermenting stable manure, which will be all the better if mixed, one-third or more of its bulk, with leaves. Put in the manure evenly, and beat it down firmly with the fork. Put six inches of light, rich soil upon the manure, and spread evenly, and put on the sashes. A thermometer should be placed in the soil, and when the heat falls to about 90°, seeds may be sown. When the plants are up, they will need daily care. The glass must be raised in the morning, and closed in the afternoon, and shutters or mats must be put on at night. Neglect in airing at the proper time will burn the plants, and leaving the sashes open too long, when the heat of the sun declines, will chill them—extremes which are to be avoided. Egg-plants, Peppers, Tomatoes, and other plants, of warm countries, are raised in this way. Hot-beds are to be started about six weeks before planting can be safely done in the open air.

Cabbages and *Cauliflower*, that have been wintered in cold-frames, may generally have the sash removed by the first of this month. Those who do not keep fall sown plants over winter, can get a very good start by sowing in the boxes described above, and placing them, at first, in a mild hot-bed. When the plants have made a proper growth, they should be transplanted to other boxes, and when established, the boxes are to be transferred to a cold-frame, and gradually hardened off, by exposure to the air whenever the temperature will allow. Enough plants for a family garden may be grown in window boxes; these should be exposed on mild days.

Cucumbers, Squashes, and other plants that do not transplant readily, may be started in hot-beds, cold-frames, or in the house, by the use of small squares of sod, which are placed grass side down, and the seeds are sown in the earth. The bits of sod containing the plants are to be set out at the proper time.

Plow or Spade the soil whenever it is dry enough. Use plenty of manure, and work fine and deep.

Asparagus and *Rhubarb Beds* may be forked over if the season permits, working in manure.

Peas.—Sow early sorts as soon as practicable.

Potatoes for the early crop should be placed in a warm room, to induce the sprouts to start. After cutting, allow the surface to dry before planting.

Onions.—Plant Sets, Top, or Button Onions, and Potato Onions, as soon as the ground, which should be rich, is ready.

General Planting.—The hardiest vegetables are carrots, beets, spinach, salsify, onions, leeks, and early turnips. The time for sowing them will depend upon locality and season. To these may be added radishes, cress, and lettuce.

Flower Garden and Lawn.

The article upon "Laying Out a Flower Garden," given last month, will be found to contain hints which need not be repeated.

Walks should be made permanent, and, if of gravel, must have a good foundation of stone. Asphalt walks are worth considering.

Lawns.—Attend to the work of preparation early. Drain, if necessary. Let the soil be deep and fertile, and use a plenty of seed—from two to five bushels to the acre. June grass (Kentucky Blue), or Red-top alone make a good lawn. Various mixtures are sold by seedsmen. Imported lawn mixtures are unsuited to our climate. See last month's notes for other suggestions.

Perennials.—Those which have been in one place for three years or more, should be taken up, divided, and reset in fresh soil.

Green-house and Window Plants.

Now that the heat of the sun increases, more frequent airings can be given, and more care in watering will be required. Many plants that have been dormant will be pushing their growth, and many of these that have been kept for the winter in the cellar may be brought out and started.

Propagation of plants for summer use in the borders, such as Verbenas, Geraniums, and the like, should now go on rapidly. The secret of success is in keeping the air of the house at a lower temperature than that of the cutting bench.

Fig. 1.—FRONT OR SOUTH-EAST ELEVATION.

A Convenient House of Medium Size.

Half-a-dozen houses, alike in plan, and differing a little in size, are now being completed in Flushing, L. I. They are designed to furnish comfortable homes for such as desire a convenient house, at a cost of about $5,000, exclusive of land. They were planned by the senior Publisher of this Journal, who has a fondness for mechanical constructions, as a pastime, especially in the way of house-building. **The Elevations** are seen in figs. 1 and 2. The houses face southward to the street, and fig. 1 shows a view from the south-east. Fig 2 gives the opposite, or north-west view. It will be seen, by fig. 2, that the rear is almost as tastefully finished as is the front. As previously remarked in these pages, people see their own houses from the rear oftener than from the front; and their own taste and self-respect should be consulted, by having the house look well on all sides—not a fine front for show to others, and a cheap look in the rear for their own habitual observation. Besides, few houses are so situated as not to be seen on every side from some point. A few dollars in window caps, etc., make all the difference.... The siding (fig. 6) is of 10-inch boards, a full inch thick, and grooved in the middle, to give the appearance of narrow boarding. The lap is 1 inch, which

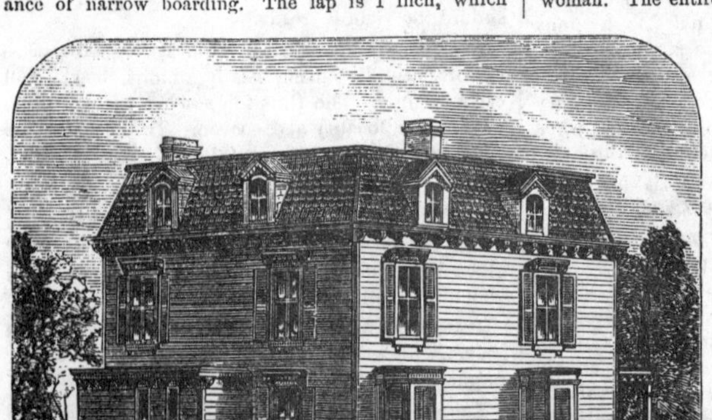

Fig. 2.—REAR OR NORTH-WEST ELEVATION.

prevents the entrance of cold air. This siding is to be greatly commended. The joints are very close, and but half as many as in ordinary clapboarding. These thick boards add greatly to the warmth and solidity of the walls. With such siding, the house is many times stronger than with ordinary clapboards, though the upright timbers be only half as large. The cost is but little greater...... Bricks, laid in mortar, with an open space on each side, fill all the walls, from the cellar to the roof. The Mansard-Roof is covered on the sides with blue slate, laid over a double thickness of felting, and on the top with heavy tin. The Mansard-roof is set more perpendicularly than has been the usual custom. This makes the rooms in the third story very commodious, and it improves the general outside appearance........ **The Cellar** is of brick, nearly five feet out of ground, with large windows. Except in very cold localities, houses are generally set too low. It is more healthful to have the living rooms well above the soil in all cases. The cellar is lighter, and more airy, and, if desired, may be plastered and used for laundry or cooking purposes.

First Story, fig. 3.—Hight to the ceiling, 10½ ft. The *Piazza, P,* supported on brick piers, extends around two sides, with banisters, and front and side steps. The *Front Hall, A,* is wide. This always gives character to a house; a narrow entrance Hall dwarfs an entire house. A side door, under the stairs, opens on to the side piazza. The *Parlor, B,* is ample for all ordinary purposes, and is much improved in convenience, pleasantness, and apparent size, by the Bay-Window. Double windows open down to the piazza floor in front. *Bk* is a bell-pull to the kitchen.—The *Living* and *Dining-Room, D,* is of good size, and is also improved materially by the bay-window. These bay-windows also add much to the outside look of the house, as is seen in fig. 2. Closets are marked at *E* and *c.* This room communicates, by doors, directly with the Parlor and Kitchen, and also, through the hall, with the front and side doors, and the cellar under the stairs. *B3h* is a bell-pull to the third story hall (*Q,* fig. 5), to call servants or others from that floor. The *Kitchen, C,* is of ample dimensions, with Pantry, *E,* and small pantry, or hall, *F,* opening out upon the rear steps. The great feature of this room is the convenience of water and washing fixtures, which should be in every house, even though the expense be cut off somewhere else. A "Victory" Cooking Range, *r,* with water-back, connects with the 30-gallon Copper Boiler, *b.* The Force-Pump-*fp,* in the corner, draws water from the reservoir through tin lined pipe, for the sink, *s,* and, when required in a dry season, it fills the supply Tank in the third story (fig. 5). A stop-cock, over the sink, also supplies hot water to it. The two stationary wash-tubs, *w w,* are supplied with hot and cold water pipes and stop-cocks, and large waste pipes. This arrangement saves all lifting of tubs and carrying out water, and furnishes hot and cold water always at hand. The ordinary fire keeps 30 gallons always hot. Where there is much washing to be done, it is equivalent to saving half the labor of one woman. The entire cost (excluding the Range, which saves the cost of a cooking stove), but including all pipes, third story tank, etc., is less than $250—involving an annual interest of less than $20. It is worth $100, or more, to every housekeeper, and will save that sum in hired help, in strength, and doctor's bills. Let all house-builders look to procuring boilers, and stationary wash-tubs. Those who have not looked into it can hardly appreciate the great advantage of them. The tubs, tank, and pipes, are placed on the south, or warmest side of the house, and kept away from the walls with double plastering behind them, which prevent winter freezing. *Ventilators* are in every room in both stories.

Second Story. *Fig.* 4.—Hight to ceiling 9 feet. The wide hall gives a convenient bedroom, or store-room, at *L.* The front chamber, *G,* has two closets, *c, c,* which are not only convenient, but with the arch thrown over between them, they give the appearance of a bay-window, and brake up the box-like look of any room. This is intended for the family bedroom. On the right side, over the bed, is, *Bk.* a bell to the kitchen; *B3h,* a bell to the 3rd story hall, to call servants in the morning; *Sk,* a speaking tube to the kitchen to talk with, or give orders to the servants or others there, and, *Sfd,* a speaking tube to the front door, opening just over the bell, to speak with any caller at night, without having to dress and go down. These little contrivances cost no great sum, and are a material help in saving woman's steps. They are built into the walls when constructing the house. *Sk* in *H* is a bell-pull to the kitchen. There are two closets *c, c,* in this room. The chamber *I,* has a closet, *c;* room for a full-sized bed at *y,* and a wash-sink at *S-hcw,* with stop-cocks over it in pipes carrying hot and cold water. The waste-pipe is large, so that, if preferred, it can be used as a chamber-slop sink. This arrangement saves all carrying of water up and down stairs, and gives constant hot and cold water on the second floor. If desired, a bath-tub can be placed in *I,* by simply connecting it with the hot and cold water pipes, and its waste-pipe with the sink-waste. The house is designed for a large family (8

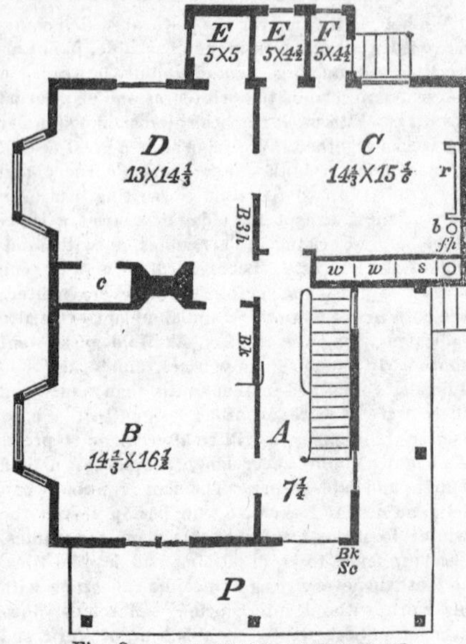

Fig. 3.—FIRST STORY—FLOOR PLAN.

sleeping rooms), but if this room (*I*) is not needed, it can be readily converted into a bath-room, store-room, etc. The stairs to the Third Story are omitted by the engraver. They are directly over the lower flight, the entrance being at the left of the door to *L,* with a half turn in the

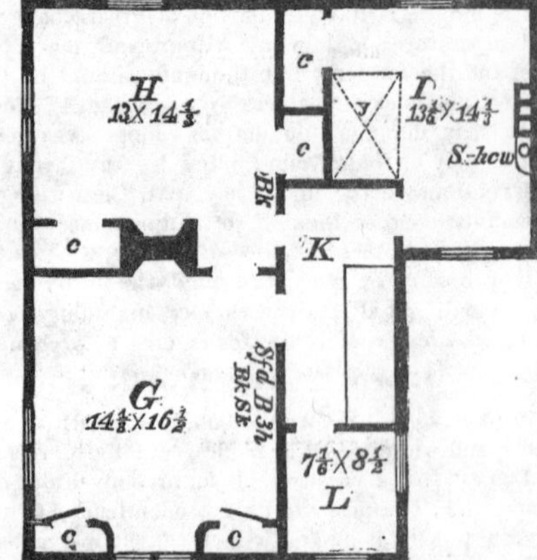

Fig. 4.—SECOND STORY—FLOOR PLAN.

steps. It will be noted in fig. 3, that the Main Hall, *A,* is well lighted by a side window at the foot of the stairs.

Third Story. *Fig.* 5.—Hight to ceiling 8½ feet. This has four finished bedrooms also, as shown on the plan. They are full size to 4 feet high, and the slightly

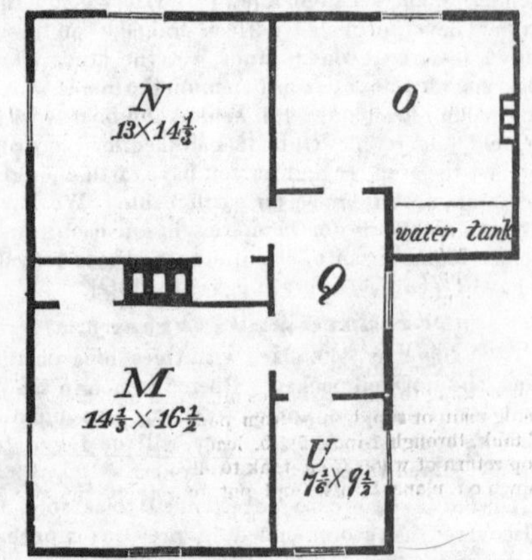

Fig. 5.—THIRD STORY—FLOOR PLAN.

inclining Mansard-roof and dormer windows, render them almost equal to the rooms in the second story. *U,* is nearly large enough for the largest sized bed. The *Water Tank* holding 450 gallons, receives the water directly from the upper roof, and when full, the surplus water flows down to the large receiver outside the house. The room *O*, is plastered and may be used as sleeping-room, or store-room.

Other Items.—Ample drainage is provided for the cellar, sinks, etc. A Grape Arbor runs back to the privy, which is flat-roofed and enclosed with an **L** of the arbor and lattice-work screens, so as to effectually hide it from view. The window-caps and cornices are more ornamental than shown in the engravings. The second cornice, at the top of the slate, adds to the beauty of the roof. The mouldings in the first story, (fig. 7,) are a new pattern, designed by S. B. Reed, architect, which secures larger ornamental mouldings with the same cost of timber, than any pattern we have before seen. It is cheap, quickly put up, and does not shrink open. The middle piece is cut from a board less than an inch thick....It will be seen that there is not an inch of waste room in the house, while there is a large amount of room in proportion to the amount of walls. Ten closets or pantries are provided.

Fig. 6.—END-VIEW OF SIDING.

All doors well grained in walnut; all woodwork, outside and in, 3 coats paint; dining-room and kitchen grained in oak; Parlor in colors; upper rooms white; stair newell, railing, and turned banisters, solid walnut. Gas-pipes to every room, from cellar to attic.

Fig. 7.—1ST STORY MOULDING—8½ in.

Cost.—This will depend upon location, etc.; but with proper economy, it may be built for about $5,000, exclusive of land, everything else included. In this case, all materials are bought of first hands, at lowest wholesale, net cash rates. The work is done by active, expert workmen, by the day, with an efficient superintendent, (Mr. John Donald), with favorable weather, etc., and the cost, all complete, is about as follows: Timber, $250; Lumber of all kinds, $760; Mason-work and material, $970; Carpenter-work, $750; digging Cellar, $15; digging and stoning privy vault, $15; Grading, 35; Painting, $265; sashes and glazing, $110; Blinds, $50; Slate and tin, $280; Plumbing, drains, and Range, $290; Stairs, $120; Stoops and Piazza railing, $165; Arbor and Privy, $60; Fencing, $100; Side-walk, $25; Hardware (nails, locks, etc.), $130; Mouldings, $85; doors, $90; 4 Italian Marble Mantels, $140; Grates and summer-pieces, $60; Gas-pipes, $40; Interest, $110; Carting and Sundries, $185. Total, **$5,100.** Land, $1,000. N. B.—Part of the houses are one foot smaller (6 inches on a room) than the plans here shown, and that is the size here estimated for. The larger size costs about $150 more. With ordinary workmen, and with materials purchased on credit, at retail prices, the cost would run from $500 to $800 more, here. A cheaper style of finish, inside and out, would materially reduce the cost. Among items of cost in these houses, we reckon: best nails at 4⅜c.; tin roof, 10c. per foot; slating, 14c; gas-pipe, 15c. per foot; timber, 2½c.; siding, planed and grooved, 42 cents per 10-inch board; bricks, $8½ per M; blinds, unpainted, 40c. per foot; parlor moulding, 8½c. per foot; mason work, $4.50 per day; tenders, $1.75; carpenters, $3.00 to $3.25, etc.

Plumbing Specifications.—As a matter of information, not generally understood, we give the specifications for the Plumbing work: "In third story, line Tank (6 ft. long, 3½ ft. wide, 2¾ ft. high), with 4 lb., per ft., sheet lead; the sides of tank lead to be "tacked": all seams to be heavily soldered, and all inlets and outlets to be flanged over and soldered. Put in 3-inch inlet, with 4-inch overflow, to connect with leader pipes......In Kitchen, furnish and put up one 30-gallon copper boiler, round, riveted head, Croton pressure, set on Lockwood's stand, properly connected with waterback in range, and a ¾ *aa* lead pipe, with ⅝ sediment brass cock, to connect at the bottom with waste from sink......By side of Boiler, furnish and put in one cast-iron sink, 30x20, to be supplied with hot and cold water, through ⅝ 2lb. lead pipe, and two brass ⅝ Bibb's cocks. Also, where shown on plans, fit up two wash tubs, with right and left cocks. Waste from wash tubs and sink, through 2-inch lead pipe, properly trapped, to connect with earthen pipe leading to drains. Cold water supply for wash tubs and sink, to come direct from pump; also, run hot water pipe from the boiler up over top of tank, to prevent accident...... Furnish and put in one 2½-inch Douglas suction and force pump, suction to be 1¼-inch 3½ lb. tin-lined lead pipe, leading to within 4 inches of bottom of cistern; the rising main or supply pipe from pump to empty over top of tank, through 1-inch 2½lb. lead pipe, with valve to stop return of water from tank to sink......Also, where shown on plans, furnish and put in one quarter circle galvanized iron sink, 27x14, supplied with hot and cold water through ⅝ 2½lb. lead pipe, with ⅝ Bibb's cocks. Waste from this sink through 1½ lead pipe, properly trapped, to connect with earthen pipe leading to drain. All pipes to be put up with lead tacks and screws, with all necessary stop-cocks and traps; trap screws to be put in all traps. All to be done in a workmanlike manner, and warranted for one year, damage by frost excepted."

HAULING OUT MANURE.—"W. G. C.," says: Commence dropping the heaps two and a half paces from one side of the land to be manured, and drop the heaps five paces apart, the whole length. Drop the next row five paces from the first, commencing half way between the first two heaps, breaking joints, as it is called, and so continue, until the whole is finished. This takes 160 heaps to the acre. If it is desired to manure pretty heavily, drop five heaps from a one-horse cart, which will take 32 loads to the acre. Six heaps from the cart take 27 loads. Seven heaps 23 loads. Eight heaps 20 loads. An ox-cart or a two-horse wagon will hold one or two heaps more. The quantity required on an acre must always depend upon the quality of the manure, the condition of the land, and the kind of crop to be raised.

SPREADING MANURE.—In spreading manure, care should be taken to scatter it evenly over the land, breaking to pieces all large and hard lumps. This should always be done immediately, or not more than half a day, before plowing, especially if the weather is dry and very windy. The manure should be plowed under, before it dries very much, or loss will accrue.

PLOWING AND HARROWING.—Never plow if it can be avoided, or go on to the ground for any purpose, when it is wet and sticky. Keep the furrows *straight*, and, if possible, reverse them at every plowing, so as to keep the land level. To fill in furrows, back-furrow pretty widely once around, and haul once around very wide; this will generally be sufficient. Harrow soon after plowing and before the lumps, if any, get dried hard; twice over with the teeth down and once with the back of the harrow, will prepare the land for ordinary crops.

DO PIGS PAY?—A correspondent in N. J. writes: "I have just footed up the proceeds of one brood sow, kept during the past year, and find I have received $406.54, and have the sow still on hand. The pigs were kept and fed in an ordinary manner during the summer on milk, and aside from the poor corn, not more than 100 bushels of ears of good corn were fed." The pigs were sold at from 5 to 11 months old.

Maple Sugar Making.—2d Article.

BY W. I. CHAMBERLAIN, HUDSON, O.

The article in February concluded with the tapping of the trees. If the day has been a good one, many of the buckets at the trees first tapped will be full, and it will be well to commence

GATHERING SAP.—Take the stone-boat sled (fig. 5, Feb.), put on three barrels or casks, a tunnel made to fit the barrel and not rock,—and a tin pail holding sixteen quarts. One pail is better than two, unless the trees are scattered far from where the sled can go. There should be a *strainer* in the tunnel. This should be of the shape shown in figure 2, and made of thin, coarse, white muslin, stitched around a wire hoop at top and bottom. The top wire should fit tightly over the top of the tunnel, and be pressed down over the outside of it about an inch, so as to hold the strainer in place. The bottom wire should be small enough to keep the strainer from touching the sides of the tunnel, else the sap cannot run through rapidly. The strainer should be so short that the bottom of it will not touch the bottom of the tunnel, for this too would obstruct the flow of sap. The strainer is to keep out from the barrels dirt or flies that may have got into the sap, and also to prevent the loose flaky ice often found in sap from clogging the tunnel. The ice may be thrown away, as it is of little value.

Fig. 1.—HOW TO EMPTY SAP.

Select a good road through the grove so as to bring the sled as near each tree as practicable. There is but one right way to empty the sap from the bucket into the pail. If you are not used to the work, probably you will set the pail on the ground, take off the bucket cover and throw it down, take the bucket from the spile with both hands and empty it, hang the bucket, stoop for the cover, and finally stoop again for the pail. If the pail does not stand on level ground, the sap will run over, and if the cover is wet, dirt will stick to it and fall into the pail. The right way is this. Stand facing the tree, with the bucket in line between it and yourself. Hold the pail in your left hand, take the cover with your right hand, and place it up under the left arm above the elbow. Now hold the pail close to the left of the bucket, grasp the rim of it firmly with your right hand, and turn the bucket gently on the spile, as on an axle, until the sap is all out. Then let the bucket back to its place, and put on the cover. (See fig. 1.) The bucket is not removed from the spile, no sap is wasted, no dirt sticks to the cover, your backbone is not bent, and one hand empties the sap far more easily and quickly than both. In gathering, if you find a large bucket only half full, and near to it a small one overflowing, change them. Much sap can be saved by a little care in changing the buckets where necessary.

Fig. 2.

When your barrels are full, and you go to empty them, be careful to drive close to the ends of the skids. These are straight poles, or 3 x 4 scantlings, fastened at the upper end to the store-trough, and at the lower end to a timber at right angles with them, and high enough to bring them just on a level with the top of the sled-rave, as shown last month in the engraving on page 50. The skids should be just long enough to let the barrels roll once and a half over and bring the bung-holes down. Or if the top of the store-trough or vat is low enough, the skids may be so short as merely to let the barrels roll half over. At any rate the skids should not be steep, not more than fifteen degrees, or the barrels will roll too hard for one man to unload easily.

Fig. 3.—DIPPER.

If there is any mud on the barrels remove it

before they are stirred, for if it gets into the sap in emptying, it cannot be strained out. Indeed, if the road is very muddy, it is well to have a tight sled bottom below the barrel rack.

About two feet from the lower end of the store-trough, a strainer-cloth should be tacked tightly to its sides and bottom by strips of lath, and should be stretched across the trough from top to bottom. The lower end of the vat is only half an inch lower than the upper end. This gives the sap a gentle motion through the strainer, permits the fine, heavy dirt to settle, strains out everything at all coarse, and drains the trough after you have stopped putting in sap. The lower store-trough (in the sugar-house) is arranged in the same way, to secure the greatest possible cleanliness.

As soon as the first sled-load of sap is in the upper vat, and has settled for a few minutes, the faucets are opened and it is run down into the boiling-pans, and as soon as it covers the bottom of these, one man should commence boiling. The most careful and capable hand should take this work,—usually the owner of the sugar-works. The wood should be well crossed in the "arch," and not permitted to pack down and keep the air and flame from circulating freely. It may pack down so that no heat shall reach the front pan. One should not be satisfied unless the sap is foaming and tumbling all over the pans. Scum (like suds), will rise and float to the edges and corners, and should be skimmed away often. It may be saved, cooled, skimmed, and settled, and put into a barrel with half its bulk of *rain-water*, and a little vinegar or "mother." In a year it will be quite good vinegar.

If the sap is inclined to boil over, a bit of lard as large as a small pea, will keep it down for two hours, and not injure the flavor of the syrup. Do not split wood in the shed near the arch without closing the door, or the dirt and chips will fly into the boiling sap.

If you have not the self-feeders (fig. 4, Feb. No.) you must use great care not to let the sap boil too low. You go out, perhaps, to help unload, or to change buckets according to the running capacity of the trees, and come back to find the sap boiled to syrup and just ready to burn. Do not, however, keep the pans too full. The less sap there is in a pan, other things being equal, the faster does it evaporate. Not merely a larger *fraction* of the quantity in the pan, but a greater absolute amount,—more gallons.

It is best to "syrup off," often. A barrel of sap makes a gallon of syrup, thick enough to strain, cool, and settle before clarifying, which, when cool, will weigh 10 lbs. When you have enough boiled in for six or eight gallons of syrup, and it is boiled to about an inch deep in each pan, slacken the fire a little, and dip all you can safely, into the front pan, and supply its place with as much cold sap. Boil that in the front pan until it will drop from the edge of the dipper in drops three-quarters of an inch broad. Then, if you have a car (fig. 5, Feb. No.), draw the pan up two inches and roll it to the front, away from the fire, lower one end two inches and dip or pour off the syrup at your leisure, and run it into a twenty-gallon cask through the strainer (fig. 2). If you have no car, slacken the fire a little, and dip off the syrup as low as practicable with a flat-edged two-quart dipper (fig. 3). Have a pail of cold sap at hand, and when you have dipped as low as it is safe to do, turn in the cold sap. Or if you wish all the syrup out, dip out all but a pailful, and then two men can easily lift off the pan, pour out the syrup at one corner, turn in a pail of sap and put the pan in its place again. When the cask is full it should be removed to the dwelling house, put on a bench and left for twelve hours to settle.

It is best to "syrup off" as often as once in ten gallons. The syrup is better, there is less risk of burning, and it boils faster if the pans are only partly full of syrup.

For clarifying, a small sheet-iron pan, similar to those at the sugar-house, is used. It is about 2 ft. long, 1½ ft. wide, and 9 in. deep. This is set on top of the cooking stove, and the syrup filled in to the depth of two inches. If proper care has been taken in gathering and boiling, the syrup, after settling in the cask, will draw off as clear, and almost as light colored as strained honey. It is common, however, to clarify it with milk or beaten eggs, or both together. Eggs make lighter-colored syrup, but injure the maple flavor; hence it is best to use milk. A pint of it will clarify ten gallons of syrup. The proper quantity of milk should be stirred into the syrup when it is first put over the stove. As soon as it begins to boil, the milk, with the impurities, will rise to the surface in a thick, dark scum. This should be skimmed off as often as it rises. Boil and skim until a gallon of the hot syrup will weigh 10¼ lbs. The scales and gallon measure should be at hand, and every mess of syrup should be brought to exactly this weight. In cooling, it shrinks so much that a gallon when cool, weighs 11 lbs., and this is standard weight for maple syrup. If it is thicker it will turn to sugar badly.

If you wish to make sugar, the syrup should boil until it "hairs," that is, drops from the edge of a dipper or spoon, and draws out into hairs three or four inches long. For cakes it should be taken off and stirred until it begins to grain and turn light colored, when it may be poured into tins of any required size and shape. For grained sugar, the stirring must be continued until the sugar is nearly dry, when it may be put in a cask with a perforated bottom to finish drying. It ought to be as dry and white as good "C" coffee sugar. In general, however, it does not pay to make sugar, except a few pounds of small cakes at the first of the season for eating. A limited quantity early in the season will bring from 25 to 35 cts. per lb. Later it will not bring more than 18 cts., and then the syrup, if nice, will buy its weight of "C" coffee sugar, which is better for ordinary cooking purposes. Nice maple syrup is far the best syrup made for buckwheat cakes; and at tea, with hot biscuit, it is better than honey. One does not tire of it so soon as of honey. But it must be of the *best quality*. There is as much difference between syrup made as I have described, and that made from sap and water caught in open wooden buckets or troughs, and boiled in kettles hung between two logs by "sweeps" and chains in the old fashion, as there is between Cauliflower and Cabbage, or Delaware and fox grapes, and people begin to appreciate this difference. Prime maple syrup, made as I have described, now brings in Northern Ohio, from $1.50 to $2 per gallon, according to the time of making. But it must be strictly first quality, and in order to have it thus, three things must be observed:

First.—*The sap must be kept clean*. . . . Second. —*It must be kept cool and sweet until it is boiled*, and in order to have this, ... Third.—It must be *gathered as soon as possible* after it runs, and boiled as *rapidly as possible*.

One can guard against sour sap in several ways. When cakes of ice form in the buckets they should not be thrown out (though there is little sweetness in them), for they keep the sap from souring as long as they remain unmelted. It even pays to gather a quantity of this ice to put into the vat if the day is warm. The most of the sap should be kept covered tightly in the *outer vat*. The one in the sugar-house is warmed by heat from the arch, and by steam from the boiling, and the sap will tend to sour. Again, after each "run," the vats, strainers and barrels should be scalded. We usually have two or three days, sometimes a week, of freezing nights and warm days, when the sap runs well. Then it rains or snows or freezes solid for as long a time. The consecutive days of sap weather before the storms and freezing, are termed a *run*. When there are indications that the run is over, barrels, vats, strainers, pails and boilers should be left sweet and clean. The vats and strainers should always be scalded. If the spiles begin to sour they should be brought in and thoroughly boiled out in water in one of the pans. The buckets can be most conveniently scalded by taking a barrel of boiling water on the sled, going through the woods, bringing twenty buckets to the sled, scalding and returning them to the trees bottom side upwards. Two men in three-quarters of a day will scald 500 buckets and spiles, and it pays over and over again for the work. Some seasons the buckets and spiles should be scalded three times, twice during the season, and once at its close. Usually, by the time the buckets need scalding the holes need reaming out, or the trees retapping. The reaming is done with a curved-lipped bit ⁵/₁₆-inch in diameter. If the trees are small and tapped every year it is not wise to bore *a second hole*. Reaming the old one answers the purpose.

Fig. 4.—BUCKET INVERTED.

If the buckets are not sour, it is well to invert them at the last gathering of the run. The bucket is not removed from the spile, but inverted on it, and left inclining from the tree at an angle of about 25 degrees (as in fig. 4). This drains it, and the frost and wind make it sweet and clean. The cover should be laid on a clean root, if placed upon the ground the dirt may freeze to it.

The sap should be gathered as soon as possible after it runs, and boiled soon and rapidly, even if it requires night boiling. Night-boiling is not so bad as it might seem. A bunk is built in one corner of the sugar-house, three feet high, with straw-bed, pillows and buffalo-robe or blankets; and two men divide the night, one boiling while the other sleeps. You will sleep soundly after gathering thirty barrels of sap.

THE PROFITS.—The expense of fitting up a "camp" of 500 trees with buckets, spiles, covers, vats, etc., including a decently good sugar-house and shed, need not much exceed $500, or $1 to the tree. In a favorable year, good trees will yield fifty cents worth of syrup, which is good interest. The buckets and fixtures will last thirty years or more, if cared for. The fuel of the kind I have described answers the purpose well, if housed, and costs little. The work can be performed by the usual force, and comes at a time when not much other profitable work can be done, at least on a dairy farm.

If a man has 500 good maple trees, growing close together, say on six or eight acres, with the other trees mostly cut out, this piece of ground will probably net him more than any other of equal size on his farm.

The Flamingo.

The Flamingo forms one of those wonderful zoölogical links, which both delight and puzzle naturalists. It may be said to be a goose that does not swim, and it is a wader with webbed feet, besides it picks up its food, holding it in the spoon-shaped end of its upper bill. In its instincts, and modes of flight, it much resembles the Wild Goose; in many of its habits, also, it resembles the crane and heron; so it really occupies an intermediate position between the waders and swimmers, besides being exceedingly curious and interesting in other respects. When Europeans first landed upon the West India islands, they found these great red birds arranged in phalanxes along the coast like ranks of red-coated soldiers. They were so tame then, that it was easy to approach them; and many years after, it is said, a sportsman might secrete himself and shoot one after another for some time, before the flock would take the alarm. Now, they are the shyest of all wary things. When they pass over the water, they fly low, but when they approach the land, up they go, away above the reach of shot. On alighting, they do so in shoal water, where they feed, standing erect, if necessary, but easily reaching the bottom with their heads. The legs are very small, and as the birds are heavy, they are able to stand in swift currents of the in-coming and receding tides. When feeding, as described, the heads are long submerged, and of course the birds would be exposed to attacks from their enemies, and especially from man, were it not that sentinels are posted at either end of the long row of feeders, which stand erect, and sound a trumpet-like alarm in case of danger. Then the sentinel leads off and all follow in Indian file, or in two lines, forming an angle, like wild geese, which they always do when on the wing. The motions of these birds are peculiarly graceful, although they are so tall and apparently unwieldy. Their heads and necks are swayed about with a gentle deliberation and ease, which is most striking when they dress their feathers.

The best known species of Flamingo are the American (*Phœnicopterus ruber*), and the European or Mediterranean (*Phœnicopterus antiquorum*). Their size and habits are very similar, the European being somewhat the larger. In color, ours is much the more intense, approaching scarlet; while the European one is of a rosy-white, except upon the wings, where the color is much deeper—sufficiently crimson to give good ground for the generic name, which means crimson-wing. These beautiful birds do not long survive in captivity, and of those from which the drawing for our engraving was made, none are now alive. They were imported by Mr. Charles Ritchie, of New York, from the Zoölogical Gardens, at Antwerp, at a cost of $350 per pair. The two larger birds in the engraving are of the European, and the smaller one of the American species. The flesh of the Flamingoes is esteemed, and among the ancients the thick, fat tongues, were regarded as the most delicate of all articles of food.

EUROPEAN AND AMERICAN FLAMINGOES.

The American Panther.

That the American Panther is widely distributed, is shown by the various names it has received. In some localities it is known as the Catamount and "Painter,"—a corruption of Panther. By the Spanish American inhabitants of Mexico and California, it is called Leon (Lion). It is Pagi in Chili, and Puma in Peru. Indeed, it is found from latitude 50° or 60° north, to the extreme point of South America. It is a true cat (*Felis concolor*), and with the exception of one found in Louisiana and farther south, is the largest of our five species. Its weight sometimes reaches 150 pounds, and it measures about seven feet from the nose to the end of the tail. It is covered with short, compact hair of a brownish-yellow on the sides, and of a dirty white on the under part of the body. In color, the panther very closely resembles the common deer, and it is said to change like that with the seasons. The tail is more than half as long as the head and body together, and has a brushy tuft at the end. The young animals are spotted and barred, but the adult ones are without markings, except a black patch on the upper lip, and dusky black upon the interior of the ears. The animal is more abundant in mountainous districts than elsewhere, and is rather shy in its habits, concealing itself among the rocks in the daytime, and prowling about at night. It is very destructive to colts and calves, and has been known to attack the full-grown animals. It climbs trees readily, will lie in wait crouched upon a limb, for deer to pass, and drop suddenly upon them. It is stated that the Panther has been known to attack and kill a man, but we have met with no authenticated account of its having done so unprovoked. It is said to be easily tamed, and in captivity its habits are much like those of a cat. In California and the adjoining parts of Mexico, Panthers are very destructive to domestic animals, and strychnine is extensively used to poison them as well as the wolves. The early Spanish Missionaries had great difficulty in sustaining themselves in Lower California, as the Panther destroyed their domestic animals. The Indians would not kill or disturb the Panther, as they had a superstitious regard for it, for the reason that they depended upon it for a good part of their food. By watching the buzzards, who always soon scent out a concealed carcass, they could discover where the Panther had hidden that portion of his prey which he was unable to devour, and regale themselves upon the fragments.

THE AMERICAN PANTHER.—(*Felis concolor.*)

For other Household Items, see "Basket" pages.

Library Steps.

Modern furniture at best, is not very strong, and soon becomes worn out even with the most careful usage. Nothing brings a chair into that rickety stage which precedes dissolution, sooner than its frequent use as a substitute for a step ladder. It is the most convenient thing near at hand, if one wishes to take something from a high shelf, reach up to arrange a window curtain, or for any similar purpose. Step ladders are not convenient to keep always at hand, and there is generally more or less trouble in arranging them. The house-furnishing stores sell what are called library steps, like those represented in the engraving. A tall stool has a strong iron rod connecting two of its legs. This rod supports two steps, which are of such a size, that, when not in use, they can be readily turned back, to occupy the space between the legs, and the affair takes up no more room than a stool. Those in the stores are made of black-walnut, with the steps and the top covered with pieces of carpeting. Such a piece of furniture, which any amateur workman can make, will be found very convenient, and save much wear and tear of chairs.

LIBRARY STEPS.

Nice Dishes that all Farmer-Folks can have.

BY MRS. "H. M. R.," MONTOUR CO., PA.

The question, What shall we do for a change? is one which often vexes the housewife, and many times is not so easily answered, especially among a certain class of farmers (and a pretty large class too), who feel that they must make the products of their own farm furnish variety for the table. Among this class, apples are the "stand-by" for fruit, and in nine families out of ten they seldom, if ever, come to the table in but two ways, viz.: between two *lard* crusts, and stewed. I give a few recipes for preparing this esteemed fruit, which, when once tried, will not, I think, be willingly abandoned....Take easy-cooking, tart apples; pare, and with a narrow-bladed knife remove the core; after they are all served in this way, place on a tin plate, the stem or blossom end down, and fill the hole, left by removing the core, with sugar; drop three or four drops of water (flavored with lemon, if you choose) on the sugar in each apple, and set in the oven and bake until done, which will be in about three-fourths of an hour, if your oven is right. Remove to an earthen plate as soon as done, which should be fifteen or twenty minutes before serving. These are excellent with meat dinners, or, eaten alone with cream and sugar, they make a very nice dessert....Some persons consider sweet apples of little use in the kitchen, but when properly prepared, I think they are nearly as good as peaches. Pare, quarter, and core, nice sweet apples, put them in an earthen or new tin dish and set in the steamer and steam until done, which can be told by trying with a fork. Serve, when cold, with cream and sugar. Do not forget to steam and can some before they are all gone, so that you can have them next spring when other fruit is scarce....APPLE PUDDING.—Set as much stale bread in the oven as will make a pint of crumbs. When it is brittle enough to roll, remove, and roll very fine. Take four medium-sized tart apples, pare, quarter, and core, cutting each quarter into four pieces by cutting it in two, both lengthwise and crosswise. Then take one quart and a gill of new milk, the *yolks* of four eggs, one cup of sugar, butter the size of an egg, melted, a little salt, and spice to your taste. Beat the butter, eggs, and sugar together, and stir in with them all the other ingredients. Bake in the same dish in which it goes to the table, unless brought to the table in small dishes. Before removing from the oven, beat the *white* of the eggs with a half cup of sugar to a stiff froth and spread over your pudding, and, when "set," remove from the oven. To be eaten, when nearly cold, with cream....APPLE JONATHAN.—Pare, quarter, and core enough tart apples for two layers over the bottom of a deep, square pie-tin, cutting each quarter in two, sprinkle with a little sugar. Then take one coffee-cupful of sour cream, one of buttermilk, a teaspoonful of saleratus, a pinch of salt, and flour enough to make a batter as stiff as will pour. Stir all well together and pour over the apples, and bake until done. Turn out of the baking tin on to a platter, and serve with cream and sugar.

The Blanchard Butter Essays.—Competitors should remember to mail their essays previous to the 10th of this month, that they may be received on or before that date.

Household Talks.

BY AUNT HATTIE.

I have been fixing up an old rocking chair that has done nursery service for many years. The cane bottom seat was almost gone, so first with a pair of scissors, I picked out the whole of it from the holes, leaving nothing but the frame. I had previously procured some strong packing twine, which I threaded backwards and forwards after the manner of cording an old-fashioned bedstead, making a very durable seat; then I made a cushion, and a cover for it, and I value it now more than ever.

I notice that my new girl wears away a broom all on one side. There is nothing annoys me so much as to be obliged to use a broom that has been so worn. I should be ashamed to have any one see Libbie's kitchen broom as it is now. I think nothing is so suggestive of extravagance and bad housekeeping as a one-sided broom. There is no necessity of having a broom worn down in this way. When sweeping, hold the broom up, almost perpendicular, and brush rather than sweep, being careful to turn the broom and keep the longest side towards the carpet or floor. A broom kept straight, will last three times as long as one allowed to wear all upon one side.

Mrs. Gilman asks me to write more about making cake. She does not mention her own skill or want of success, but writes: "Some of my neighbors make shocking cake! What is the reason that cake is not only heavy but oily at times?"—I remember that mother used to say, the way to have light cake is to beat the whites of the eggs to a perfect froth, and add them lightly, the last thing. That was very well in those days, when eggs were only 5 to 8 cents per dozen, and we put in nine or ten to an ordinary sized cake, but now, when the farmers get 35, and the grocers 40 cents for them, it is evident that cake must be made light in some other way. I use only one egg to one loaf of cake, but I am careful to obtain good baking powder, or pure cream of tartar, and good soda. I think one great secret of light cake making, is this keeping the ingredients in a state of foam, during the mixing. For instance, beat the sugar, butter, and the yolks of the eggs until they turn white and creamy. Then add milk, if used, and some of the flour, beat and stir one way all the time until the ingredients are mixed, and bake immediately. If cream of tartar and soda are used, mix the former with the flour, and add the soda dissolved in a tablespoonful of warm water, the last thing before the white of the eggs. I imagine that cake is oily, at times, for the reason that the proportions are not properly measured. The sugar in all cases should be as much as, or more than twice the quantity of butter, and the flour should be equal in quantity to the butter and sugar together, thus: 1 part of butter, 2 parts of sugar, and 3 parts of flour. If these rules are observed, and the baking powder is good, you cannot fail to make good cake. I make a very nice cake using only one egg, as follows: Take one small teacupful of butter, warmed, two and a half teacupfuls of sugar, and the yolk of the egg; beat well together, then add one-half teacupful of milk; add gradually, beating one way, three teacupfuls of flour; flavor with very fine strips of citron or candied lemon peel. Finally add three or four teaspoonfuls of baking powder, the white of the egg, and bake in a buttered pan lined with white paper.

I manage my washing in this way. The evening previous to washing-day I put on the boiler and two gallons of soft water, allowing about a quarter of a pound of sal-soda, and a quarter of a pound of soap sliced; let them boil together until the soap and soda are dissolved. I then put the fine white clothes into one tub, and the coarsest ones into another. To the water in the boiler I add enough cold water to make the whole luke warm, and pour it over the clothes, and cover the tubs with a blanket. In the morning I add a pailful of hot water to the fine clothes and rub them well from this. They then go through another rub in fresh water, are next boiled twenty minutes, sudsed, rinsed in blue water and hung up to dry. The coarse clothes receive the same treatment. My colored clothes are washed, rinsed, and starched before hanging out.

In starching my muslins and the shirt bosoms and wristbands, I use boiled starch, being careful to rub in well the starch, that is, after the shirt bosom has been dipped and wrung out as dry as possible, it is well rubbed and patted between the hands. This is a secret known to all good laundry women, and it is almost impossible to prevent the iron from sticking unless this precaution is observed. I cannot make Edward's collars stiff enough with boiled starch, so I always use cold starch for them. Take two teaspoonfuls of starch, and perhaps a half teacupful of water, or a little more. Have the starch thoroughly mixed before dipping in a collar, wring out, and rub and pat with the hand, spread on a clean towel, and when all are done, roll up the towel and iron in about an hour.

Hints on Cooking, Etc.

Cooking Parsnips and Cabbage.—"Lucy Lamb," says: Wash the parsnips, scrape, boil tender, and then slice and brown on a griddle, with butter to prevent sticking. Carrots are good, cooked in the same way....Boil cabbage tender in clear water; drain dry, add salt and a good piece of butter put on in little bits; cut in small squares for convenience in serving. Another method, requiring only half as much time in cooking, is to chop the cabbage fine before cooking; boil in clear water forty or fifty minutes, and serve as above. It is nearly equal to cauliflower.

Raised Cake without Eggs.—By Lucy Lamb.—Eggs are scarce and expensive, and we invent recipes which require few or none. Stir together a coffee-cupful of light sugar, and half a cupful of butter; add a pint of warm water, half a cupful of yeast, and flour enough to make as thick as ordinary fruit cake. Rise over night. When very light, add a little mace, cinnamon, allspice, and nutmeg, and a cup of chopped raisins. Put in the pan, let rise until light; then bake.

Marion's Cake.—Stir to a cream a teacup of butter and two of sugar; then add four eggs beaten to a froth, one small grated nutmeg, and a pint of flour. Stir until just before it is baked. Bake in cups, about 20 minutes.

Corn Cake.—By Mrs. R. E. Griffith, Chester Co., Pa.—1 pint of buttermilk, 1 pint of corn meal, 2 beaten eggs, 1 teaspoonful of soda; beat well together and bake in shallow tin pans. Increase quantities for more than four or five persons.

Cookies.—By Mrs. L. A. G. 1 large cup of butter; 2 cups of sugar; 3 well beaten eggs; ½ a cup of sour cream; 1 level teaspoonful of soda; cinnamon or carraway seeds, and flour enough to roll.

J. & P. COATS'

BEST SIX-CORD

IS NOW THE

ONLY

Thread put up for the American market which is

SIX-CORD IN ALL NUMBERS,

From No. 8 to No. 100 inclusive.

For Hand and Machine.

FOR SALE BY

All Dealers in Dry Goods and Notions.

India Rubber Gloves

For Gardening, Housework, etc., etc. A perfect protection for the hands, making them soft, smooth, and snowy white. A certain cure for Salt-Rheum, Chapped Hands, etc. Ladies' short, $1.50; Gauntlets, $1.75 per pair. Gents' short, $1.75; Gauntlets, $2.00 per pair. Sent by mail on receipt of price, by GOODYEAR'S I. R. GLOVE M'F'G CO., No. 205 Broadway, New York, Manufacturers of all kinds of Rubber Goods.

FARMERS,

Ask your Merchants for the

"CHAMPION SHOE."

See June Number *American Agriculturist*, page 237.

A. BALLARD & SON, 32 & 34 Vesey St., New York.

ON REMITTANCE OF $42.00

We will send to a Club of Ten Smokers, warranted genuine Meerschaum Bowls of medium size in cases with Weichsel stems and Amber bits (worth retail $6.00) for $5.00 each, allowing to canvassers $8 in cash from the full amount of $50.

POLLAK & SON, Manufacturers,
Letter Box 5846. 27 John St., New York.

MACY HAS IT.

Every new thing, **Macy** has it.
If you order anything, **Macy** has it.
Any advantage in price, **Macy** has it
We buy and sell for cash— **Macy** has it.
Every new book, **Macy** has it.
Everything in Lace Goods, **Macy** has it.
Hosiery, Gloves, Underwear, **Macy** has it.
Small Wares, Trimmings, &c. **Macy** has it.
All at popular prices— **Macy** always has it.
Cash orders, by Mail, or Express.

R. H. MACY, 14th St. & 6th Ave., New York.

Samples Free.
The Great Improvement
IN
ROOFING.

SEPTEMBER 1st, 1870.

Having just completed our new manufactory containing new and improved machinery, and all the appliances found necessary during our experience of nearly thirteen years in the manufacture of Roofing Materials, we desire to furnish samples to Builders, Roofers, General Merchants, and others, of our

IMPROVED ROOFING MATERIAL,

which is ENTIRELY DIFFERENT FROM ANY OTHER, and is TEN TIMES STRONGER than any Composition Roofing in use.

An examination of this material will satisfy competent judges that it is in every respect a substantial and durable fabric, made upon correct principles, perfectly adapted to the purpose, and totally unlike the numberless cheap and flimsy articles heretofore sold for Roofing purposes; while owing to our extensive manufacturing facilities we can supply it for about the same price.

We also manufacture from the fibrous mineral asbestos a

FIBROUS ROOF COATING,

prepared ready for use, which can be applied with a brush, and forms an *Elastic Water-proof Felting*, on any surface; by use of this Coating, old and leaky Roofs can be made serviceable for many years.

These materials can be readily applied by any one. Descriptive Pamphlets, Testimonials, Prices, etc., by mail.

The Roof of the Agriculturist Building is coated with these materials, which have so far proven satisfactory.

EDS. AMERICAN AGRICULTURIST.

H. W. JOHNS, 78 William St., New York.

Manufacturer of Asbestos Roofing, Roof Coating, and Cement Roofing and Sheathing Felts, Preservative Paints, etc. Asbestos, crude, crushed, and ground.

Established in 1858.

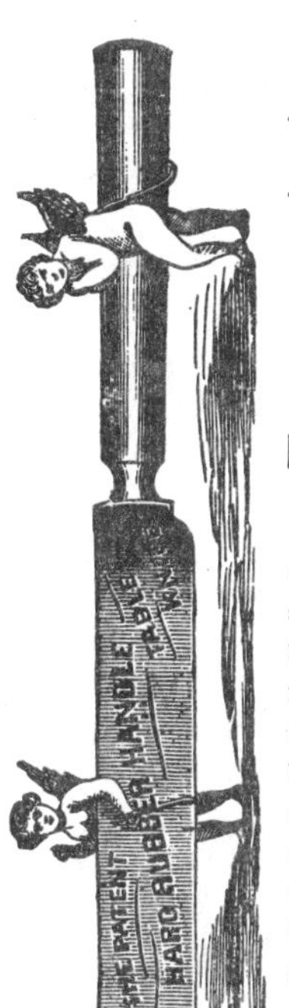

THE
MERITS
OF THE
Patent Hard Rubber Handle
TABLE KNIFE

are fully established. It is not affected by HOT WATER. It is less expensive than *Ivory*. HUNDREDS of THOUSANDS have been made, and the HANDLES of none have come off. Sold by all dealers in Cutlery, and made exclusively by the Meriden Cutlery Co., who also manufacture the Patent Solid Steel Table Knife, the Heavily SILVER-PLATED Handle and Blade of one piece, and all other kinds of TABLE CUTLERY.

Meriden Cutlery Co.,
45 Beekman Street,
New York.

IVES' PATENT LAMPS.

"The next best thing to daylight yet discovered."—*New-York Tribune.*

"They are, without hesitation, the best in the world."—*American Institute.*

THE SILVER LAMP WICK—The best and cheapest ever introduced. Does not smoke. Requires no trimming. Lasts longer than any other wick. Does not incrustate, and therefore always burns with a clear, silvery flame.

The Folding Pocket Lanterns,

VERY LIGHT, STRONG and DURABLE. Can be folded and carried in the pocket or traveling bag.

Nearly 100,000 already sold.

THE IVES PATENT LAMP CO.,
Sole Agents for the United States,
37 Barclay-st., and 42 Park-place, New York.

IMPROVED MONITOR
CLOTHES WRINGER.

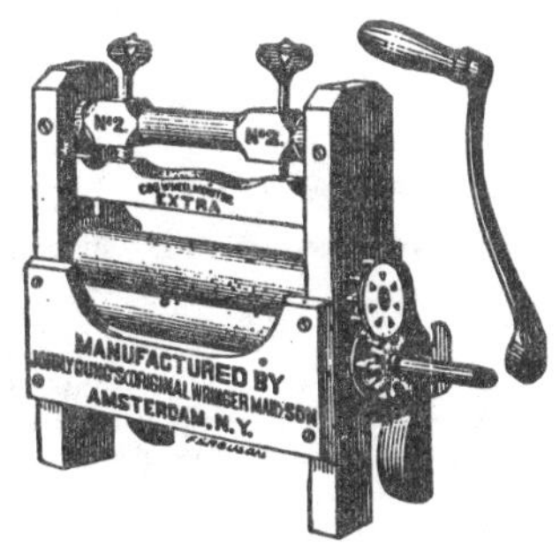

The only Wooden-frame Clothes Wringer with Moulton Patent Immovable all White Rubber Roller. Warranted to outlast any other Roller in use. Sold everywhere.

RUSSELL & ERWIN M'F'G CO.,
45 & 47 Chambers St., New York, General Agents.

Manufactured by

JOHN YOUNG'S SON, Amsterdam, N. Y.

$30 to $50 per week sure.—CANvassers!

Look at this for your own benefit.

"THE UNIVERSAL FAMILY SCALE is worthy the special attention of all housekeepers. It is a new and desirable article for use in families; is the most complete thing of the kind we ever saw, and should be in every house in the nation." It has a face like a clock, and is so simple that a child can use it."—*New-York Independent.*"

Canvassers wanted to whom liberal inducements will be given, and territory guaranteed by certificate. For pamphlet and terms, address G. W. LEONARD, Gen'l Ag't,
P. O. Box 2,833. 32 Cortlandt St., New York.

AGRICULTURAL IMPLEMENTS, MACHINERY, etc.

Our large **Catalogue** of the above is a handsome volume of about **300 pages**; containing nearly **600 illustrations** of the **newest** and **best** for Farm and Household use, and is sent post-paid by mail on receipt of **$1**; *but we will refund this on receipt of the first order for our goods to the amount of* $5. No Progressive Farmer ought to begin another year's work without a copy of this book and a careful perusal of it before entering on the coming winter will benefit every one who has need of Tools or Machinery for Agricultural Purposes. Address all letters to

R. H. ALLEN & CO.,
189 and 191 Water St. P. O. Box 376.
New York City.

N. B.—If you want anything for use on your farm send a stamp to us, and we will either write you the desired information, or send you a Special Circular, of which we issue a large number.

Prindle's Agricultural Steamers & Farmers' Boilers.

For cooking food for stock and a great variety of other purposes. All sizes constantly in stock.

They are perfectly safe, simple, and popular.

Over **1,000** now in use by 100 different trades.

The large descriptive 12-page pamphlet, with prices and illustrations, will be sent gratuitously on receipt of stamp.

The great Prize Essays on Cooking Food for stock, written by eminent and practical men, will be sent post-paid for 20 cts.

R. H. ALLEN & CO.,
189 & 191 Water St. P. O. Box 376.
New York City.

N. B.—The fullest assortment of AGRICULTURAL IMPLEMENTS and MACHINES, SEEDS and FERTILIZERS in the City, may be found at our Warehouses, and open to the inspection of all.

Anti-Friction Lever Horse Powers.

(See pages 64 and 65 of our large Catalogue.)

Two sizes and two styles, from $175 to $265. Various other sizes and styles, as low as $100. Every variety of Agricultural Tools, &c., in our large Warehouses, 189 and 191 Water St., for exhibition and sale.

R. H. ALLEN & CO.,
P. O. Box 376,
New York City.

THE
GREAT WANT
SUPPLIED.
A New Water Pipe!

A cheap, healthful, durable, and flexible metal pipe for conducting water into dwellings, to be used for drinking and cooking. This pipe can be had at **20 per cent less in cost** than **Lead Pipe** of the same strength, and **10 per cent less than Galvanized Iron Pipe** of the same bore, and transportation will cost but one-half. Satisfactory testimony can be given that water conducted by this pipe is much superior, by a great percentage in purity, to that carried by either lead or galvanized iron pipe. This pipe can be worked and soldered by Plumbers the same as lead pipe. Please send for circulars.

NEW YORK LEAD CO.
63 and 65 Centre St., New York.

FRANK MILLER'S

Leather Preservative and Water-Proof Oil Blacking for *Boots and Shoes.* Frank Miller's prepared *Harness Oil* Blacking, for Harness, Carriage Tops, &c.; by the use of these articles one-half may be added to the durability of leather. For sale in nearly every city and town in the United States and Canadas.

FRANK MILLER & CO., 18 & 20 Cedar St., New York.

HAWKES' PATENT FOUNTAIN PEN.—No Inkstand required. One filling writes 12 hours. Send stamp for Circular. Also other styles of Gold Pens. Sent by mail. Pens repaired for 50 cts. GEO. F. HAWKES, Manufacturer, 64 Nassau street, New York.

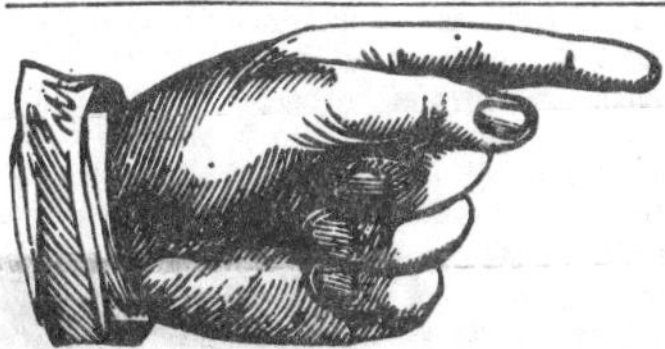

A TRIAL Trip.
3 Months for
30
Cents.

ALL YE WHO LOVE **Fruits, Flowers, Gardening, Designs of Cottages, Ornamental Trees, Shrubs,** and embellishment of home grounds, take a trial trip for 3 months or a year, with

THE HORTICULTURIST.

Splendid Premiums of Lilies, Grapes, Gladiolus, and Books given to all who get up Clubs or bring new Subscribers. Every trial trip subscriber is entitled to a fine Engraving, *Gathering Fruits and Flowers.*

HENRY T. WILLIAMS, Proprietor,
5 Beekman St., New York.
Send stamp for Illustrated Premium List.

containing a great variety of Items, including many good Hints and Suggestions which we throw into smaller type and condensed form, for want of space elsewhere.

Maple Sugar Items.—Mr. Chamberlain, author of the articles on Maple Sugar Making in the present and last months' issue says: "In the first sentence of the third paragraph of the article in January, the last clause should read "and two sheets of 12x12 inch tin." Recently a patented metalic sap spout has been brought to my notice, manufactured at Chagrin Falls, Ohio. I think it will prove superior in several respects to the wooden one described in my January article. I shall give 200 of them a careful and exact test through the entire season of 1870, and if they prove as I think they will, shall use them exclusively. I forgot to say, that maple syrup may be canned like fruit, and retain its flavor perfectly any length of time. Fruit cans in plenty are empty in time to can the syrup."

Unanswered Letters.—The large number of letters to Mr. Judd, requiring his personal attention, have recently become so numerous, that he finds it absolutely impossible to give them prompt consideration, and on this account he begs his friends and correspondents to excuse what might seem to be discourtesy or inattention. The hours in a day, and the days in a week are limited, as are human strength and endurance.

Olm Bros., of Springfield, Mass., make a specialty of sending green-house and bedding plants by mail, and offer all the novelties. Their recent catalogue presents a new feature—a lithographed design for laying out grounds, which may be used for a large place or a small one, and will afford useful suggestions to those intending to make improvements.

Stables, Out-buildings, and Fences, is the title of a new architectural work which reaches us just as we go to press. The author is Geo. E. Harney, and it is published by Geo. E. Woodward. The illustrations of farm buildings, rustic work, stable fittings, and the like, are numerous, and we doubt not the work will supply a long felt deficiency in our architectural literature. Price, by mail, $10.

Norway Oats.—What Oats to Plant.—Last fall, we asked for the experience of those of the readers of the *American Agriculturist* who had raised the Norway Oat, or had tried it in comparison with other sorts. We have received a number of responses, and proposed to print them this month, but they are crowded out. The writers who have had good seed, with a few exceptions, seem greatly in favor of the Norway. The yield in some instances has been enormous. The straw uniformly extolled for stiffness, length, and leafiness. The best success seems to have been where little seed was drilled in, on good land. The plants, when not crowded, make many stalks. In procuring seed for planting, we surely would try some Norway, even if they failed last year. Other varieties should be tested also, but of the new kinds, these take the lead. They are *not* white, but of various tints of lighter or darker gray, or rather smoke color, never black.

To Save a Poisoned Dog.—"H. H. H." recommends the administration of flour and warm water, as the only effective emetic he has tried, having, by its use, saved a valuable dog which had taken poison.

Winter Butter in Virginia.—"J. F. B." writes from Greenville Co., Va., that he is milking eight very indifferent cows, and averages only 28 pounds of butter per week, from October to February, feeding corn fodder, and a pint of corn-meal, and a quart of ship-stuff, daily, to each cow. It makes him think of the cows he used to milk in Vermont, and he thinks if he had such now, he could do very well at butter making in that milder climate. This is true; and farther South it would be even better, for winter pasturage is often abundant, and might be easily provided. One great need of the South is good neat cattle. Ayrshires and Devons are especially recommended. Buy thorough-bred bulls, and gradually bring the native stock up.

Fumigating Trees.—Who has had any experience in fumigating trees in the open air, either for the purpose of killing insects, or for destroying fungi?—We are induced to make this query by an examination of a very ingenious fumigator, made by Badouin & Fteley, and which was advertised last year in our columns. If out-door fumigation is really useful, this implement will apparently accomplish it.

Winter and Spring Barley.—"C. H. G." There is no difference between them, except that the winter barley has, by being repeatedly sown in the autumn, become hardy and stands the winter.

Gypsum in Tennessee.—W. T. Waters.—As long as plaster costs you $35 per ton, or even $25 per ton, it will probably not pay you to use it as a manure.

Petroleum for Roofs.—"Will not petroleum make roofs more inflammable?"—If the roof catches fire, the petroleum would favor combustion; but a roof treated with petroleum is not as likely to catch fire from sparks. It fills up the pores, and the wood becomes hard, firm, and smooth.

Lime Kilns.—An article on kilns in September 1867, gives some valuable details with engravings.

Forest Trees from Cuttings.—"A. W. W.," Muscotah, Kansas. The only trees likely to succeed from cuttings, except in the hands of an experienced propagator, are the Willows and Cotton-woods. Of these the cuttings are best taken before the sap starts in the spring. The article on Hedging, in February, gives directions for planting them. Other trees are best raised from seed.

Flower Seeds Gratis.—Mr. Chas. D. Copeland, of Lima, Livingston Co. N. Y., writes that he has a considerable surplus seed of "Fancy Pinks" and "Sweet Williams," embracing German, French, Italian, China, and Japan varieties, and he will be happy to send a free paper of the seeds, mixed, to any subscriber of the *American Agriculturist* who will furnish him their address, with a Post-office stamp, for the postage. He will find it a larger job than he has looked for, but he is willing to risk it, and we vary from our general rule not to publish such offers. One of our subscribers formerly offered to send a descriptive circular of a certain breed of animals. After distributing 13,000, at a cost of 10 cents each, he was glad to beg us to withdraw the offer.

Trouble with a Seed Drill.—A subscriber of the *Agriculturist*, in Georgia, writes that he has bought a wheat drill, that it sows dry wheat well, but it will not sow wheat that has been treated with a solution of blue vitriol. Neither would it do so after the wheat had been rolled in ashes to dry it. If the difficulty cannot be overcome, he says the drill is of no use to him. He does not mention the name of the drill. With a "cup" drill, we have never had any difficulty in sowing vitrioled wheat. The only remedy we can think of is to use a stronger solution of vitriol. Turn over the wheat repeatedly, or run it through a fanning-mill until it becomes dry enough to sow. The vitriol will not injure the wheat if kept dry for several weeks before sowing. Ashes or lime should not be mixed with the wheat immediately, as they will neutralize the vitriol. But after the application has destroyed the fungus on the wheat, the ashes or lime would probably do no harm.

Timber in Southern Indiana.—"A Subscriber" has 75 acres of heavy timber two miles from a Railway station, where it is worth $2 per cord. Shall he clear it or not?—No. We think, if he holds on, it will rise in value so as to pay not less than 12 per cent interest per annum for several years, and if the trees are second growth, they are gaining 3 to 8 per cent per annum.

Manuring Corn.—"L. W. G." Barreville Pa., proposes to plant thin sward land limed, with corn, and intends "to put one handful of dry cow dung without straw on each hill." He is in doubt, whether to mix plaster or lime in considerable quantities with the manure. Of course he should use plaster, for thus, any ready formed ammonia present, and liable to escape would be saved, while lime would drive it off if not immediately buried. Besides the soil has probably lime enough.

Improvements in Field Rollers. Mr. N. G. Jenkins, of Brookside, N. J., suggests some improvements upon the plan of making rollers, described in the January number. His suggestions are excellent, but he, perhaps, does not remember that we describe one that any farmer may make, with only a little help from the smith. "I would make the drums not less than 3 feet in diameter, which gives lighter draught, as well as greater weight. In place of the 1¼ plank in the centre, I would substitute a piece of iron, ⅝ x 4 in., passing through the inner timbers of frame-work, with nuts upon either end; this would allow the drums to run closer together, say not more than 1 inch apart. In place of the journals I substitute a rod of 1-inch round iron, running the entire length of both drums and frame. The rod remaining stationary, I use cast boxes, running through the head of each drum, with flanges on the same, and bolt through the heads. The rod can be drawn out at any time for repairs, or other purposes. I use old tire from wagons, and hoop the drums at each end, and put dowels in the centres of the drums, to keep them in place, and also arrange a seat for the driver over the centre of the roller, which is both easy and convenient."

The Monte-Bello Apple.—Thos. Gregg, Secretary of the Warsaw, Ill., Horticultural Society, writes: "Permit me to thank Mr. Charles Downing for calling attention to so valuable a new apple as the Monte-Bello, noticed in your January number. The sketch given does not, however, do justice to it. It is regarded by the members of the Warsaw Horticultural Society—which has the honor of bringing it into notice—as one of the very handsomest, as it is one of the very best fall apples known; and as the few trees in existence are hardy, and good bearers, it is deemed a great acquisition. It is a remarkable fact in horticultural history that a variety so valuable should have existed in a community for over twenty years almost unknown. Now that our society has brought it to notice, it is exciting much attention, not only abroad, but at home, where its merits should have been recognized before."

Bots in Horses.—A subscriber who believes that "a great many horses are killed by the bot grubs in this section" wishes a remedy. It is a remarkable fact that though every quack horse doctor in the land attributes all sorts of evil and fatal effects to the bot grub, no educated veterinarian believes that they do much harm. The best books on the diseases of the horse say, they do little or no harm; and the man of all others who made the different kinds of bot flies and grubs his especial study (Bracy Clark V. S.) thinks, they do good, rather than harm. One thing is certain—there is only one thing to be done—that is let them alone. They are now nearly full grown and will detatch themselves and quietly pass out in the course of the season.

What is a Ton of Manure?—Peter Henderson says: I must have had over a hundred letters during the past year, asking me how many bushels of manure go to make a ton or how many tons are in a *cord* of manure. These letters still come asking the same questions, exhausting not only my stationary but my patience in answering. One would think that the slightest reflection would show that the condition of the article—wet or dry—solid or light—would so affect the weight, that no comparison between measurement and weight could be given. A little exercise of judgment, will make it apparent, that weight only, is the fair test of value; for if it takes 50 bushels to be a ton, costing $2, it is likely to be of no more value than 25 bushels would be, having the same weight; for the one article would be strawy and loose, the other moist and solid.

Van Buren's Golden Dwarf Peach and Horse Plum.—J. Beachy, Preston Co., W. Va. The Dwarf Peach is worked on the common peach stock, and retains its dwarf character. The Horse Plum is a purple variety, also known as Early Damson, but we cannot tell if it is your common plum.

Eumelan Grape.—Mr. F. R. Elliott, in a recent letter, says: "I notice you are offering the Eumelan Grape among your premiums as compensation for labor in obtaining subscribers. Perhaps no grape of very recent introduction better merits approval. Like the Delaware, it is a sort with which in quality of fruit we may hope to tone up and improve public taste, while realizing a profit in its sale. I have watched the growth of the vines for two years, and they compare favorably in all respects with any and all other varieties. My valued friend, Prof. J. P. Kirtland, fruited it the past season, with, I think, as good or better promise of value, as any other young vine of whatever sort. The question comes to me almost daily, 'What grape do you advise to plant?' and I reply, plant mainly of those you have heretofore known to prove profitable in your section and like soils; but of the new black grapes don't fail to plant some of Eumelan, as a grape of promise. I have known several acres of it planted the past year resulting in uniform and satisfactory growth, giving good cheer to the enterprizing planters, who doubtless congratulate themselves on their foresight."

Cabbage Insect.—"R. B. K." Sprinkling the young plants with ashes and plaster keeps off the "black louse." We suppose the "large white grub" referred to is the larva of the May-bug. No remedy short of digging it out and killing is known for the White-grub.

AMERICAN AGRICULTURIST

FOR THE

Farm, Garden, and Household.

"AGRICULTURE IS THE MOST HEALTHFUL, MOST USEFUL, AND MOST NOBLE EMPLOYMENT OF MAN."—WASHINGTON.

VOLUME XXIX.—No. 4. NEW YORK, APRIL, 1870. NEW SERIES—No. 279.

ESSEX SWINE.—*Drawn from Photographs and Engraved for the American Agriculturist.*

The Essex are at present attracting much attention from those interested in improving their swine.

The animals above represented are upon the farm of Mr. L. A. Chase, of the *American Agriculturist*, at Northampton, Mass., and are only in breeding condition. The two young sows were bred by Joseph Harris, Esq., ("Walks and Talks,") Moreton Farm, Rochester, N. Y., and the old boar, represented in two positions, is from the herd of Edwin Thorne, Esq., Washington Hollow, N. Y. The influence exercised by man upon the form and character of animals is nowhere more strikingly shown than in the pig. The form and peculiarities which adapt the animal to a wild life in which it has not only to provide its own food, but to protect itself and its young from enemies, are quite different from those required in a state of domestication, where the the animal's whole energies are to be concentrated upon turning vegetable food into pork. By selection and crossings, breeds have been established in which not only aptitude to fatten has become a fixed character, but the amount of bone and useless parts reduced to a minimum. The offal in the Essex is only about ten per cent of the live weight.

For the mere purpose of making pork we would not recommend the pure bred Essex. They are too fine and delicate. Their value consists in their capacity of improving the large breeds, or in fact, any kind of common pigs. For this purpose they must be bred *pure*. What a farmer needs to improve his stock is *thorough-bred* males. And in pigs there is no breed more thoroughly established than the Essex.

TOMATOES.

The Unrivalled Prize, "Gen. Grant."

This Tomato introduced by us last season, has again proved the best in all respects, taking for the third season the first and second prizes at the Mass. Hort'l Soc. Exhibitions, in competition with all the leading sorts. We have received numerous testimonials from all quarters, but for want of space are obliged to omit publishing them. Price, post-paid, per packet, 25 cents.

EARLY WYMAN CABBAGE.

Introduced by us. The best and most profitable early market-cabbage grown. It heads early, is of large size and first quality, and brings the highest price of any early cabbage brought into the Boston market, selling readily last season by the thousand, at $18 per hundred. It took the first prize at the last and several previous exhibitions of the Mass. Hort. Soc. Per packet, post-paid, 25 cents.

For further descriptions and a beautiful colored Lithograph of above Cabbage, see Jan. No. Agriculturist, page 21, or

Twenty-fourth Edition of this popular work, which has met with so much favor in the past, is now ready. It has been re-written and improved, printed with new type, and on fine paper; illustrated with a beautiful Lithograph, and many other fine engravings from nature. It contains full description and the culture of over 1,500 leading varieties of Flowers and Vegetables; also descriptive list of the novelties of the present season; to which is added a collection of 200 choice French Hybrid Gladiolus. This work, we feel confident, will compare favorably with any similar one.

From Levi Bartlett, Warner, N. H.

"I have received a copy of your superbly gotten-up Amateur Cultivator's Guide. I think it far ahead of anything of the kind ever before issued from the American press."

Sent to any address upon receipt of 25 cents for paper cover, and 50 cents for tastefully bound in cloth.

WASHBURN & CO., Boston, Mass.

Three Best Squashes!

The Hubbard is the sweetest, dryest, and richest table squash for winter, the *American* Turban for fall, and the Boston Marrow the best pie squash. Seed sent post-paid to any address, at 10 cents a package; Hubbard and Turban 25 cents an ounce, and Boston Marrow 20 cents an ounce. I take pride in having been the original introducer of the Hubbard squash. My Illustrated Catalogue of Garden and Flower seed sent gratis to all.

JAMES J. H. GREGORY, Marblehead, Mass.

THE CASABA MELON.

"One of the specimens weighed twelve pounds; and if there is a more sugary, tender-fleshed, and altogether good melon than this, we would like to see it."—*Editors American Agriculturist, Oct., 1868.*

"Every one who ate of them have pronounced them the finest melons they ever tasted; they weighed from ten to sixteen pounds each. They suffered less from drought than any melons on the farm."—J. M. Phillips, Newtown, Md.

"I am especially pleased with the Casaba Melon on my ground; they have borne abundantly, are of very large size, and superior quality."—T. C. Wallace, M. D., Cambridge, N.Y.

"The earlier years of my life, which I spent in Constantinople, Turkey, in the midst of the profuse abundance of melons, which used to overstock the market of that city, we would now and then come across the Casaba. It was a luxury, not solely for its rarity in that market, but for its own size and excellence; though as large as the largest water-melon, it was also of really exquisite flavor. The cause of its rarity was stated to be the fact that it would not grow in excellence except in the province of Casaba, in Asia Minor." —*Extract from a letter from W. D. Dwight, Principal of the West Point School.*

This melon can be grown quite as easily as any of the other varieties of muskmelon.

The seed for sale in sealed packets, 25 cents each, or six for $1; also the following superior varieties:

	pkt.	oz.	lb.
California Golden Orange, ℔ pkt.,	10 c.;	℔ oz. 20c.;	℔ ℔., $2.50
Turks-cap, extra,	10	20	1.50
Pineapple,	10	20	1.25
Water-melon Mo'nt'n Sweet	10	10	1.00

By Mail, at above prices. For other varieties, see Dreer's Garden Calendar for 1870, mailed on receipt of a stamp.

HENRY A DREER, Seed Grower,
714 Chestnut St., Philadelphia, Pa.

American Sage Seed.

True Broad-leaved Sage Seed, grown by the famous Danvers growers who raise Sage by the acre. *Warranted growth of 1869.* Price—10 cts. per package; 35 cts. per oz.; $1 per ¼ lb.; $3.50 per lb. *Sent by mail, post-paid, with full directions for cultivation. Catalogues free.*

JAMES J. H. GREGORY, Marblehead, Mass.

Gregory's Annual Catalogue of Choice Garden and Flower Seeds.

Having in former years introduced to the public the Hubbard Squash, American Turban Squash, Marblehead Mammoth Cabbage, Mexican Sweet Corn, Brown's New Dwarf Marrowfat Pea, Boston Curled Lettuce, and other new and valuable vegetables, with the return of another season I am again prepared to supply the public with Vegetable and Flower seeds of the purest quality. My Annual Catalogue, containing a list not only of all novelties, but also of the standard vegetables of the garden (over one hundred of which are of my own growing) and this season for the first time a carefully selected list of flower seeds will be forwarded *gratis* to all. Sent without request to my customers of last season. All seed purchased of me *I warrant to be fresh and true to name, and that it shall reach the purchaser.* Should it fail in either of these respects I will fill the order over without additional charge.

JAMES J. H. GREGORY, Marblehead, Mass.

SEED STORE AT YOUR DOOR.

GARDEN & FLOWER SEEDS

Sent by mail, *postage paid.* Send for priced Catalogue.

Farm & Garden Implements, and Fertilizers,

Of every variety, at *very low prices.*

J. VANDERBILT & BROTHERS,
23 Fulton Street, New York.

EARLY WYMAN CABBAGE.—The earliest Boston Market variety. Per packet, 25 cents, mailed.

IMPROVED EARLY PARIS CAULIFLOWER.—One of the finest grown by the Market Gardeners about Boston. 25 cents per packet, mailed.

NEW EARLY PEAS.—Supreme, per half-pint packet, $1.00; Alpha, small packet, $1.00; Cook's Favorite, per half-pint packet, 75 cents. Sent by mail.

EVERGREEN PODDED BEANS, a new variety of pole beans from the Pacific Coast. As a string-bean, and for pickling, we consider it one of the best recently tested. Per packet, 25 cents, mailed.

SURPRISE OATS, ripen three weeks earlier than any other variety, and weigh forty-five pounds to a measured bushel. For full description we refer to our New Catalogue. Stock limited. 50 cents per quart, postage paid.

CURTIS' & COBB'S

New Illustrated Seed Catalogue, and Flower & Kitchen Garden Directory.

The SEVENTEENTH EDITION of our popular and comprehensive Catalogue is now ready, and will be mailed to all applicants enclosing us Twenty-five cents. Regular customers supplied without charge.

Address CURTIS & COBB,
348 Washington St., Boston, Mass.

BOSTON MARKET CELERY.

I can supply best seed of the genuine Boston Market Celery, the variety described on page 420 of the Nov. (1869) Agriculturist. Per package, 15 cts.; per ounce, 50 cts.; by mail, post-paid. *Catalogues free.*

JAMES J. H. GREGORY, Marblehead, Mass.

BEAUTIFUL FLOWERS AND CHOICE VEGETABLES.

MY NEW ILLUSTRATED CATALOGUE OF CHOICE FLOWER AND VEGETABLE SEEDS for 1870, is published and ready to send out. It contains a list of nearly everything desirable in the Flower and Vegetable Garden. All the best varieties of Beans, Beets, Cabbage, Cucumbers, Melons, Radish, etc., etc., 15 varieties of Tomatoes, embracing several new sorts. Read my advertisement about NEW TOMATOES in the Agriculturist for February, page 70. The three varieties described there will be sent for 35 cents; *not* **50** *cents*, as the printers made it read. My Catalogue tells how to get the Agriculturist, Our Young Folks, The Childrens' Hour, Arthur's Home Magazine, The Independent, Phrenological Journal, and several other valuable Magazines, *at the lowest Club price.* Catalogue mailed free to all who apply.

Address J. F. MENDENHALL, Carmel, Ind.

WARRANTED GARDEN SEEDS.

CHOICE FLOWER SEEDS, NEW SEED POTATOES, AND SEED OATS, ETC.

By Mail, to any Post-office in the United States. Our Illustrated Descriptive Priced Catalogue for **1870**, is issued, and will be mailed to any address on receipt of 10 cents.

EDW'D. J. EVANS & CO., York, Pa.

Conover's Colossal Asparagus!

SPLENDID ROOTS!

Grown singly in drills in rich soil and with good culture. One year old. $3 per 100; $25 per 1,000.

Send for Circular. R. M. WELLES,
Towanda, Pa.

Phinney's Early Water-Melon, per packet..........25 cents.
Rising Sun Tomato, " " ....25 "
Early Wyman Cabbage, " " ....25 "
Conover's Colossal Asparagus Seed, per packet....50 "
Early Wakefield Cabbage, per ounce......................$1.00
Fine roots of Conover's Colossal Asparagus, at $3.00 per hundred.

Our New Seed Catalogue, containing the names and prices of all the **New Vegetable** and **Field Seeds**, with a description of the same, is ready for mailing to all applicants. Address

R. H. ALLEN & CO., P. O. Box 376,
189 & 191 Water Street, New York.

Fresh Garden, Flower, Fruit, *Herb, Tree, Shrub* and *Evergreen Seeds* (25 sorts of either, $1.00) prepaid by mail. Agents wanted.

Priced Catalogues gratis. Trade List. Seeds on commission. Small Fruits. Fresh Onion Seed, $4 per lb.

B. M. WATSON,
Old Colony Nurseries and Seed Warehouse,
Established 1842. Plymouth, Mass.

PURE CANE SEED.

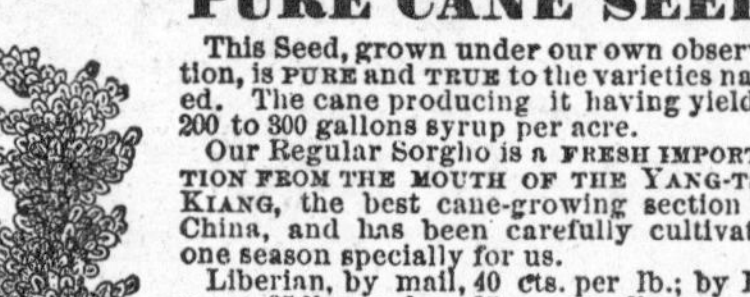

This Seed, grown under our own observation, is PURE and TRUE to the varieties named. The cane producing it having yielded 200 to 300 gallons syrup per acre.

Our Regular Sorgho is a FRESH IMPORTATION FROM THE MOUTH OF THE YANG-TSE-KIANG, the best cane-growing section in China, and has been carefully cultivated one season specially for us.

Liberian, by mail, 40 cts. per lb.; by Express, 25 lbs. or less, 25 cts. per lb.; over 25 lbs., 15 cts. per lb. Regular Sorgho, Oomseeana, Neeazana, by mail, 50 cts. per lb.; by Express, 25 lbs. or less, 30 cts. per lb.; over 25 lbs., 20 cts. per lb.

Packages included. Send for Circular.

BLYMYER, NORTON & CO.,
Manufacturers of **Cincinnati, O.**
VICTOR CANE MILL and COOK'S EVAPORATOR.

Alsike Clover.

Fine for mowing and pasturage, or for bees. Price 10 cts. per oz.; 75 cts. per lb., by mail, post-paid; also $7 per peck, and $25 per bushel of 60 lbs., by Express. *Catalogues free.*

JAMES J. H. GREGORY, Marblehead, Mass.

New Tomatoes.

ALGER.—This new variety has the same potato-like foliage as Keyes', but the fruit is larger, of good market size, early and very productive, 15 cts. per package.

GEN. GRANT.—Remarkably solid, round, flat in shape, handsome, and of excellent quality. Received the first premium for the two past years at the Annual Exhibition of the Mass. Hort. Society, 15 cents per package.

CRIMSON CLUSTER.—Early, grows in large clusters bearing handsome fruit, oftentimes elegantly spotted with gold, 15 cents per package.

MAMMOTH CLUSTER.—Very large, round, crimson, 15 cents per package.

BOSTON MARKET.—The result of most careful selection by the Boston market-men for a series of years; large, flat, round, solid; enormously productive, 15 cts. per package.

EARLY ORANGEFIELD.—An English sort, yields its fruit in large clusters. This Tomato is of a peculiarly rich and sweet flavor, and excellent as fruit for dessert, 15 cts. per package.

Also the following excellent varieties at 10 cts. per package. Early York, Dwarf Scotch, Keyes' Early Prolific, Yellow Fig, Maupay's Superior, Cherry, Large Yellow, Feejee, Cook's Favorite, Lester's Perfected, Large Smooth Red, Tomato De Laye, Tilden, New Mexican, Strawberry, or Ground Cherry.

All of the above are of my own raising, each grown isolated, scattered over three square miles of territory. Catalogues free.

JAMES J. H. GREGORY, Marblehead, Mass.

Improve Your Seed.

"The Best is the Cheapest."

SANFORD CORN.

The experience of the past season fully confirms all that has been said in its favor. In many instances, being planted in the same field and having in all respects the same chance, *it has ripened from two to three weeks earlier, and yielded double the amount of other corn.* Testimonials from reliable farmers in nearly every State endorse it as being *the best field corn.* One quart by mail post-paid, 75c. Two quarts, $1.25. One peck, $3. One peck by Express, $2. One bushel, $5. Send for Descriptive Circular. Address

S. B. FANNING, Jamesport, Long Island, N. Y.

Seed Potatoes and Oats.

Genuine Early Rose for seed at $4 per Bbl. Four lbs. by mail, @75c. Ramsdell's Norway Oats, genuine, pure, and perfectly free from foul seeds, @$3.00 per pushel (32 lbs.).

The above will be delivered to Express free of charge.

Address E. W. GRISWOLD, Centrebrook, Conn.

AMERICAN AGRICULTURIST.

NEW YORK, APRIL, 1870.

However much winter we may have in March, we are sure to have spring in April. At the time of writing, our summer-like winter is tardily asserting itself, and making spring dates like apples of Sodom—or broken promises. Farmers are likely to have their generalship and management sorely taxed, for a late spring puts one's power of every kind to the test. If a farmer cannot multiply men and teams to suit the tardiness of the season and his own convenience, he must manage his work to accomplish the most possible with the fewest steps of men and teams, and in the shortest time. A late spring does not insure a late June, but we may calculate on rather an early one—no harm will be done whether it comes or not. We are too apt to look over the fence, and plan work by our neighbor's, and take somebody's else estimate of the weather, etc. Men seldom blunder in getting ahead with work. The common failing with American farmers is, that they cannot manage men except as foremen, doing half as much again work as any man they can hire. If such a one makes farming pay, it is by the labor of his hands rather than by "the sweat of his brow," which we are inclined to interpret as meaning *head-work.*—"By the sweat of thy brow thou shalt eat bread." One of the best farmers we know, says, sometimes he fairly longs to do a day's work, but he can't afford the time. Now and then he takes hold just to show the men that he knows exactly how he wants the work done; but his time is worth double or triple what it will cost him to labor in the field. We train our boys to work, that they may know how to do everything well; but they should also be trained to manage hands, and plan work. It requires both to make an accomplished farmer. No farm work can be done without hands, tools, and teams, and it is the farmer's first business to provide these, then to make good use of them.

Hints About Work.

Live-stock.—As the weather grows warmer, vermin increase in numbers on live-stock of all kinds. The long coats upon horses, cattle, and sheep, favor their increase, and when once they have a foothold, only vigilance in the thorough application of remedies will rid one's stables. Washing with carbolic soap we have found effectual. Those who have not, must have used the solution too weak. The curry-comb, or card and brush, are efficient also, and do all kinds of animals much good. Spring is a trying season with most stock, and all should have plenty of feed and water, with all the salt they want. Let cattle and horses drink all they will, unless heated, or very thirsty.

Orchard and Nursery.

Preparing to Plant.—The land should have been prepared last autumn, but many will do it this season. In the hurry to have the soil ready for the trees, do not be tempted to plow while it is wet. It should crumble as the furrow is turned, and be in a condition to produce a good corn crop.

Draining will often be needed, but this can be done after the trees are set.

Pruning before Setting.—See that all roots, mutilated in digging, have a clean cut, and cut back the top one-third or one-half of last year's growth, according to the amount of roots that remain.

Trees in bad order, from delay in transportation, must be treated according to their condition. If shriveled, bury them, tops and all. If they have heated, and the buds started, cut back severely.

Trees that cannot be well planted.—It sometimes happens that the land is not ready, or other work presses so, that the trees, though at hand, cannot be set in a proper manner. Rather than hurry the planting, it is better to set them out in good soil, in nursery rows, and allow them to grow there for one season. They will then be ready for planting in the orchard early in autumn.

Root Grafts.—Set out in rows far enough apart to work with the cultivator, and a foot apart in the row. Bring the soil close in contact with the roots.

Nursery Stock.—Head back before the buds push, as cutting when in leaf will check the growth.

Budded Stocks.—Cut off above the bud those where the budding has been successful. Apples and pears, upon which the buds have failed, may be stock-grafted close to the root.

Seeds of apple and pear should be sown early in rich soil, to get a growth before hot weather. Take Peach stones from the seed-beds or heaps, sift the earth from them, and plant in nursery rows.

Fruit Garden

The amateur is much puzzled with descriptions of new varieties of fruits, each claiming to be better than those that are in general cultivation. Among all fruits there are some which, though not the best in all points, have the most desirable quality of being reliable. While we would have all who can afford to do so, to test the new kinds, we advise those about to plant for family use to take for their main dependence the well tested sorts.

Grape-Vines.—The Concord is the most generally known variety. The Eumelan promises to be a favorite. Delaware, Iona, Salem, and others, are excellent where they will succeed. Varieties that have been laid down are to be taken up and tied to the trellis. Plant good one-year-old, or at most, two-year-old vines. It is common to see vines an inch and more in diameter for sale. Such are not worth planting. If it is desired to propagate a variety by layers, open a trench a few inches deep, and lay down a cane of last year's growth. This may be fastened in place by pegs, but it is not to be covered with earth until the shoots have pushed. Cuttings of those varieties that start without artificial heat may be put out.

Strawberries.—Plant as directed last month—the earlier the better. Go over beds that were covered with straw last fall, and part the straw over the plant, but leave it to mulch the soil.

Blackberries.—Set early. See last month. The Kittatinny is the best generally tested variety.

Raspberries.—The tender varieties are to be uncovered. The Black Caps do not sucker, and the fruit is liked by most people. The McCormick (called Mammoth Cluster by some) is one of the best.

Kitchen Garden.

Seed Sowing is to be done according to locality. Garden vegetables are divided into hardy and tender. Onions, leeks, beets, carrots, spinach, cabbage and its relatives, lettuce, parsley, parsnips, peas, radish, cress, salsify and turnip are hardy, and may be sown when the soil can be well worked. Beans, melons, cucumber, squash, tomato, okra, nasturtium and sweet corn are tender, and need the the same conditions of soil and temperature as required by Indian corn.

Seed-Beds are used for all plants that require transplanting, or that it is most convenient to raise in this way. These should be of light and rich soil, placed in a sheltered part of the garden.

Varieties.—We usually give this month a list of a few of the leading sorts as a guide to the inexperienced. There are others equally good with those we name; our object is to save the novice the perplexities which a crowded catalogue presents.

Novelties.—Each year the seedsmen offer new varieties. Some of them prove good, and some inferior to old sorts. It is well to try a few new things every year, if one can afford it.

Beans.—Early Valentine is the standard bush sort. Black Wax is excellent. Plant in drills, 2 feet apart. Among pole beans, the Giant Wax is best for snaps, and Large Lima for shelling.

Beets.—Bassano is earliest, but Early Blood Turnip is better. Dewing's, Hatch's and Egyptian are new and highly praised. Drills a foot apart.

Broccoli.—White and Purple Cape are good. Treat the same as cabbages.

Cabbage.—Early Wakefield and Little Pixie are good early. Early Wyman, new and large. Winningstadt is best for light soils and medium late. Marblehead Drumhead, Fottler's, Flat Dutch and others are good late. Early, Blue Savoy and Drumhead. Sow in seed-bed or cold-frame. Set plants out from cold-frame, 16 inches apart in rows 2 feet apart, as early as the soil can be worked.

Carrot.—Sow same as beets. Early Horn.

Cauliflower.—Half Early Paris, of which the Boston Market is a strain, is the best for general culture. Treatment, as for cabbage.

Celery.—Dwarf White Solid, Boston Market. Some of the red sorts, such as Incomparable, and Dwarf Crimson, are fine. Sow in seed-bed. Turnip-rooted is for soups and salads.

Corn.—Crosby's Early, Farmer's Club, Mammoth Sweet and Stowell are all good. Mexican, though black, is the sweetest of all.

Peppers.—Squash, for pickling, Sweet Mountain, for stuffing. Monstrous is a new sort. Need to be started under glass.

Potatoes.—We have seen nothing equal to Early Rose for garden culture.

Radish.—Early Scarlet Turnip, Short-top Long Scarlet, Olive Shaped. Sow early and at intervals of 10 days wherever there is room.

Rhubarb.—Manure old beds.

Salsify.—Sow and cultivate the same as beets.

Spinach.—Sow in 18-inch drills. Round-leaved is the most used. The Lettuce-leaved commended. New Zealand is excellent in summer, as is the Perpetual Spinach Beet.

Sweet Potatoes.—The tubers should be put in hot-beds the middle of the month. Cover the manure with 2 inches of good compost, and lay the tubers close together. When the buds start, cover them with an inch of compost. Nansemond, the standard sort, Southern Queen, new and good.

Squash.—Summer Crook-Neck is the best bush variety. Boston Marrow and Hubbard, for late.

Turnip.—Flat Dutch, for early. Red and White-top, for late. Of the Ruta-baga sorts, the Sweet German and White French are best for family use.

Flower Garden and Lawn.

In laying out a place the roads and paths should receive the first attention, and then the lawns and flower-beds.

Lawns.—Sufficient information on forming new lawns is given in previous months. Old lawns should be top-dressed with good compost, guano, and bone-dust, a good phosphate, or ashes. Where the grass is taken off, some fertilizer must be used to keep the turf in good condition. Sow seed on thin spots, and roll as soon as the frost is out.

Trees and Shrubs.—Transplant all deciduous ones before they start. Those taken from the woods must be trimmed severely to make them succeed.

Perennials.—Those that have grown three or four years in one spot need to be taken up, divided and set in a new place before growth starts.

Annuals.—Sow hardy sorts as soon as the ground can be worked. Tender ones must be left until later, or sown under glass. The number is so large that we must refer to the catalogues for novelties.

Bedding Plants must not be put out until the soil is well warmed and chilly nights are over.

Roses require a rich, well drained soil. For summer blooming the Tea and China roses are best, but the Remontants, which usually bloom in spring only, are the finest. Get them on their own roots.

Green-house and Window Plants.

The plants should be prepared for turning out of doors by the admission of air every pleasant day. Be prepared to heat up during cold, damp weather.

Water will be needed more frequently than during winter, and window plants often become too dry.

Propagate such bedding plants as will be needed, before the sun gets too warm, and pot off as fast as they strike.

Fig. 1.—FRONT, OR SOUTH-EAST, ELEVATION.

A Convenient Country or Village House,

Having the "Modern Improvements" of Gas, Hot and Cold Water, Bath-room, Water-closet, Warm Air Furnace, Speaking Tubes, Bells, Dumb-waiters, etc.—(all of which, except the Gas, are available in Country Houses generally).—Cost, $7,000 to $8,500.

The demand upon us for hints and suggestions on building and improving houses, is becoming very great. This is hardly to be wondered at, when we remember that there are in this country some Forty Millions of people, all of whom need some kind of shelter in the form of a house. We will try to do more than hitherto, towards supplying practical information. Perhaps the best way to do this is, to take up successive plans and describe them, weaving into the descriptions various suggestions. Some of these plans will furnish exact models for those having no experience in building, and no favorite plans

Fig. 2.—REAR, OR NORTH-WEST, ELEVATION.

of their own. Usually, however, these descriptions will serve merely as suggestions......Our Senior Publisher has for several years kept one or more houses in course of erection, not as a source of profit—for they are sold at *cost*—but partly as a matter of pleasant pastime or source of pleasure in exercising a natural mechanical taste, and partly, to gain information that may be useful to our readers, by practical *experience* in working out plans. In March of last year, we published a description of one of his houses costing $12,000. These houses were further improved upon in two others, costing about $13,000 each, having more piazzas, circle head or arched doors and windows in the first story, and one foot larger each way. Last month we described those, alike in plan, but varying a little in size, which cost $5,100 to $5,500. These prices are exclusive of the ground. We now give plans of those costing a little over $8,000. Mr. J. is also thinking over some plans (for testing by actual trial) of houses to cost respectively $1,200 to $1,500 each; about $2,000 each; about $3,000 each; and about $4,000 each.

Fig. 1 shows the **Front,** or South-east, **Elevation;** this stands on a corner, but back from the streets, with the Piazza extending parallel with both the streets.

Fig. 2 shows the **Rear,** or North-west, **Elevation.** For notes on kind of siding, roofing, hight from ground, filling in walls, painting, etc., see notes last month, pages 28-9. Gas Pipes extend to every room and hall, from the basement to the 3d story. Color, very light gray.

Fig. 3. Basement.—Hight in clear, 8½ feet. The walls are of Brick, with hollow in middle, and rise about 4½ feet above the ground surface. *Aa*, the Kitchen, is of ample size, having: *r*, a 9-inch "Victory Range" with water-back, and by its side a pot and shelf closet, *Cc;* a Sink, *u*,—with Suction Pump, *p*, drawing water from outside reservoir through a tin-lined lead pipe; a Pantry, *Cc*, and a Dumb-Waiter, *dw*, to the butler's pantry above; Speaking Tubes, *s,s*, one to dining-room, and one to family chamber (*F*, fig. 5)......*Bb* is the Laundry, having Stationary Wash Trays (or tubs), *t,t,t*, supplied with hot and cold water stop cocks, chain-plugs, and outlet pipes with stench traps; a Suction and Force pump, *n*, which draws cold water from the reservoir for the wash-trays, and when needed in dry weather, forces water to the Tank (figure 6). The pump is supplied with an air chamber, and a valve to prevent drawing water from the tank, as shiftless "help" would be inclined to do. *Z* is a 40 gallon Copper Boiler, put here instead of in Kitchen, for heating. This is of course connected with the range water-back, and has waste-pipe and stop-cock at the bottom, to draw off any sediment that may collect at the bottom in the course of time, if the water be not absolutely pure. It is of course kept full by a pipe from the tank. *Ee* is a milk or food cellar, plastered and finished off like *Aa*, *Bb*, and *Hh*. *Dd* is a general or vegetable cellar, and *Ff* a coal cellar. *Oo* is a 22-inch "Oriental" Base-burning Furnace, with door opening into *Ff*, and man-hole door in *Hh*, for supplying water to vapor pans. A door from the main Hall, *Hh*, opens back into the basement Area and Stairs. A Closet, *c*, is put under the stairs—the rule being to put closets wherever they can be worked in, throughout the house, which all housekeepers will value. The tin pipes conveying warm air to the 2d story, are carried up through the partitions, with iron laths on each side. These are made by cutting strips of sheet-iron 2 inches wide, bending the two edges over to hold the mortar, and nailing them upon the studding. This prevents any possibility of firing the wood-work. Too much care cannot be given to keeping all hot air pipes nearly an inch away from any wood-work, and lining all wood near them with tin, for absolute safety.

Fig. 4. First Story.—Hight in the clear 11 feet. A *Piazza*, 8 feet wide, extends around two sides, and another (Area 12½x6 feet) is enclosed in sash—a cool resort in hot weather......The *Main Hall*, *H*, is sufficiently wide for a house of this size, and with the double doors on each side gives a very much more spacious look to the whole house, on entering, than if it were only the usual width of 6 feet. The *Vestibule*, *V*, is shut off by glass doors......The size of the main *Parlor*, *P*, was adopted after much measuring and inquiring among housekeepers of good taste and judgment. The front double windows open down to the floor of the Piazza. The chimney has a statuary or white marble mantle, grate and summer piece. *r*, near a square, here as elsewhere throughout the houses, indicates a register and warm air pipe from furnace. *Bk* is a bell-pull to the kitchen. In this room, as in others, double ventilators, one at the baseboard, and one with cords placed near the ceiling, connect with a flue in the wall. (When heating up a cold room, close the upper and open the lower ventilator, and the warm air rising to the ceiling soon forces the cold air out through the lower ventilator. When desirable to remove the heated, rarified, or impure air from above, the order of opening is reversed.)......*R* is the *Reception Room*, or Smaller Parlor, or "Living Room," with pantry or closet, *p*, Italian Marble Mantel, and Grate, etc., and warm air register, *r*. *Bk* is a bell to kitchen......The *Dining-Room*, *D*, has pantries or dish-closets, *p,p*, at the ends, with an arch thrown over, and side brackets or pen-

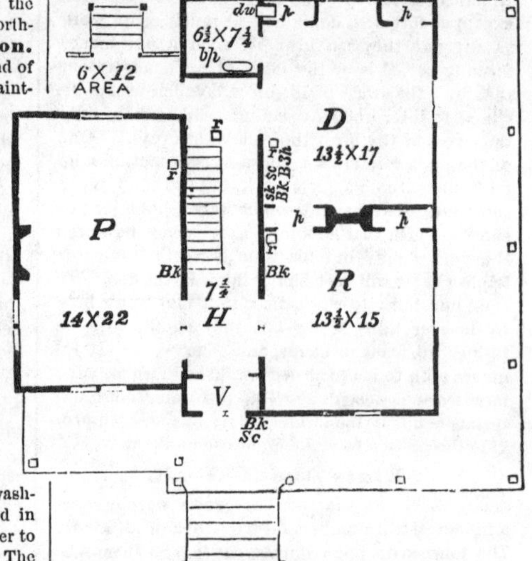

Fig. 4.—FIRST STORY—FLOOR PLAN.

dants. This is ornamented; it breaks up the box-like look of the room, and has the appearance of a Bay-window. The *Butler's pantry*, *bp*, has an eleptical plated copper basin or wash sink, supplied with hot and cold water cocks with elevated, curved discharge pipe, and waste pipe and trap below. This is convenient for washing silver, and China or other dishes not to be sent below. The Dumb Waiter, *dw*, opens into this pantry, so that food can be brought up from the kitchen and prepared for the table out of sight of the dining-room; and soiled dishes are taken here to be sent down. (The elevator is suspended in the center by weight and cord over a pulley at the ceiling, for which there is room, as the shelves do not need to rise more than 4½ or 5 feet above the floor.) *Sk* is a speaking tube to kitchen; *Sc* one to the family chamber; *Bk*, a bell-pull to kitchen; *B3h*, a bell-pull to 3d story hall to call down servants or others when desired.

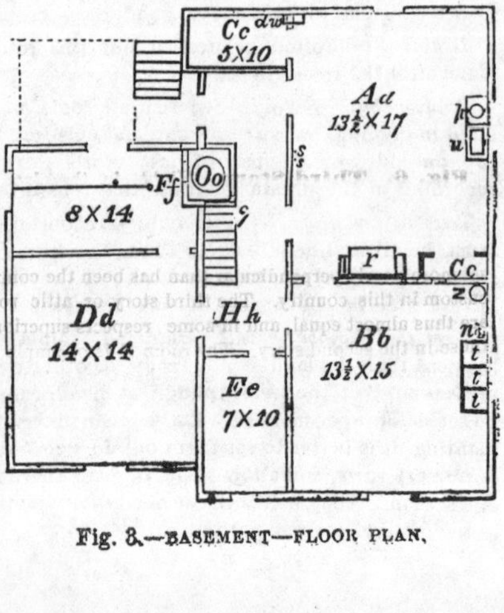

Fig. 3.—BASEMENT—FLOOR PLAN.

Fig. 5. Second Story.—Hight in clear, 9½ feet. The *Family Chamber*, *F*, opens into a dressing-room, or bed-room, *O*, which will take in a full-sized bed by the side of the door. (This room, *O*, may be used as a sewing-room, or a child's room, or a sick-room.) The Marble wash-stands, *w*, *w*, in *F*, and *G*, are placed in arched recesses, with an arched light cornice over the top, which has a pretty appearance. The rounded Marble table extends slightly into the room, and the closet beneath is cased, with a door. Marble pieces against the wall protect it from water. These basins are supplied with plated self-acting supply pipes for hot and cold water, with plated plugs, and stench traps. This arrangement of wash-stands in recesses, devised by Mr. Judd, has several advantages: The wash basins are out of the way, giving much more space in the room; one set of supply and waste pipes answers for two basins; the pipes run down by the chimney and are out of the way of frost, and the arched and corniced recesses are quite ornamental. In this room, *F*, there are, also, *Sk*, a speaking tube to the Kitchen; *Sd*, one to the Dining-room; *Sfd*, one to the front door, opening over the bell-pull (*Sc*, fig. 4), for conversing with callers at night; *Bk*, a bell-pull to Kitchen, and *B3h*, a bell-pull to the 3d Story hall to call servants in the morning, if need be. As remarked last month, these little conveniences, put in at small cost when building, save tens of thousands of steps, and much hallooing through the hall. The Housekeeper, though an invalid in her room, can thus call to other rooms and speak with the occupants through the tubes. So, also, directions may be conveyed to the kitchen and dining-rooms, before dressing in the morning. (A piece of flexible 1-inch rubber tube, attached to the speaking tubes, and extending on to the bed, will enable a sick person to talk with those in the kitchen or dining-room, though unable to rise from the pillow.) The *Chamber*, *G*, has the double closet and arch, above described; bell, *Bk*, to kitchen, register, *r*, etc.... The Chambers *M* and *N*, are convenient bed-rooms, with closets, and a warm-air register, *r*, in *N*.... The *Bath-Room*, *S*, has a wash-stand, and bathing-tub, *b*, both supplied with hot and cold water cocks, waste-pipes, and stench traps; also, a warm-air register, *r*, and gas pipe. The Water-closet, *w*, with patent basins and water arrangements, has a double cover, or seat, on hinges, so that on raising both, the broad-top basin answers for the reception of slops, and also as a urinal. It is supplied with water from the Tank in the hall above. The waste-pipe is iron, 4 inches in diameter, extending down the corner to the cellar, and out through the wall below frost, into the glazed pipe, 6-inch drain, running into the out-door privy vault. The floor under *w*, is cased with lead turned up 4 inches all

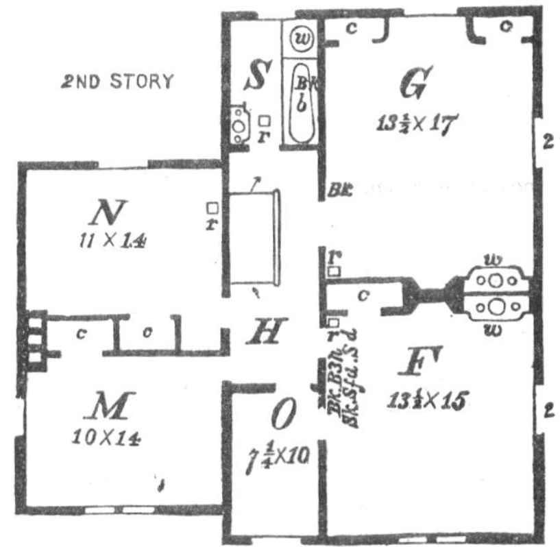

Fig. 5.—SECOND STORY—FLOOR PLAN

round, to catch any possible drip from the water-closet apparatus. A bell, *Bk*, communicates with the kitchen. The door has sash, with translucent or ground glass, to admit light to the hall, *H*. (Both flights of stairs have a landing and turn, four steps from the top, which eases the ascent, and lowers the floor of the Bath and Tank-rooms two and a half or three feet below the hall floors.)

Fig. 6. Third Story.—Hight, in the clear, 8½ feet. All rooms in this story are finished with white plaster coat. The outer walls rise perpendicularly 3½ feet, and then incline inward with the Mansard roof, which is set more nearly perpendicular than has been the common custom in this country. The third story or attic rooms are thus almost equal, and in some respects superior, to those in the second story. The room *Z* is of ample size for a servant's room, leaving the others for general use. *T* is left undivided, and will answer for a childrens' exercise and play room, and for drying clothes on rainy days, or when too damp to dry them in the laundry. There is a very pleasant outlook from the dormer-windows. (The houses here described stand on ground 70 feet above tide-water.) The tank is 7 feet long, 5 feet wide, and 3 feet deep, giving a capacity of 785 gallons. The water from all the upper roof runs into this, and when it is full the surplus discharges through a 5-inch leader into the general reservoir. If emptied in a dry season, it is refilled by the force pump in the laundry. This, of course, keeps the the boiler (*Z*, fig. 3) always full and gives both hot and cold water to the second story rooms. A safety pipe from the top of the boiler passes up over the top of the tank with a goose-neck, and when the fire is brisk, as in a cold day, some hot water will flow over into it, which alone would prevent freezing, though with a considerable body of water, and with double lath and plastered walls around it, and with the warm air constantly rising up through the open stairways (the walnut railing and banisters

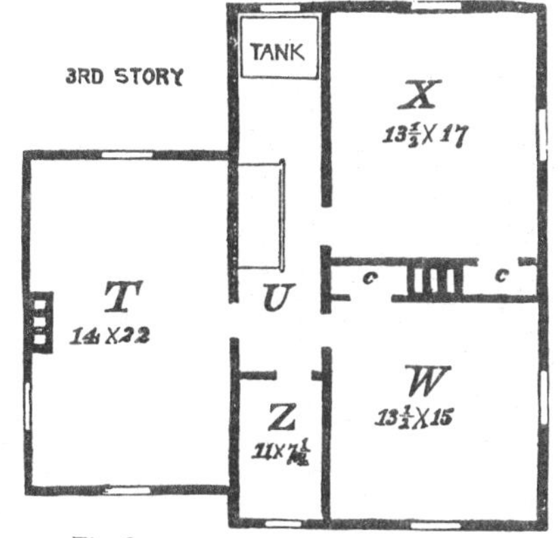

Fig. 6.—THIRD STORY—FLOOR PLAN.

extend into the third story hall) there is no danger of frost. The tank is made with a strong frame, covered inside with matched plank and lined with 4 lbs. per foot sheet-lead, strongly soldered at the joints, and the center of each side supported by plumbers' "tacks," that is, a circular segment of lead attached to the lining and let into the wood. A film soon coats the lead and prevents its solution in the rain-water, though this is of no consequence, as all water for cooking and drinking is drawn through tin lined pipes at *u*, fig. 3. Those, fearful of contact of water with lead, can arrange in the tank a gate or valve worked by a floating ball and lever, so as to turn all the water directly from the roof into the cistern when the tank is full. No spring or well water, which filters through the soil and necessarily dissolves out more or less of mineral salts and organic material, can be so pure as the "heaven-distilled" rain-water from a clean upper roof. We have used none other for many years, and would use no other. When accustomed to it, it is far more agreeable and more salubrious than any compound water drawn through the soil. If the roof be exposed to leaves and dust, a simple filter of washed sand, or sand and charcoal, attached to the cistern will remove these impurities. If many leaves are likely to fall on the roof, a wire netting should be placed over the entrance to the conducting pipe, and the accumulation of leaves be occasionally removed. (The dormer-windows give ready access to the gutters and roof.)

Other Items.—See last month, page 29, for various items, as grape arbor, siding, moulding (used in first and second stories in this house) for painting, etc., etc. There are 16 pantries or closets—an important thing to be provided in any house. One of the upper rooms may be used for a store-room, if desired, as there are in all 9 rooms that may be used as bed chambers.... As a matter of due credit and a stimulus to perfection in workmanship, we give the names of the chief artisans of these and the house described last month: General Superintendent, as well as architect, and a hard worker himself, JOHN DONALD; Mason-work, Hendrickson Jarvis, and Geo. W. Lewis (1 house); Painters, Torrington & Brown; Plumber and Gas-fitter, Henry Lewis; Roof-work, and putting up Furnaces, Benj. G. Field; Sash and Blinds, Henry Christie; Stairs, etc., John L. Smith—all of Flushing; Doors, by Little, Fowler & Fleet, of New York City.

COST.—We give the actual cost of these houses, which may be taken as a comparative criterion. (These were built with special economy in purchase of material, and work, as noted last month, but probably timber, etc., will range below the rates of New York and vicinity, and skilled labor averages lower, in most of the country. For ordinary country and village residences of the same size and conveniences, a less expensive style of finish, outside and in, would often be adopted, so that the cost would range all the way from $7,000 to $9,000): Timber, $345; lumber of all kinds, $1,105; mason work and material, $1,547; carpenter work, $1,103; digging cellar, $20; digging and stoning privy vault, $25; grading, $40; cistern, $35; painting, $350; sash and glazing, $130; blinds, $85; roof, slate and tin, including gutters, and window and chimney zinc and tin, $450; plumbing, with marble and wood-work, $490; range, with water-back connection, $60; outside drains, $50; furnace, with setting, piping, registers, etc., $300; stairs, $260; piazza steps, newels, banisters and railing, $150; arbor, privy, and lattice screens, $80; fencing, $125; sidewalks, $40; hardware, nails, locks, hinges, sash-weights, bells, etc., etc., $288; mouldings, $130; doors, kiln-dried, $140; marble mantels, $200; summer pieces and grates, $80; gas piping, $60; ventilators, $25; cartage, and many sundries, $293; average interest on outlay while building, $180. **Total Cost** of House, exclusive of land, **$8,186.**

Dog and Sheep Power for Churning.

Something has excited the interest of our subscribers in "dog-powers"—that is, in contrivances for utilizing the power of dogs for churning, and, perhaps, other light work—as we judge from the numerous inquiries lately received. This is a good symptom, it shows that there are some people who have waked up to the need of alleviating the drudgery of woman's toil. Where there is much churning to be done, a dog-power is truly a labor-saving device. There are several different kinds, the best, perhaps, is a "tread-power," like the ordinary one, or two-horse tread-powers. These, however, are rather costly, and can only be made by experi-

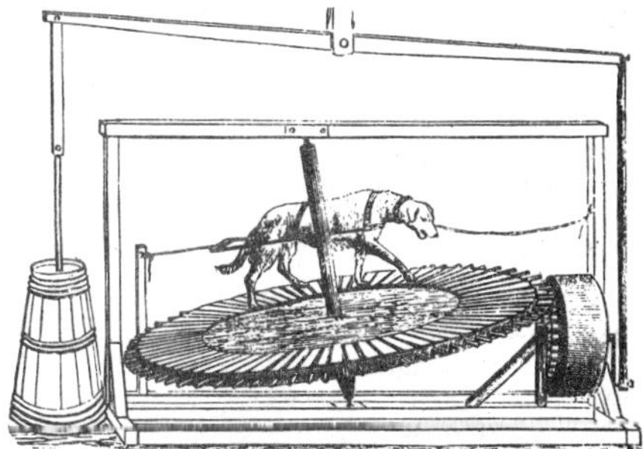

Fig. 1.—PLATFORM DOG-POWER.

Fig. 2.—WHEEL DOG-POWER.

enced mechanics. There are forms, however, which may easily be made by any one familiar with the use of tools. Two of these we represent in the accompanying engravings. They are worked upon very different principles. The revolving platform, fig. 1, is set at such an angle that, though the weight of the dog operates to favor the turning, it is, after all, by his strength of draft that the machine is effective. The animal must be harnessed in some simple way and attached to some fixed object. The harness figured has no advantage over the simpler one by which the arctic sledge-dogs are attached to the vehicles they draw. This is by means of a broad collar-band, and a small rope or thong passing from it between the legs and held in position by a belly-band. The platform power is made as light as possible, consistent with strength. There is a frame made, supported upon the shaft by means of cross-beams pinned to it, braces beneath, and a rim in which pins are set to act as cogs, meshing in with flat ones in a drum. The churn may be operated by a crank and walking-beam, as shown, or, if a rotary churn, by a band running upon the drum. A track must be made for the dog by nailing radiating cleats upon the platform. Any arrangement to lessen friction, like friction-wheels, and iron sockets and bearings, will be of essential service. The platforms are usually 7 or 8 feet in diameter. The drum should be a little swelled in the center to prevent the band running off.

The wheel-power is a little more difficult to make, but has some advantages. It is operated by the weight of the animal, his labor being exerted precisely as in running up hill—as a squirrel runs in his cage. A heavy block is sometimes suspended from the axle to hang down behind the dog, and cross-bars may be nailed to the arms or spokes to prevent his jumping through. The wheel is either made only wide enough for one dog to run in, or wide enough for two to run abreast. Friction rollers, in this case, are also very useful. The wheel is hung in a frame, that will not shake with its motion, which is sometimes quite irregular. It is made with bent rims, fastened to the spokes, and boards are nailed to these.

The greater the diameter, the easier and slower will the wheel turn. Eight feet is about the right size, and the wheel should be banded with common hoop iron nailed on—regular hoops not being necessary. A three-quarter-inch iron rod makes the bent axle, and this should be keyed fast in the wheel, while the ends should run in metal boxes, which may be oiled.

Dogs, sheep, and goats, are used in these and similar "powers." The last are rather light but active and hardy, and the exercise does them good, especially if they are kept stabled.

A Boat for Getting Out Muck.

Digging muck is work that may be done at almost any season, provided drainage can be got. Where the water of the swamps cannot well be drawn off, the work is usually deferred until very dry weather, which is not necessary, as a little contrivance will make the job an easy one. "S. R.," of Ashtabula Co., Ohio, writes: "We have been drawing out muck upon gravelly ground, hauling it out of the swamps in its wet state by means of a boat, made as follows: Take five one-inch boards, one foot wide, and eleven feet long. The side boards are sloped at the forward ends to five inches, and at the back end to eight inches. Three cross-pieces are put in, the bottom boards are bent to them and nailed. The boat is drawn into the swamp by hand, and drawn out by a team attached to a long chain." We think more cross-pieces would be desirable to give strength, but it is clearly unnecessary, and in fact undesirable, to have

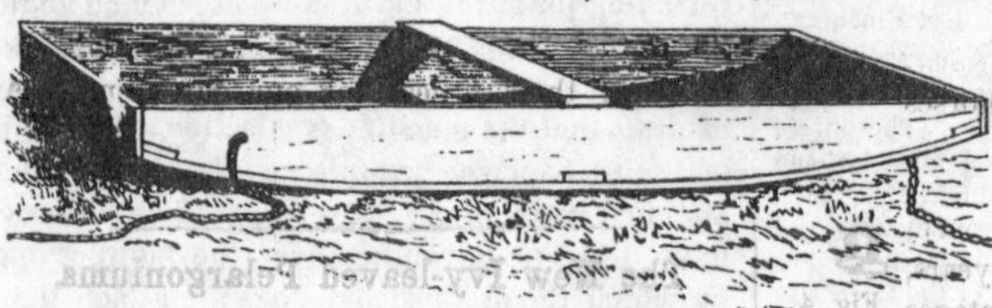

BOAT FOR GETTING OUT MUCK.

the boat water-tight, for the water must have a chance to run out. Five five-inch cross-pieces, and three-quarter-inch oak boards would make a stiff, good bottom, which would wear a long time even if hauled a good ways over anything but a stony or gravelly soil. The engraving shows hand ropes near the stern for hauling the boat backwards, and the chain at the bow.

Three-Story Barns.

Three-story barns, or "three-deckers," as they are sometimes called, when conveniently arranged, are decidedly the most economical, both of material in building, and of labor and

Fig. 1.—ELEVATION OF BARN.

care in management. The accompanying plan has been prepared in response to several requests for a barn-plan to accommodate a small farm, and not to cost more than $1,500 to $2,000.

It is rarely or never worth while to attempt to build a three-story barn upon level ground, but where a descent can be used having a slope of 18 inches in 10 feet for a space of 50 or 80 feet, it will do very well. A bridge, or a walled approach to the barn floor is often dangerous. Access by a self-sustaining sodded earth bank, sloping off gradually to the general surface, is decidedly preferable. A good cellar is seldom to be had without considerable digging, and the wall against the bank must be a substantial one of concrete, or of stone, or brick, well laid in cement, and guarded from the action of water by surface channels and underdrains. The cellar ought to be at least 9 feet high, the floor grouted and cemented water-tight, and should be accessible from the south. Being used for manure, convenience of loading carts requires it to be reasonably high. The feeding or cattle floor is not necessarily so high; 7 feet in the clear is as low as one ever ought to be, and 8 feet is about right. This floor should be accessible from each end, and well lighted. The openings for the manure to be thrown into the cellar, and for the liquid manure to flow through, must be where the liquid will not rot the beams. The floor should be laid of well-seasoned inch-and-a-half plank, merely tacked down at first if not perfectly dry, especially if the barn can stand through one summer before it is used. In this case the floor can be re-laid permanently in autumn, after this extra seasoning, and the seams caulked and pitched. The thrashing floor should be not less than 12 feet wide, the doors opening nearly the full width, and 10 feet high. From the sill to the plate cannot well be less than 14 feet, and the barn should be framed to dispense with the great cross-beams so much in the way of the horse-forks (see figure 5). The side beams, connecting the inner posts with the outer frame, should be level with the top of the great door. We commend a feature which we have long known to work well, namely, laying a corn floor upon the tie-beams of the roof. In this barn such a floor would afford 18 × 40 feet of space for spreading out corn to cure, in the hottest place to be found. Such a loft will hold easily 1,000 bushels of corn in the ear. The corn is lifted in tubs attached to the common horse-fork rigging. The corn sheller is placed here, and the shelled corn run down by a shute.

In building such a barn economically, it is expected of course that the farmer will do a good deal of the labor with his own men and teams, at times when other work does not press. He will dig the cellar and grade the ground for the approach and for the barn-yard. He will haul all the stones, sand, cement, and lime, for the wall, and, perhaps, mix the mortar and lay the walls himself. He will cut and hew the timber, haul to the saw-mill and back again, and assist in the framing and raising. It depends, therefore, a good deal on the part of the country in which the barn is to be built, what timber, as well as what foundation, can be economically used. The barn can be built near New York with bought materials and hired labor, for $2,000.

Description of the Barn. — The barn is 30 × 40 feet inside measure, and the plans are drawn to a scale of $^1/_{16}$ of an inch to the foot. Fig. 2 is a plan of the main floor. On the left, space is taken for the shop and the grain room.

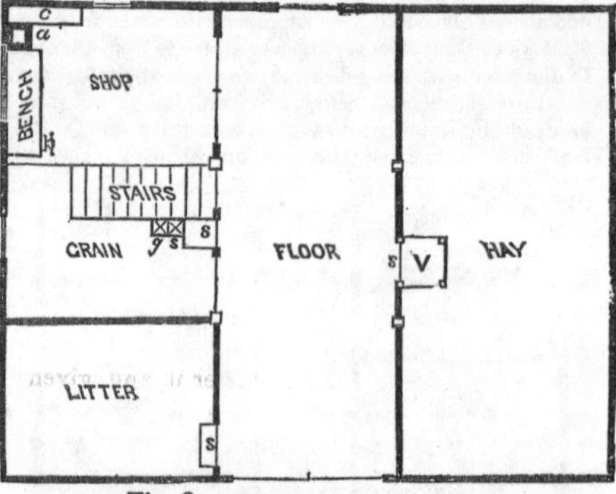

Fig. 2.—PLAN OF MAIN FLOOR.

The former, a room 10 × 14, has a large double window and a single one. The double doors make it possible to run a wagon or carriage into the shop, for painting or other repairs. There is a carpenter's bench and a closet for tools.

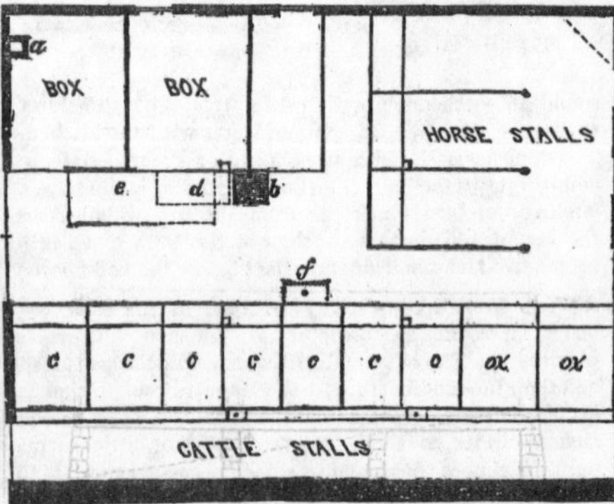

Fig. 3.—PLAN OF CATTLE FLOOR.

The chimney passes through this room, and a fire can be made if necessary. The grain room should be supplied with bins, and there should be two or three shutes for different kinds of grain or meal. These are shown at *g s*, fig. 2, and are intended to be of canvass, after passing through the floor, so that they may be turned to one side and out of the way. These rooms need not be more than 7 or 7½ feet high. From the thrashing floor, two shutes, or trap doors communicate with the floor below—one near the stairs for cut hay, etc., one near the litter bay, through which bedding may be thrown down at the rear of the cows, while the ventilator is also used as a shute, and through it, long hay is

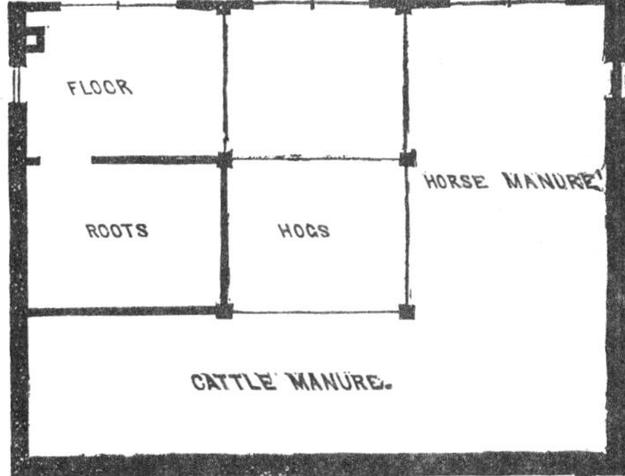

Fig. 4.—PLAN OF BARN-CELLAR.

thrown down, falling in the passage-way between the horse stables and the cow mangers. This ventilator (*V*) is 3 × 3 feet square, and extends from the cattle floor out through the roof of the barn.

The plan of the cattle floor (fig. 3) needs little explanation. The stairs at *e* are hinged and may be turned up and fastened, so that a cart can be driven, or backed under them, to dump a load of green fodder upon the floor, or one of muck to be thrown through the trap door *f* to the pigs. The feed-box *d* is movable. Near the end of the barn, where the horse and ox stalls are, the shed for wagons, carts, tools, etc., is supposed to be placed. At *b* a hydrant and water-trough is placed. One corner of the horse stable may be partitioned off for nice harness, etc., if desired. In the cellar plan, fig. 4, a root cellar is provided, also a floor where steaming apparatus may be set up. Here the "working hogs" are to be kept, and either shut off from the manure, or allowed to range over it, and given the range of the barn-yard besides, if that be desirable. This cellar is accessible to carts or wagons through three 8-foot doors, and it is light-

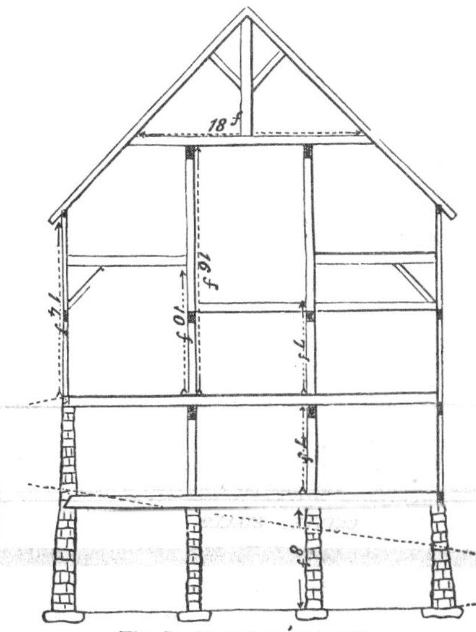

Fig. 5.—SECTION OF BARN.

ed by windows in these doors, and by others in the ends. The ground of the barn-yard slopes towards the cellar, and the water is collected in a tank to be pumped through a hose over the manure whenever it gets too dry, or too hot. Fig. 5 represents a cross-section of the barn, and shows the framing and slope of the ground.

Wooden Shoes for Horses.

Much labor would be saved in gathering the hay crop upon salt marshes and on reclaimed swamps, if horses could be used for mowing, raking, and carting. It often happens that the farmer has several acres of this soft land, where the crop costs all it is worth to gather it. He must mow, rake, and move the grass by hand, or let it rot upon the ground. We recently saw a wooden shoe that completely remedied this difficulty. It was made of stout oak board, one inch in thickness, 10 inches long, and 8 broad, and rounded at the corners. Fig. 1 shows the bottom of the shoe. A cleat is fastened across each end to prevent it from splitting, and to give additional strength to the shoe; this should be fastened either with stout screws 1¾ inches long, or with wrought iron nails driven through and clinched upon the upper side. An iron strap is fastened across the middle of the shoe to receive the shanks of the strap going over the hoof of the horse, which are held in place by screws and nuts. Fig. 2 shows the upper side of the shoe. The horse should be rough shod, and places should be cut into which the toe and heel corks will snugly fit.

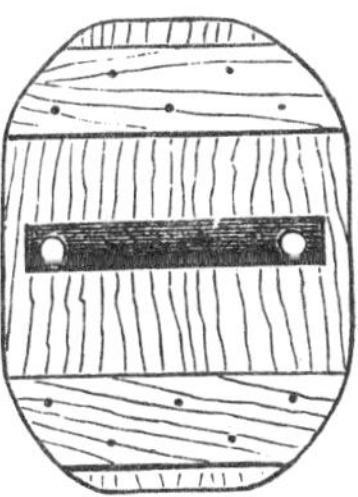

Fig. 1.—UNDER SIDE.

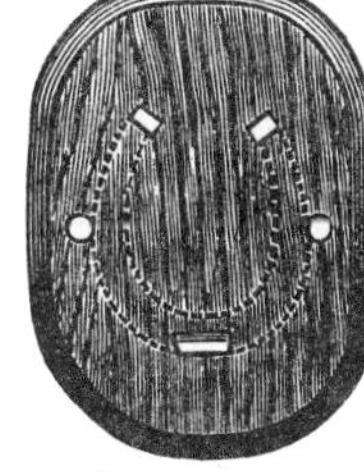

Fig. 2.—TOP OF SHOE.

In fig. 3 the hoof is shown as it stands upon the shoe. An iron strap (fig. 4) is fitted to the hoof, and the shanks pass through the plank and are fastened with a nut and screw. This shoe is so simple in its construction that any one accustomed to the use of carpenters' tools can make the wooden part of it, and a blacksmith can furnish the iron straps and screws. A common monkey-wrench will be needed to put on the nuts, and to tighten them, if they get loose.

Fig. 3.—SHOE FITTED.

A set of shoes made of good white oak will last a great many years. It will be seen that the shoe enlarges the lower surface of the hoof about four times. It is found in practice that a light horse, weighing, say 900 lbs., shod in this way, can go upon any soft land, where a man could walk, with safety. If the horses are very heavy, or the land very soft, the shoes must be enlarged. These articles had been in use upon the farm where we saw them, some ten years, and so manifest were their advantages that they had been adopted by all the farmers in the neighborhood who had occasion for them. They were in use by all the owners of a large reclaimed salt marsh, and the facility they afforded for gathering the crop, had added very much to the value of the land. To owners of marsh lands these shoes will be invaluable.

Fig. 4.

The Ivy-leaved Toad-Flax.

The little Ivy-leaved Toad-Flax, *Linaria Cymbalaria*, is a native of the south of Europe, but has become perfectly naturalized in England, and is entirely hardy with us. Unlike the Toad-Flax we are most familiar with—the trouble-

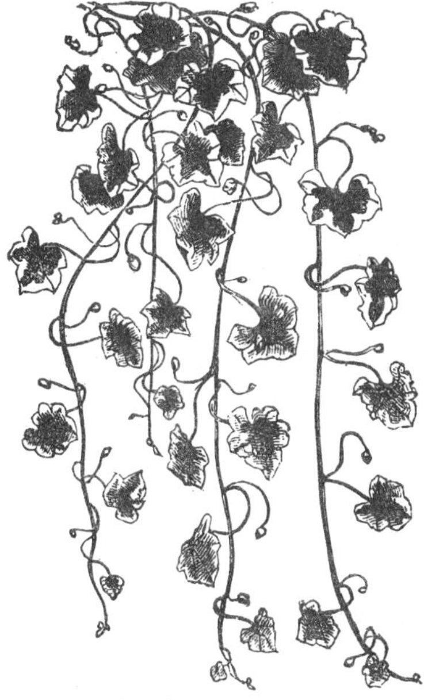

VARIEGATED IVY-LEAVED TOAD-FLAX.

some weed known as Butter and Eggs and Ranstead Weed—this is a delicate trailer, which has long slender stems, neatly lobed leaves, which are dark green above, and purplish below, and very small lilac flowers. It shows to the best advantage when growing in positions where its branches can hang down, and where it grows wild it prefers crevices in rocks, old walls, and such situations. This habit adapts it for growing in suspended vases or baskets, as well as upon rock work, for which purposes it is quite popular under the names of Kennilworth Ivy, and Coliseum Ivy—though it is hardly necessary to say that it is not an Ivy. When grown in the border it makes a neat little mass; the long leaf-stalks become much entangled, and the plant is more compact and less graceful than when the branches have a chance to trail. Every season brings some novelties in the way of variegated leaves, and this year this well-known old plant is offered with leaves which are blotched with green upon a yellow ground. The engraving represents the variegated plant which differs from the common form only in its blotched leaves. Should the plant hold its variegation, and prove reasonably hardy, it will become very popular.

We are cautious about recommending plants with variegated foliage until they have been tested in out-door culture. Neither the descriptions of the European journals, nor the appearance of the plants in the green-houses, will allow us to judge of their fitness for our gardening. A large number, if not the majority of variegated leaves, become green before the summer is over.

The New Ivy-leaved Pelargoniums.

The Old Ivy-leaved Pelargonium (*Pelargonium peltatum*), or Geranium, as it is most commonly called, is a much neglected plant. When well grown it makes a fine plant for a hanging basket and is handsome for both its foliage and flowers. Within a few years new varieties have appeared which show a marked improvement.

(☞ *For other Household Items, see "Basket" pages.*)

A Shaded Garden Seat.

The engraving of a garden seat with a tent-like roof is from an article of English manufacture, that attracted much attention at the Paris Exposition. We do not know that it is sold in this country, but we introduce it here as a suggestion to the ingenious. A shaded seat of some kind would be found very useful by those who have croquet grounds, as well as by those who wish to enjoy sewing or reading in the open air, and cannot avail themselves of the shade of trees. The top of the one figured is made to incline backwards or forwards, as may be required. It would not take much contrivance to fit up a light awning that would be quite as useful if not so elegant as this. Where the seat is to remain stationary, the awning may be supported by poles driven into the ground.

Scouring Knives, etc.—Mrs. "O. A. H.," writes that for five years she has used water-lime for scouring knives, forks, tins, and the like. She says: "I have a box with a partition and keep the lime in one part and the cloths in the other. I wet a small cloth a little and dip it in the lime, and after the articles are well washed and wiped, I rub them until the spots are removed. Then I take a larger, dry cloth, dip it in the lime, and rub the articles until polished to suit me. Wipe off the dust from the knives and forks with a dry cloth, and they are ready to put away."

"Sun-light Oil"—A Household Humbug.

Most of the "Humbugs" exposed in the column especially devoted to rascality, affect only the pocket. Here is one which, worse than the highwayman, demands both your money *and* your life; and we notice it in the Household Department, as in this case mothers need especial warning. We have several times seen circulars proposing to sell recipes for preparing "Sun-light Oil," which should give a better light and be cheaper than any other illuminating oil. Now we have, from one in Wisconsin, who has invested $2 in the thing, the recipe itself, which reads as follows: "To make one gallon, take 3 quarts of BENZINE, 1 oz. Pulverized Alum, 1¼ oz. Alcohol, 2 oz. Cream of Tartar, 2 oz. Salsoda, 1 pint of Potatoes (cut fine), 2 tablespoonfuls of fine Salt, 2 drachms Oil of Sassafras, 4 drachms Gum Camphor. Dissolve the Alum in the Alcohol as much as possible; then add the Gum Camphor, stir for a few minutes; then add to one pint of the Benzine, stir it well for ten minutes; then add all the other ingredients, except the Benzine, stir well until it foams, then add the remainder of the Benzine; leave it open and exposed to the air; shake it occasionally, and in ten hours time it will be fit for use, although it should stand, if convenient, for 48 hours before using."

If this thing were not so wicked, we should ridicule its absurdity. It is an evident attempt to induce people to believe that, with the various additions and the prescribed mixings, the Benzine can be converted into something safe to be used for illuminating purposes. The Oil of Sassafras and Camphor are added to make the compound smell differently from pure Benzine, they are combustible and so far are not foolish; but the other ingredients, Alum, Soda, Cream of Tartar, Salt, and Potatoes! are all sheer nonsense. They have not the slightest effect in making the Benzine other than Benzine, and the person who proposed to add them is a fool if he did not know it, and a rascal if he did. These recipes are hawked about the country by agents, who by pleading the cheapness of the light doubtless sell a large number. We do not know of any law that exactly meets the case, though by any sensible justice an agent of this kind would be punished as a common nuisance. Recollect that this Sun-light Oil is Benzine, and as far as danger is concerned, nothing but Benzine—unsafe, explosive, deadly. Better that your children should have no other light than tallow dips, or go to bed at dark, than run the risk of sudden death, or what is worse, being maimed for life by this "Sun-light Oil." If you will use it, make your will, insure your life, and ask your minister to get ready a sermon on the "Mysterious Dispensations of Providence."

To Restore Faded Black Lace.—In many a housekeeper's bureau drawers, lie old black lace veils, edgings and head dress, faded and rusty, yet not worn out. By a simple process they can be quickly restored to their pristine beauty. Strain off some black tea from its leaves (having made it too strong to drink), let it cool until milk warm, pour over the lace, and let it stand several hours, then squeeze it very gently, dipping it frequently into the tea, until it shows that the dirt has been extracted.

How to have a Variety in Food.

There are two ways of having variety: one, that of heaping upon one's table a great many kinds of food at one meal, so many that it is impossible to do more than to taste of each one; the other to have but a few kinds at one time, but so varied from meal to meal that the same thing does not appear upon the table very often. The latter seems to me to be the better way on the score of economy, which the larger portion of farmers' wives probably find it necessary to study to a greater or less extent. Besides this, a great jumble of food is injurious to the digestion. I will explain what I mean by the best kind of variety.

Bread, either wheat or rye, should be a constant article of diet, and should be made fresh three times a week. Occasionally raised biscuit, or soda biscuits, can be substituted, or some form of corn bread, or Graham bread, gems, or some kind of muffins. The plain white bread will probably hold its place as the standard kind for a long time to come, and its substitutes should be so judiciously introduced as to make all kinds seem ever fresh and new. Meat ought, also, to be varied, both in its kind and in the manner of cooking it. Beef is acknowledged to be superior to other meats, but it should be cooked in different ways. Roast beef and well-broiled steaks will always stand first in general estimation; but boiled beef, beef stew, and beef soup, help to variety, keep up the appetite for the other kinds, and are more economical than the constant use of such pieces as are needed for good roasts and broils. Many families are in the habit of having a nice piece of roast beef for their Sunday's dinner, and bringing it on cold every day thereafter, until it is vanquished by the force of sheer persistence. Such a piece of beef can be used in many ways. Was it quite rare, to suit the tastes of some members of the family? Then for Monday's dinner cut off some slices of the rare part, lay them on the gridiron over some very hot coals, let them brown upon both sides, and bring them directly to the table with a little butter upon each slice.

The more thoroughly cooked part can be made into several kinds of hash; coarse hash, with or without potatoes; fine hash, with or without potatoes, warmed in the mass or made into small thin cakes, and browned in a little hot lard; fine hash, with potatoes, spread upon a piece of pastry rolled thin, then rolled up tight and baked; or fine hash, without potatoes, put upon toast. A savory stew can also be made of bits of cold beef by the addition of two or three onions, as many tomatoes, or some canned tomato, and a sufficient amount of seasoning. A pie can be made of cold beef, either with common pastry for crust, or layers of mashed potatoes, or boiled rice, alternated with the meat.

A good many farmers' families are out of the reach of a market or a butcher's wagon, and are obliged to subsist, as to meat, the year round upon ham, salt pork, and chickens. The ham is always fried, the pork is always boiled or fried, swimming in grease, and the chickens are cooked in some one way. The ham might be boiled or broiled, or cut in small bits and made into dumplings, the crust for them being like soda biscuit, and steamed or baked. Cold pieces of ham are nice cut in small mouthfuls, and warmed with eggs; they are also very palatable fried in batter, like veal cutlets.

The methods of cooking chickens and pork are numerous, and most housekeepers know them, but fall into the habit of cooking them in some stereotyped way, so that they never seem to have a variety. It is the same with vegetables. The first care is, in the summer and fall, to make provision for winter; this should be plentiful and varied, both in vegetables and fruits. Dried or canned green corn, and canned tomatoes, and the many kinds of sour and sweet pickles, are great additions to the winter dinner table; but my opinion is that they are better appreciated if the same kind does not appear upon the table oftener than once a week. Potatoes seem an essential part of every dinner, and one can make such a rotation of turnips, parsnips, onions, cabbages, beans, corn, beets, etc., according to the nature of the dinner, as never to get tired of any one vegetable.

In the matter of canned fruit, preserves, cakes, pies, and puddings, the same rule for variety should be observed. A housekeeper should not fall into the common error of making year after year the same kind of preserves, the same kind of cake, and the same kind of pies.

It is a good plan to keep a written recipe book, and to add to its contents occasionally such recipes as are known to be good, and then to use them all in their own time, not settling down upon a very few. By canning fruit with but little sugar, the winter routine of pies can be very much relieved. Cherry, raspberry, currant, peach, and blackberry pies, can take their turns with the mince, pumpkin, and apple, usually supposed to be exclusively winter pies, and there is little difference between them and those made of the fresh summer fruits. Apples, which very few farmers are without, are capable of being made into most delicious sauces, jellies, preserves, pies, and puddings; and yet they are often as bad as wasted by being made into miserable, lumpy apple sauce, with the flavor all washed out of them by the admixture of too much water, and into hard, tough, indigestible pies.

Citron and Raisins. — A Maryland Housekeeper asks us to tell her how she can prepare green citron to be used in cakes, and how to make raisins from her grapes. The citron used in cake, etc., is the preserved rind of a fruit like a large lemon. The citron melon cannot be used as a substitute, as it has no aromatic quality. None of our American grapes will make raisins.

English Seed Biscuits.—1 ℔ of flour, ¼ ℔ of sugar, ¼ ℔ of butter, ½ oz. of caraway seeds, 3 eggs. Roll out, cut round, and bake in a moderate oven.

Cinnamon Cakes.—By Mrs. L. A. G. 1 cup of sugar; ¾ of a cup of molasses; 1 cup of butter; 1½ tablespoonfuls of ground cinnamon; 2 level teaspoonfuls of soda, dissolved in 6 large tablespoonfuls of warm water; stir well, and add flour enough to allow to roll quite thin; cut out with a biscuit cutter and bake in a quick oven.

The Northern Hare.—*Lepus Americanus.*

The true rabbits live gregariously, and form burrows, in which they rear their young. None of our rabbits have these habits, and the animals which, in this country, are popularly called rabbits, are strictly hares. We have, in North America, some thirteen species of hares, all of which are solitary in their habits, and instead of burrowing, make forms or nests of grass upon the ground, upon which they sit. The most common, after the Gray or Brown Rabbit, is the Northern Hare, also called White Rabbit. It is found from Virginia to Canada, and as far west as the plains of the Missouri. It is considerably larger than the common rabbit, has ears larger in proportion to the size of the head, and a longer hind foot. The feet are so thickly covered with hair, that the animal makes but little impression upon the snow. In summer, the color of the animal is of a reddish or cinnamon brown above, and white beneath the body; the short tail is sooty brown above, and dull grayish beneath. In winter, the pelage changes to white, but when the hair is parted, lead color and cinnamon color are seen below. The fur, at all seasons, has a rough and shaggy look, on which account the skins are of little value. The animal weighs from three to six and a half pounds. This hare inhabits dense swamps in winter, but in summer finds a retreat on higher ground. It is very seldom seen in the day-time, as it does its foraging during the evening, and night. It forms well worn paths or runs, which it is said to follow for years, and many are taken by means of snares or traps, set in these runs. When pursued, it runs, or rather leaps, with great speed, and endeavors to escape the hunter by avoiding the open ground, doubling and turning among the thickest woods and undergrowth. It is much more fierce than the common rabbit, and when captured, bites and scratches with considerable energy. Numbers of this hare are sent to the New York market every winter, but as they are not highly esteemed, they sell for a low price. The flesh is dry and hard, and much inferior to that of the common gray rabbit. It has been stated that the hare was introduced from England into Canada, and from thence spread over a large part of the United States. This is an error, as the European hare differs from ours in many respects. It is larger, with ears longer in proportion to the size of the head, and it does not change its color with the seasons.

THE NORTHERN HARE—(*Lepus Americanus.*)

Sultan Fowls.

The Sultans were introduced from Turkey into England about the year 1854. They are an exceedingly beautiful and attractive breed, closely resembling the Polands in many of their characteristics, yet differing from them essentially in others. Their plumage is of the purest white, and they are abundantly feathered on every part, as shown by our excellent engraving.

A PAIR OF SULTAN FOWLS.

The comb is forked, at least it is in two parts, which, in the case of the cock from which the engraving was taken, so resembled a crescent, that every one was struck with the appropriateness of the combination of crescent and turban for a Turkish fowl. They have full Polish crests, and muffs or beards in both sexes. The body feathers are soft and abundant; the hackle flowing; the sickle feathers well developed; the legs fully feathered to the tips of the toes; the feathers of the hocks having stiff quills and extending back in a line with the thighs, which is called being "vulture hocked." The hens are persistent layers, rarely or never wanting to sit. They lay large, white eggs, and many of them. The breed is reputed to be easily kept, and hardy, easily confined, and useful as well as ornamental. The chickens feather very young, and hence are delicate, as is usually the case with breeds upon which the feathers come before the young bird has built up its frame sufficiently to sustain the draft upon it without becoming too much weakened. Several importations of these fowls have been made within a year or more, and naturally have attracted much attention at the exhibitions where they have been shown. The stock of Sultans in this country is small, but they have proved themselves hardy in our climate, and useful as layers. They are less than medium-sized fowls, but have plump bodies, and, if abundant, would be excellent for the table.

Osier Willow as a Hedge.

We recently saw a successful hedge grown from the Osier willow. It was strengthened at the top and in the middle by interweaving the branches, without severing them from the stock. These living ligatures became stronger every year and added to the stiffness of the fence. It was kept trimmed on the top at the hight of six feet. It turned cattle perfectly in its fourth year, and is constantly growing stronger. It was kept at its required hight by cutting annually a crop of boughs from the top for basket willow. This is a novelty in the way of growing willows, but we do not see why Osiers may not be grown just as well six feet from the ground, as upon its surface. The wands are quite as vigorous, and of as good quality. The double office this willow serves, treated in this way, would induce some, perhaps, to cultivate it, who otherwise would not think of it. The owner of this hedge peeled his wands by hand, and sold them in the market for ten cents

containing a great variety of Items, including many good Hints and Suggestions which we throw into smaller type and condensed form, for want of space elsewhere.

The Torrey Botanical Club.—The botanists of New York and vicinity have for some years been associated under the above name. With the present year they commenced the publication of a Monthly Bulletin of matters mainly relating to the local Flora. They desire to be in communication with all who are interested in studying plants within 30 miles of New York City. These and others can procure the Bulletin for a year by enclosing $1 to Wm. H. Leggett, 224 East 10th-st., N. Y.

Another Vine Protector.—If bugs are not kept away from cucumber, melon, and similar vines, it will not be for lack of contrivances to effect it. The latest one we have seen was in the stock of R. H. Allen & Co., and is shown in the engraving. It is made of mosquito net, four triangular pieces of which form the sides; the edges or corners are of two thin strips of wood,

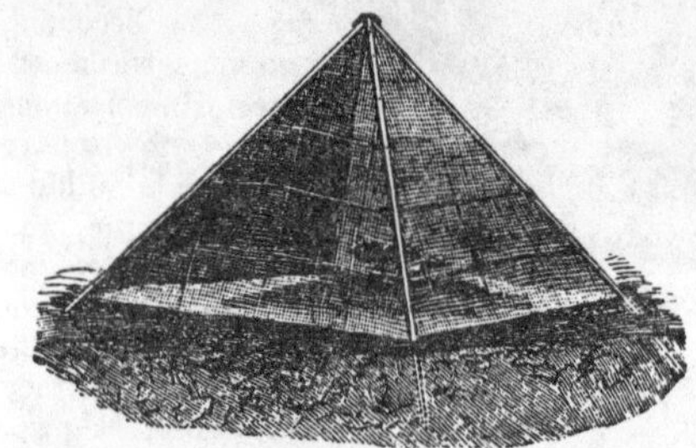

VINE PROTECTOR.

nailed together with the netting enclosed between them. At the top are two small bits of leather, each one of which is tacked to two of the wooden supports in such a manner as to form a hinge. The protector, when not in use, shuts up and occupies a very small space. It is claimed that 24 dozen can be put into a flour barrel. They are easily made, but are sold by the dozen very much cheaper than one can afford to make them.

How to Improve a Meadow.—"S.," Seymour, Conn.—Harrow it thoroughly, sow on some fresh grass seed this spring. Pile or compost your manure, putting in a half bushel of bone-dust to the ton of manure, turn it two or three times, and apply it to the meadow at the rate of 20 tons per acre. Or, if more convenient, use good Peruvian guano or superphosphate of lime, broadcast, at the rate of 400 lbs. to the acre, during rainy weather this spring.

Good Pigs.—Porter Frisbee, Secretary of the Delaware County (N. Y) Agricultural Society, writes, that at their annual meeting they had a good show of grains, seeds, dressed meats, etc. Five spring pigs were shown—three of them, 10 months old, weighed, dressed, 446, 354 and 332 lbs.; and two pigs, of 9 months old, weighed, dressed, 371 and 338 lbs.

Bark Louse.—D. B. Boyer, Montgomery Co., Pa. Yours is a bad case of the Oyster-shell Bark-louse. The eggs are under the scales, and will hatch in June. If this tree is in an orchard we should cut it down and burn it; and give close attention to the other trees upon which you say it is "starting." We do not commend Mr. Norris' plan of painting trees "from stem to stern," but your tree is good for nothing else but to burn, or devote to some such desperate experiment.

Floral Exhibition at Charleston, S. C.—An exhibition will be held by the Charleston Co. Ag'l and Hort. Soc., from May 1st to 10th, and liberal premiums are offered for the leading classes of flowering plants as well as for baskets, bouquets, strawberries, etc. H. B. Holbeck is Secretary.

Corn vs. Wheat in Virginia.—"H. G.," Augusta county, Va., writes: "My farm is a sandy loam, excellent for corn, but rather too light for a maximum crop of wheat. I can raise from 40 to 60 bushels of corn per acre, while the same land rarely yields over 20 bushels of wheat, and often less. If we attempt to push it beyond this, it falls down and does not fill well; but it brings fine clover." We can see no reason why such land cannot be made to yield 30 or 35 bushels of wheat per acre. Underdrain where needed, make the land clean, so that there shall be no weeds or grass in the wheat—raise more clover and keep more sheep. If this does not give stiffer straw and a well filled ear, try lime.

Grafting the Hickory.—Hickory nuts vary much in their wild state, but the difficulties of propagation have prevented the multiplication of the choicer specimens. Mr. David B. Dickinson, having tried various methods of grafting without success, hit upon the following, which allows him to propagate the hickory with satisfactory results: Early in the Spring he grafts the stock, which should be small, very near the surface, or, if possible, below it. The graft is waxed in the usual manner, and then the earth is heaped around to cover all of the cion except the upper end. Mr. D. has grafted trees now coming into bearing, from which he brings us specimens of the nuts.

How Crops Feed—A Treatise on the Atmosphere and the Soil as Related to the Nutrition of Agricultural Plants, with illustrations; by Samuel W. Johnson, M. A., etc. New York: Orange Judd & Co. Professor Johnson's work, "How Crops Grow," is devoted to the chemical composition of the plant and the changes which take place during its life. Not only has that work received the highest commendations in this country, but it has been republished in England under the editorship of two of the professors of the Royal Agricultural College, and it is being translated into German under the auspices of Liebig. As a companion to "How Crops Grow," we have a work by the same author entitled "How Crops Feed," in which are given the latest researches upon the relations of the atmosphere to plants, and the origin and composition of soils and the offices they perform in the nutrition of crops. It is sufficient to say of the present work that it is marked by the same thoroughness and clearness that characterized its predecessor, and is, like that, the best work upon its subject in the language. As the author says in his preface, "His office has been to digest the cumbrous mass of evidence in which the truths of vegetable nutrition lie buried out of the reach of the ordinary inquirer, and to set them forth in proper order and plain dress for their legitimate and sober uses." Those who would study the principles which underlie all correct agricultural practice can have no better aid than these works will afford them. A third of this series is in preparation, which will treat upon manures and fertilizers. Price, $2, post-paid.

Rosebugs on Grape-Vines.—J. McCoy, West Va.—We know of no better way to treat the rosebug than to shake them off early in the morning, while they are dormant, catch them on a cloth and put them in the fire. The netting you speak of might do for a vine or two, but it would be too expensive on a large scale.

Cropping an Orchard.—"J. B. H.," Rancocas, N. J.—The soil of an orchard 16 years old should be given up to the trees, and instead of taking anything off by planting soiling crops, it would be much better to put fertilizers on to it. We know that it is the custom to mow or otherwise crop the orchard, and we know it is also the custom, in places where the practice prevails, to complain that fruit trees do not do well.

Peaches in Indiana.—W. W. Borden, Clark county, Ind., states: "The night of Feb. 20th killed the entire peach crop in this vicinity. Such is the report from the high lands (the Knobs), by those extensively engaged in peach culture. In the valley the result is about the same. The peach buds were very much swollen, and the lilac buds were bursting. The thermometer indicated 3° below zero."

"Shot Land" in Kentucky.—"J. F. F." writes: "I have about twenty acres of what we call 'Shot land.' If it were all together I would probably put it in some kind of grass, but it is scattered over my farm. The soil consists of pebbles from the size of a large buckshot down to that of a pin's head, nearly round, hard, but it can be cut with a sharp knife, exhibiting a dark iron-ore appearance on the inside. They are in some places very thick on the ground; and the thicker they are, the less the land produces. It appears to be unable to retain ordinary manure."—Probably the best course would be to plow and seed to red clover encouraged by plaster, and calculate to turn under two or three crops of buckwheat, clover, or corn before it would begin to look like good soil or bear other crops.

Inland Fisheries.—It is gratifying to notice the steady progress made in some States in stocking streams and ponds with fine kinds of fish. All the New England States have commissioners who are active in investigating and devising means for multiplying and introducing fish into their streams, lakes and ponds. We have received from the Massachusetts commissioners their report to the Legislature for the year 1869, which is an instructive document to those interested in this subject, as it details both successes and failures. The names of the New England commissioners are given, and we print one for each State for the benefit of persons in other States desirous of inducing their legislatures to take action in fostering inland fisheries: Chas. G. Atkins, Augusta, Me.; W. A. Sanborn, Weirs, N. H.; A. D. Hager, Proctorsville, Vt.; Theo. Lyman, Brookline, Mass.; W. M. Hudson, Hartford, Conn.; Alfred A. Reed, Apponaug, R.I.

How to Apply Manure.—E. Taylor, Michigan, asks: "What shall we do with our barn-yard manure? George Geddes says plowing it under 10 or 11 inches deep is the poorest way to apply it. This has been our plan ever since I can remember. We sow wheat after corn, then clover two years, and then corn again, and so on *ad infinitum*. Now, what shall we do with the manure?" Do as you have been doing, if it does well, and never mind the agricultural doctors. If the corn does not get the benefit of the manure the wheat will. But *do* you plow 10 or 11 inches deep? Measure and see if the furrow slice is over 8 inches thick. Most people think they plow 2 or 3 inches deeper than they really do.

Plaster on Clover.—"H. G.," Augusta county, Va., writes: "We all use plaster here on clover, and it increases the yield very much—in fact, I think it doubles it in most cases."

Raising Lambs by Hand.—Use the milk of a new milch cow, and let the cow be well fed, so that the milk may be as rich as possible. Put an India-rubber nipple on a glass bottle; and be sure that the bottle is thoroughly cleansed with *boiling* water every day.

Clover for Pigs.—A correspondent writes: "With a good clover pasture, I can easily make a full-bred pig weigh 400 lbs. with 22 bushels of corn."

Button-hole Bouquets.—Some time ago we gave engravings of some button-hole bouquet holders made of glass tube. A writer in a recent number of the Gardeners' Chronicle (Eng.) gives a plan which will be acceptable to those who cannot conveniently procure the tube holders. Some moss is tied to the stems of the flowers and thoroughly wetted; then the stems are surrounded by a piece of oiled silk of the shape and twice the size shown by the dotted lines in the accompanying diagram. This is bound by means of a thread, and will prevent evaporation from the moss.

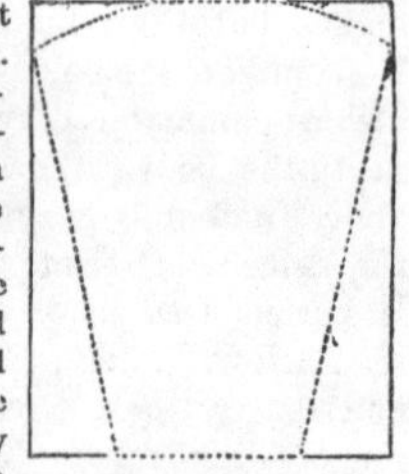

Allen's Catalogue.—Among the numerous American and foreign illustrated catalogues we have from time to time received, none equals in completeness and fullness of illustration the volume recently issued by R. H. Allen & Co. It embraces tools, vehicles and machines used by the Northern farmer and by the planters of the South, of Cuba and South America, by nurserymen, gardeners and fruit raisers, by brick-makers, carpenters and others; implements for draining, for ice cutting and handling, and a great variety of household articles, dairy implements, apple-parers, etc., etc. It includes scales, pumps and hose, edge tools, bells, and even thorough-bred stock—and almost everything is pictured. The price asked is $1, and this is refunded with the first order. This offer to refund has given rise to some exhibitions of shrewdness almost past credence. For instance, a person in Belmont, Ohio, sends for the book and $1 worth of sundry seeds, saying; "I see you propose to refund the price of the book on receipt of the first order; and, as you propose to pay postage on the book, you may send the *amount of postage* in peas, box the book and seeds and send by express." The order was filled.

Burning Bones.—"A. J. R." Rees Corners, Md.—Bone ashes are useful, certainly; but there is a heavy loss of value in burning them. A bone mill is a pretty heavy affair—Stewart's, one of the best, costs $1,000. Bones may be slowly reduced by being broken with a sledge, and then mixed with fresh wood-ashes kept moist for two or three months. When overhauled the heap should be mixed with earth, and the still hard bones thrown out to be worked over again.

Climbing Fern.—C. C. Schenck informs us that this plant, figured in Jan. last, grows abundantly upon the Cumberland Table-lands in Tenessee. It is known there as "Wild Pea-vine," and the starveling cattle feed upon it in the absence of better food.

AMERICAN AGRICULTURIST

FOR THE

Farm, Garden, and Household.

"AGRICULTURE IS THE MOST HEALTHFUL, MOST USEFUL, AND MOST NOBLE EMPLOYMENT OF MAN."—WASHINGTON.

VOLUME XXIX.—No. 5. NEW YORK, MAY, 1870. NEW SERIES—No. 280.

DEVON CATTLE FROM THE HERD OF HON. E. H. HYDE.—*Drawn and Engraved for the American Agriculturist.*

The above engravings are portraits of animals owned by Hon. E. H. Hyde of Stafford, Conn., and were taken from photographs. The cows were in full milk and the bull in fair working order only. It is impossible to show in any engraving the beauty of this breed which owes so much to its rich, almost cherry-red color and white horns. It is a snug, tightly knit race, very different from any other, showing no indication of intermixture of blood in its origin—and hence by many claimed, with good reason too, to be the original breed of Great Britain. Modern breeders have perhaps introduced a dash of Shorthorn blood to give greater aptitude to fatten and earlier maturity. The Devons are medium-sized cattle, the bulls often rather low in stature; the cows of fair size giving 14 to 20 quarts of rich milk, valuable for both butter and cheese making; the working oxen, both pure Devons and grades, are among the very best in the world, for everything except very slow heavy work. They are spry, intelligent, handy and trusty, fast walkers, and we have known them fair trotters; and they make, when fattened young, the very best beef of our markets. Devons are rather slow in coming to full size and maturity, but they fatten easily, and last in full vigor as cows and oxen until 12 to 18 years old. The cows are docile and quiet, and the steers easily broken. Upon the whole, the Devon probably combines all good points and valuable qualities to a degree not approached by any other breed. These cattle are hardy at the South, and their activity adapts them to pick up a fair living where Shorthorns or Herefords would starve. It is not very unusual to find among the cows deep milkers, giving over 20 quarts of milk, and in quality of the milk takes high rank next to that of the Jerseys.

WOODEN BIRD HOUSES

at 50 cents for the single house, (like the cut,) and 75 cents for the double house. The single house is about 8 inches long by 6 inches wide and high.

They are cheap, yet attractive ornaments to a country home, and very desirable articles of sale for country merchants, to whom we offer large discounts.

R. H. ALLEN & CO.,
P. O. Box 376,
New York.

CAHOON'S BROADCAST SEED SOWER.

Hand Machine. Sent for $10, on receipt of the money.

R. H. ALLEN & CO.,
P. O. Box 376, New York.

DRAIN TILE.

We offer a complete assortment of **Drain Pipe** and **Drain Tile**, thoroughly made, and uniformly hard burned. Adapted to the following uses:

SEWERAGE of houses, stables, manufactories, etc.

THOROUGH UNDERDRAINING OF LAND, CONVEYING of water from springs, for use in Farm Houses, Fish Ponds, Fountains, Cheese Factories, etc., etc.

DRAIN TILE MACHINES furnished to order. Send for Circulars to **C. W. BOYNTON & CO.,**
Woodbridge, N. J.

FOR Drain Tile, Pipe and Ditching Machines and Dies applicable to any Tile Machine, to make sockets to connect the same as gas and water pipes, address Lake Erie Stove and Tile Machine Works, Willoughby, Lake Co., O.

Machines delivered at your R. R. station at shop prices.

A perfect protection to the SHEEP, AND Speedy CLIPPING

Send for Descriptive Circular to EDWIN LYON, Butler, Pa.
Or to R. H. ALLEN & CO., New York.

Send for Specimen to R. H. ALLEN & CO., P. O. Box 376, New York, Sole Agents. Price, **$5.** Discount to Trade.

HOLBROOK'S SWIVEL PLOWS leave no "dead furrows" nor "ridges," turn furrow-slices flat on level land, work equally well on side-hill, save considerable time, pulverize thoroughly, will not clog, and equalize the work of team. Eight sizes, from 1 horse to 4 cattle.

Address F. F. HOLBROOK & SMALL, Boston, Mass.

THE PLANET DRILL.

For Garden Seed or Guano, the best; the most simple, compact, largest, easiest regulated, lightest, cheapest; no gearing, no slides; sows in open sight, and *evenly, whether full or not,* all seeds ordinarily sown; also salad, salsify, broom-corn, nursery seed, &c.; spreads fertilizers in the row at any rate, *without loss from winds.* No. 1, 5 lbs. seed, $12; No. 2, 12 qts. of seed or 25 lbs. guano, $20.00. Manufactured by S. L. ALLEN & CO., Forest Building, Philadelphia, Pa.

Hexamer's Prong-Hoe.

Is the most useful implement for Farm and Garden. No one can afford to do without it. Price $1.50. Send for Circular.

Address REISIG & HEXAMER, New Castle,
Westchester Co., N. Y.
Or, B. K. BLISS & SON,
41 Park Row, New York.

SELF-ACTING GATES.

The American Gate Co. of Cleveland, O., are the sole manufacturers of Nicholson's *Improved* Self-acting Gates. These gates are beyond competition—the best ever made. They are recommended by the "Scientific American," the "American Agriculturist," Gen. Horace Capron, Commissioner of Agriculture; and many others. Send for Circular.

Address M. G. BROWNE,
Secretary Am. Ga. Co., Cleveland, O.

EVERY FARMER WANTS IT.—Fits on hoe-handles. Saves half the time, and does all the work of counting and dropping the grain. No. 1, 75 cents; No. 2, $1.00; per doz., $6.00 and $7.00. Try it. Sent by Express on receipt of price. Send for Circular.

HARPER & CO., Patentees,
45 Dey St., New York.

DRAIN TILE MACHINES.

The best in use. Warranted to make the most tile with the least labor. The tile are delivered on boards. For complete instructions how to make tile, cost of works, machines, &c., address

JAS. W. PENFIELD, Cleveland, Ohio.

Holbrook's Regulator Seed Drill.

A perfect hand-machine for sowing with *regularity and in proper quantity.* Beet, Carrot, Onion, Parsnip, Spinach, Sorghum, Turnips, Peas, Beans, etc. It is very simple, compact, durable, easy to operate, and *shows the seed as it drops,* to prevent any mistake or failure. Price $12.00. Address

F. F. HOLBROOK & SMALL, Boston, Mass.

FRUIT AND MEAT PRESERVING HOUSE.—*Rees' Patent.*—The best and cheapest ever invented. See Report of the National Pomological Society, Sept. 18, 1869. Send for Circulars. Houses built on reasonable terms.

Address REES & HOUGHTON, Box P.,
Philadelphia, Pa.

STEEL HORSE HOE.

(Page 35 of our large catalogue.)

at $15, or other styles of **Horse Hoes** and **Cultivators,** at various prices.

Send stamp for Circular to

R. H. ALLEN & CO.,
P. O. Box 376, New York.

Seymour's Grain Drills and Fertilizer Sowers.

Special Circulars of these, *the best* Drills in use, sent on receipt of stamp, by the Sole Agents,

R. H. ALLEN & CO.,
P. O. Box 376, New York.

Rustic Flower Stand No. 1.

8 ft. high and 4 ft. diameter. Manufactured by the
MILLER IRON CO.,
Providence, R. I. Manufacturers of Ornamental Iron Work for Gardens, Lawns, Parks, Cemeteries, etc. Illustrated Catalogue sent free.

LET NO ONE

STAND BETWEEN US.

Four Ton Hay Scales, $75. 1,250 lb. Portable Platform Scales on Wheels, $25. 2,500 lb. Portable Platform Scales, $40. Order directly of the Manufacturers and save two or three Commissions.

Send for **Free Price List, No. 196.**

EDWARD F. JONES,
Binghampton, N. Y.

AQUARIA CEMENT.

The best article discovered for Glazing Graperies and Green-houses, having been used for the past Ten Years with great success, which those will testify to that have Houses Glazed with it. Manufactured only by

E. D. DAVIS, No. 52 Fulton St., New York.

("SEE ADVERTISEMENT.")

The Oriental Base-burning Air-heating FURNACES.

("Referred to in this edition,") Portable and for Brick setting, sizes suitable for buildings of every description, are manufactured and sold by PERRY & CO.,
250 Water Street, New York City.

Descriptive Catalogues furnished free.

CELEBRATED MOHAWK VALLEY STEEL CLIPPER PLOW. NISHWITZ'S PULVERIZING HARROW, SHARES' COULTER HARROW, HARRINGTON'S COMBINED SEED SOWER and CULTIVATOR, COMSTOCK'S COMBINED SEED SOWER and WEEDER, CAHOON'S BROAD-CAST HAND SEED SOWER. BONE MEAL for feeding cattle and poultry.

Send for 1870 Agricultural Almanac.

GRIFFING & CO.,
58 & 60 Cortlandt St., New York.

THE MELLISH Fruit Basket, for berries, peaches, etc. Handsomest, cheapest, and most durable in use. Manufactured by BAIRD, ROPER & CO., Norfolk, Va. Send for circulars to A. D. Hopping & Wilson, Gen. Agts.,
218 & 220 Washington St.,
New York,
where a large supply is on hand.

WEED KILLING MADE EASY, by Allen's Weeding Hoe. This favorite Weeder, for ease and efficiency of operation is superior to all others. Circulars and goods furnished by

CLEMENT & HAWKES MF'G CO.,
Northampton, Mass.

POST HOLE BORER.—The only good one. Pays cost ($5) in one day. Address for cut, proof, &c., IRA HART, Clarksburg, West Virginia.

SEEDS.

Our new ILLUSTRATED CATALOGUE, with Description and Prices of SEEDS, of the new and standard GRAINS, GRASSES, VEGETABLES, TREE, FRUIT and FLOWER SEEDS, will be sent to all applicants on receipt of stamp.

R. H. ALLEN & CO.,
P. O. Box 376, New York.

LAWN MOWER.

(Pages 51 and 52 of our large catalogue.)

The two smaller sizes at $25 and $35, and the larger machines up to $310.

Send stamp for Circular to

R. H. ALLEN & CO.,
P. O. Box 376, New York.

AGENTS WANTED.

We wish to get energetic agents in every section of the United States and Canada, where we are not now represented, to sell

The most simple and durable double thread Sewing Machine ever offered to the public.

This Machine is first-class in every respect, and at the same time furnished

At a price within the reach of all.

Licensed by Wheeler & Wilson, Grover & Baker, and Singer & Co. We will give parties with sufficient capital the exclusive agency of an entire State. To persons seeking a profitable business, we offer unparalleled inducements. The attention of Sewing Machine Agents is especially solicited.

For terms address, GOLD MEDAL SEWING MACHINE CO., **334** Washington St., Boston, Mass.

HINKLEY KNITTING MACHINE

For Family use—simple, cheap, reliable. Knits everything. AGENTS WANTED. Circular and sample stocking FREE. Address HINKLEY KNITTING MACHINE CO., Bath, Me., or 176 Broadway, New-York.

Bloede's Ink and Stain Erasive Pencil.

It easily removes ink spots, fruit stains and iron mould from the skin, linen, wood, bone or ivory. Invaluable as an article for the laundry, counting house, writing-desk or dressing-case. For sale by Druggists, Stationers, etc. Sample sent free by mail on receipt of **25 cts.** VICTOR BLOEDE, Derby, Conn.

CHEESE-MAKERS AND DAIRYMEN,
ATTENTION!

If you would have a quality of ANNATTO always reliable and uniform in color and strength, buy only that known as **large "F,"** and be sure you get the Genuine.

Always on hand by WARD, SOUTHERLAND & CO.,
Wholesale Druggists, 130 William Street, New York.

Also, an assortment of the other brands that are best known.

BROOM Machinery of all kinds; also, Broom-corn Planters & Seed Scrapers, both hand and horse-power; also, Broom-corn Seed. WM. STEELE, Agt., New Hartford, Ct.

1,000 CHILDRENS' CARRIAGES, from $4 to $25, by LEWIS P. TIBBALS, 478 Broadway, New York.

Carriages made to order and repaired.

WATER-PROOF PAPER FOR OUT, AND inside of buildings. C. J. FAY, Camden, N. J.

The DOLLAR Sun.

CHAS. A. DANA, EDITOR.

The cheapest, smartest, and best New York newspaper. Everybody likes it. Three editions: DAILY, **$6;** SEMI-WEEKLY, **$2;** and WEEKLY, **$1** a year. ALL THE NEWS at half-price. Full reports of markets, agriculture, Farmers' and Fruit Growers' Clubs, and a complete story in every Weekly and Semi-Weekly number. A present of valuable plants and vines to every subscriber; inducements to canvassers unsurpassed. $1,000 Life Insurances, Grand Pianos, Mowing Machines, Parlor Organs, Sewing Machines, &c., among the premiums. Specimens and lists free. Send a Dollar and try it. I. W. ENGLAND, Publisher, New York.

No Choir Should be without it!

THE AMERICAN TUNE BOOK.

Third Edition Ready.

A Collection of all the widely popular Church Tunes, Anthems, and Set Pieces which have formed the foundation of our American Church Music for the past fifty years. Containing 1,000 choice pieces selected by 500 Teachers and Choir Leaders. Price, $1.50. $13.50 per dozen. A specimen copy will be sent by mail to any address, post-paid, on receipt of price. DITSON & CO., Boston and New York.

ARE YOU GOING TO PAINT THIS SPRING? Then look at the advertisement of the AVERILL PAINT CO., on last cover page.

BACK VOLUMES OF THE
AMERICAN AGRICULTURIST,

From the Sixteenth to the Twenty-eighth. Price of each bound volume, post-paid, $2.50. Address
ORANGE JUDD & CO., 245 Broadway, New York.

AMERICAN AGRICULTURIST.

NEW YORK, MAY, 1870.

This month brings active work, usually over much, and as every man's most precious property is health, this should not be wantonly jeopardized. The fact is, we continually risk too much and often suffer for it most terribly—as is the experience of almost every family in the land. The laws of health in both man and animals cannot be disregarded or violated without a penalty, and it is much more agreeable to keep one's self, one's family, one's laborers, and one's stock in good health than it is to recover health if lost. Steady, regular, hard work is conducive to health, but it must be sustained by good food and shelter. It is a very important matter to most farmers to know how, and to economise the labor of animals. We are very apt to tax them severely one week and let them stand still another, and this irregularity and its accompanying exposures often cause disease. Turning unused horses and oxen to pasture is a remedy, but this is accompanied by inconviences, unless the animals are not to be used at all. Grooming is a partial cure, and takes the place to a certain degree of both exercise and food. After very hard work for some weeks, horses and oxen need rest and time to recuperate. Liberal feeding is essential, but no severe tax upon the system can be compensated for by food alone. Entire rest is equally important. This should not be continuous, however, but it should be given like the feed at regular intervals and in connection with moderate labor or exercise. Mere fat gained in the stall is of little use before the plow, and it brings a certain delicacy or tendency to disease.

Plan for a crop of corn for cutting and feeding green to milch cows and hogs; for roots, mangels, carrots and parsnips for feed—the last two to be put in this month; for some crops to be sold for cash if you are situated so that this is possible. Some crop may almost always be raised which is nearly as good as money. This in some sections is tobacco, in others fruit, in others broom-corn, in others flaxseed, in others cotton, in others castor-oil beans, etc., etc. As a rule we think some such crop is as advisable as it is usually profitable, if grown understandingly and not in a way, or so largely, as to impoverish the farm. Many food crops may be raised on the same principle, *i. e.* not for feeding out on the place, but for immediate sale.

Hints About Work.

Orchard and Nursery.

Planting is to be pushed forward as fast as consistent with good work. Trees may be kept back by heeling them in and shading them.

Young Orchards may have root crops between the rows. When these crops are hoed, cut up the weeds and mellow the surface around the trees. Corn shades the trees too much late in the season, and prevents thorough ripening of the wood.

Mulch.—Covering the surface around the trees with bog hay, coarse manure, or other available litter, will help the trees greatly to resist the drought.

Nursery Trees.—Budded stocks need to have all buds removed, except the inserted one. Rub them off as soon as they start.

Seed-beds.—Success in raising evergreen and most other forest-trees from seed, depends upon shading, proper moisture, and a free circulation of air. A lattice of lath makes a convenient shade.

Insects.—The annual fight must be kept up. Wherever a tent-caterpillar's nest is to be seen, there is a challenge to combat. The insect will get the best of it if it is allowed time. Make it somebody's business to destroy every nest, and let that somebody be yourself, if you wish it done faithfully. Jarring the trees and catching the insects upon a sheet or other convenient receptacle, is the only certain mode of warfare with the curculio yet known. Begin early in the season, and early in the morning. All washes and "invigorators" plague the inventor more than they do the curculio.

Fruit Garden.

Strawberries.—It does not seem to be generally understood that strawberry plants must have a season's growth before they will bear. Plants are best set in the spring, the earlier the better, unless one strikes layers in pots, in which case they may be set at any time. Where the beds have been covered with straw, this is parted over the plants and left on until after fruiting. What weeds make their way through the straw may be pulled. Beds that have not been mulched should be thoroughly hoed over and a mulch put on before the fruit attains much size. Liquid manure and wash from the house will increase the size of the fruit.

Grape-vines.—Upon young vines set this spring but one bud should be allowed to grow. Rub out all but the most vigorous one. Two buds may be allowed to grow upon vines planted last year. Established vines will push buds where they are not wanted; rub off these, and save pruning. When the young shoots are large enough to show the little clusters, which are only buds, and not, as many suppose, grapes, the end of the shoot is pinched off. Some pinch at the leaf above the last or uppermost bunch, and others leave two or even three leaves. This, when done thus early, avoids the check which is given to the vine by the old method of summer pruning after the fruit is set. Layers may be made as directed last month.

Blackberries and *Raspberries* are to be tied to stakes or other supports. A strong wire stretched between posts answers well in garden culture.

Marketing.—Those who send fruit to market should decide in season upon the kind of baskets, crates, or other packages they will need, and procure them in advance. Have them distinctly marked. Berries sent from Southern to Northern markets should not be too ripe when picked.

Kitchen Garden.

A list of the leading varieties as well as some of the promising novelties, was given in last month's notes. In most places the hardy vegetables for the first crop were sown last month.

Succession Crops of these should be put in at intervals if it is desired to prolong the season of peas, salad plants, radishes, etc.

Asparagus is to be cut with a sharp knife, and when marketed it is to be made up into bunches 9 inches long and 4 or 5 inches in diameter. An engraving of a bunching frame was given last May. Beds only two years set should be cut sparingly.

Beans.—Plant bush sorts, and when the soil is well warmed, Limas. If a cold rain comes on, the Limas are apt to rot in the ground. If Limas have been started under glass, put them out when the weather becomes settled.

Beets and *Carrots.*—Run a weeding hoe of some kind between the rows as soon as they can be distinguished. As soon as large enough, weed and thin. A few days' neglect will often be of great injury to carrots.

Cabbages and *Cauliflowers.*—Frequent hoeing is very beneficial to these. If cut-worms attack the plants they must be dug out. Sow seeds for later crops in open ground in a well-prepared seed-bed.

Celery.—Sow seeds in a well prepared seed-bed. They are slow in germinating.

Cucumbers.—If there are cold-frames or hot-beds from which other plants have been removed, they may be profitably devoted to cucumbers. Plants which have been started in-doors or under glass, may be put out when cold nights are over; they will need sheltering for a while at night; two boards nailed together roof-shape will answer. Sow seed in the open ground as soon as it is warm enough. Use plenty of seed to allow for losses. See vine-shield figured last month.

Egg-Plants.—Nothing is gained by putting these out before the weather has become warm and settled. Give them a rich warm spot.

Herbs.—These may be sown in a seed-bed and transplanted, or sown where they are to grow and thinned. In market gardens they are transplanted to the ground from which early cabbages, etc., have been taken. Sage, Thyme, Sweet-Marjoram, and Summer Savory, are the kinds generally grown.

Martynia.—The tender pods are much esteemed for pickles. Sow where they are to grow, or transplant to 3 feet apart each way.

Melons.—Treat the same as directed for cucumbers.

Onions.—Seed in cold localities may yet be sown. When the plants are up sufficiently to allow the rows to be distinguished, run a weeder or hoe through them. Weed as soon as the rows show that it is needed. A moderate dressing of salt is beneficial. Onions from sets and potato-onions are to be kept well cultivated.

Parsley.—Sow in soil as free from weeds as possible, as it is long in starting.

Parsnips.—Hoe and weed as soon as large enough, and thin to about 10 inches.

Potatoes.—Keep the weeds under. A little ashes and plaster at hoeing will help them.

Flower-Garden and Lawn.

Lawns.—To have a fine close turf, it is essential that the ground should be rich, the seed thickly sown, and the grass be mown frequently. Weeds for the most part flourish best in a soil too poor for the grass to make a sufficiently strong growth to crowd them out. Frequent mowing not only keeps the turf thick, but it destroys the annual weeds. Good hand mowers may now be had for $25 and $30. It is best to mow often and leave the cut grass to act as a fertilizer. Bone-dust and ashes are excellent as a top-dressing.

Margins around beds or along walks are to be kept neat by the use of an edging knife, which is much like a meat chopping-knife with a long handle. The outline of beds can be preserved by driving down wooden pins an inch square, their heads being below the level of the turf.

Grass may be sometimes weeded to advantage very early in the month. Clover and grass seed sown upon them, and they may be top-dressed with plaster, ashes, superphosphate, etc.

Annuals in Ribbon Gardening.—Those who cannot afford the necessary bedding plants, may produce pleasing effects with annuals. These are best sown in a seed-bed and transplanted to the show-bed, taking care to have in reserve, plants to replace any which should not come true to color. No annual is more useful for this purpose than Drummond's Phlox, which can be had from pure white to deep purple. Where different species are used, the hights and times of flowering must be considered. These are given in the Seed Catalogues.

Bulbs.—The principal bulbs to be planted in spring are Lilies, Mexican Tiger-flower, Gladiolus, Jacobean Lilies, Amaryllis, and Tuberoses. All but the Tuberoses may be put out in the border; they should be started in pots in a warm room or under glass.

Green-house and Window Plants.

Window plants usually go out of doors altogether, but it is less the custom than formerly to turn everything out of the green-house. It is but little more trouble to properly care for the plants in the house than out of doors, and the liability to injury is much less. By proper shading of the glass and ventilation, plants can be kept in good condition.

Shelter.—Plants out of doors should be sheltered from the winds and in part from the sun. Camellias especially need shade.

Plunging—setting the pot up to the rim in soil—is done to avoid the necessity of frequent watering and for ornamenting the border. Put coal ashes in the bottom of the hole in which the plant is set, to prevent worms from gaining access to the pot.

Fuchsias bloom much better in partial shade than in the sun. In this country they do poorly as bedding plants. Well-grown specimens make a fine show upon a veranda where they have some shade.

Our Gray Rabbit.—(*Lepus sylvaticus.*)

This little animal so closely resembles its cousin, the rabbit of Europe (*Lepus cuniculus*), that it passes by its name in this country. Our rabbit is, however, very distinct, particularly in its habits. It does not burrow like its congener, which is the parent of our tame rabbits, but occasionally takes refuge in deserted or occupied burrows, thus often falling a prey to foxes and perhaps other carnivorous animals. The rabbit is about 16 inches long to the tail; it is larger at the West than at the East or South. It is something over two feet long from the nose to the tips of the hind feet, when stretched out. The head is roundish and the ears short. The fur is soft and fine; the skin is rather thin and delicate; the color dark gray, tinted with yellowish-brown above, changing into gray and ash-color upon the sides and rump, and white upon the belly. The tail is very short, dark above and white below. It is very timid, but not suspicious; is easily trapped or shot from its "form," the hollow where it habitually sits, or, started by dogs, and shot upon the jump. The flesh is delicate and good eating, but not particularly rich or "gamey." The rabbit multiplies with such astonishing rapidity that notwithstanding its many enemies it often increases so in numbers as to do farmers, nurserymen and gardeners, great damage. The worst thing it does is to gnaw the bark of young fruit trees near the ground, often girdling them entirely. This is usually done in winter when a lack of other food occurs; and the remedies are to encase the trees with birch bark, sheathing felt, or tarred paper, or to put some offensive substance upon the bark of the tree; sprinkling it with blood is found to prevent their attacks; and it is said that a mixture of assafœtida and soap will have the same effect. The food of this and its related species is exclusively vegetable. Sweet fruit they are very fond of, and box traps are usually baited with sweet apples. Clover and other sweet grasses, cabbages and turnips, grain of all sorts, and many garden vegetables, are their food; their shyness, acute sense of hearing and sight, their instinctive tendency to remain still until discovered, and their rapid flight when once started, are their means of defence. So far as we know they never take the offensive except in their occasional fights with one another, or with other species of rabbits or hares, to which they manifest strong repugnance.

THE GRAY RABBIT.—(*Lepus sylvaticus.*)

The American Dipper, or Water Ouzel.

(*Hydrobata Mexicana.*)

The American Dipper is found along the Rocky Mountains, from British America to Mexico. It belongs to the same sub-family as the Thrushes, the Bluebird, and the common Robin, and resembles them in many points of structure, though in habits it is widely different. The bird, the general form of which is shown in the engraving, is lead-colored above, paler beneath, with the head and neck of a clove color, or sooty brown. The Dipper is remarkable for seeking its food under water; it wades or dives into the water, and remains beneath the surface for a considerable time. There is a similar species found in the mountainous countries of Europe, the habits of which have been more closely watched than have those of ours. Observers there state that the bird progresses beneath the water by clinging to the stones at the bottom, by means of its long curved claws, and by the use of its short wings. Its food is mussels and other fresh water mollusks. The European species has the popular reputation of going beneath the water for the purpose of feeding upon the salmon spawn, and is hence regarded as an enemy by the fishermen. An examination of the stomach of the bird has shown no foundation for this belief. The song of the European bird is very powerful and pleasing, but we do not find that observers have noted the song of our species. The nest of the European species is a curiously constructed dome, often placed behind a sheet of falling water or in some locality where it could be found only with great difficulty. The nest of the American Dipper, though it has been diligently sought for, has remained unknown to ornithologists until the past summer, when it was discovered by the naturalists of the U. S. Geological Survey of Colorado and New Mexico. Mr. Henry W. Elliott, of Washington, D. C., who furnishes the sketch for the engraving, says: "While we were in camp near Berthoud's Pass, a member of the party was fortunate enough to discover one of these nests, and to secure it and its builders. I was with him, and made the accompanying sketch. The nest is now in the Smithsonian Museum. It was placed on a rocky shelf mossed and grassed over, just about four or five feet above the stream; it was built of layers of moss, so laid one over the other, as to give it the shape of the crown to an ordinary 'Derby' hat. The moss was kept fresh and growing, from a habit which the bird had of shaking the water over it from its plumage before going in on to the eggs. The eggs were four in number, and of a pale white."

THE AMERICAN DIPPER, OR WATER OUZEL.—(*Hydrobata Mexicana.*)

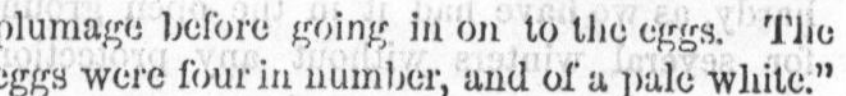

LARGE ONIONS.—A writer upon onion culture in England says the best way to get large onions, is to tramp and roll beds firmly; the seed is then to be sown on the compact surface and covered with a rich compost the usual depth.

Evergreens are best moved this month. Success depends upon keeping the roots from drying.

THE VARIEGATED GRASSY CALAMUS.—(*Acorus gramineus.*)

The Grassy Calamus.

Almost every one is acquainted with the Sweet Flag or Calamus of our swamps, and which is prized by many for the pungent aroma of its creeping root-stocks. The Grassy Calamus from China, is very much like a miniature reproduction of the one with which we are familiar. It has similar long and narrow, two-edged leaves. The flowers, as in the common one, are minute and densely crowded into a spike (or *spadix*) which issues from the edge of a flattened stem that is quite like the leaves in appearance. The species in its natural state has the leaves entirely green; but a striped variety has been introduced from Japan within a few years, and is now not rare. The engraving represents the plant somewhat under the real size, and shows the manner the clusters of leaves are given off. The clumps all have a tendency to assume a curved or cresent-like form. The plant is generally grown in greenhouses, and is a useful one to place in rustic baskets and vases, as it endures the dry atmosphere of the dwelling. It is, however, quite hardy, as we have had it in the open ground for several winters without any protection.

EXPERIENCE WITH A WILLOW HEDGE.—Mr. Geo. M. Chase, Freeborn, Minn., writes: "In the spring of 1868 I built a few rods of willow fence, which has succeeded so well, I think others should try it. First I dug a ditch 18 inches deep, and 1 foot wide at the bottom. Then I cut the willow into pieces 5 feet long, taking care not to wound them more than was necessary. They were placed in the ditch 4 inches apart, and nailed to a board 4 inches from the top, using clinch nails. The soil was thrown back and pressed gently against each piece. Digging a ditch to set the fence in, has two advantages: first it gives you a chance to build a straight fence out of material more or less crooked, and secondly it loosens the soil, giving the roots a better chance to start and furnish the nourishment now so much needed. It was quite dry for nearly three weeks after the fence was built, but I saved the life of nearly every piece by mulching with coarse manure. This makes a picket fence 3½ feet high, which nothing has ever attempted to pass, and it grows better every year. In this soil a man can dig the ditch, prepare his trees, and put up 5 rods per day. The necessity of resorting to some plan to secure cheap and durable fencing, is almost imperative in this sparsely timbered country. I ought to have said that the willow (White Holland is the kind I used) never sprouts from the roots, no matter how close you plow or how many roots you may cut off."

SILLS' HYBRID MELON.—Mr. Gregory, of Marblehead, Mass., exhibited this variety at Boston last fall. It is small, white, and netted. The flesh is of a deep salmon color, of fine texture, rich, sweet, and good flavored. Mr. G. thinks it too sweet to suit some tastes. The one we tested was certainly sweet enough.

LAYING OUT A VINEYARD OR ORCHARD.—Mr. W. W. Smith, of Napa Co., California, gives the following method of laying out a plot for planting with vines or trees which would seem to be easy of execution: "Set two flag stakes as if about to lay out the ground with a horse and plow. Then take a wheel-barrow and fasten a stake of the proper hight to the center of the box or bed, and just behind the wheel. Get the three stakes exactly in a line and start across the field, pushing the wheel-barrow before you and keeping your eyes steadily upon the stakes. After a little practice one can, in this way, strike out a row almost as 'straight as a line,' and the wheel makes a mark sufficiently plain. By running across in both directions, the intersections of the marks, or 'checks,' will indicate the proper stations for the vines or trees. By putting stakes in the 'checks' before commencing to dig the holes, a person can, with a little care, set his vineyard or orchard in perfectly straight rows."

VALLOTA PURPUREA.

An Easily Grown Bulb—The Vallota.

The Vallota is a plant so easily managed, and withal, so showy, that we wonder we so seldom see it employed as a garden ornament. There seems no reason why it should not be as common among the spring bulbs as the Gladiolus. Perhaps it is because the books put it down as a green-house plant that so few are to be found in gardens. Seeing some bulbs in the collection of B. K. Bliss & Son, we were reminded to say a word in behalf of the Vallota. It is from the Cape of Good Hope, and has been in cultivation for a century. Its leaves do not die down, as do those of many bulbs, hence, after the flowering season, it cannot be packed away like a Gladiolus, but must be kept at rest in earth, though dry. The leaves, which are nearly flat, spread in a fan-like manner, and are of a dark, rich green. The flower-stalk is a foot or 18 inches high, and bears in August a cluster of brilliant, scarlet lily-like flowers, of the shape shown in the engraving. The plant may be turned out in the open ground after frosts are over, and in autumn be taken up and potted, and kept in a green-house or dry cellar. The bulbs produce an abundance of offsets by means of which the plant may be readily multiplied. Recently, botanists have placed this in the genus Amaryllis, and it is found in some works as *Amaryllis speciosa.* It is, however, best known by the florists and dealers as *Vallota purpurea*—though *purpurea* is a misnomer, as the flowers are scarlet. At all events, Vallota will serve for the popular name.

Fish and Muck Compost.

BY AN OLD SEED GROWER.

To prepare fish and muck compost, commence with three shovels of swamp muck, and spread it on the ground in a circle, if for a small heap of a few barrels of fish pomace, or in a long heap for a large quantity; then beat the fish fine and scatter one shovelful over the muck, and so continue alternately until all is mixed, leaving the heap cone-shaped. In about a week the heap will begin to heat, and should be turned and mixed, commencing at one side and making all fine with the back of the shovel. In a week or ten days more, it should be turned again. In three weeks it will be fine and fit for use. It may be kept until wanted to be used, but will require further turning if it continues to heat. The muck should be damp when mixed, or it will not heat sufficiently. Peruvian guano and muck, or earth, should be mixed in the proportion of six to one of guano. It does not heat, but requires the same turning and mixing as fish and muck. A handful of either in the hill is about the quantity generally used, but of the fish compost, more is required than of the guano—as much as can be held in the hand with the palm uppermost and the fingers spread. If thrown into the hole in a heap, it should be spread before being covered, to avoid the danger of destroying the seed, which never should be planted directly upon it.

Fig. 1.—IMPLEMENT FOR COVERING CORN.

Planting Indian Corn.

It is very desirable to have corn so planted that it can be worked both ways, and every good farmer prides himself in having his rows straight and true. The distance the hills should be apart is determined by the size of the corn, and the strength of the soil; very small corn ought not to be in hills, but in continuous rows. Hills of ordinary field corn should stand from $3^1|_2$ to $4^1|_2$ feet apart each way, and it looks best to have the rows of hills cross exactly at right angles. So important is this, that we think it well to measure the angle at starting when large fields are marked out for corn, and where the fences or other boundaries are not exactly at right angles. This is done thus: establish one true base line, as *A-B*, fig. 2, for the first row upon one side; then measure 30 feet upon it, with a 50-foot tape-line, between two points, *c* and *d*; next take 40 feet upon the tape from the point *c*, and draw part of a circle where you suppose a line, meeting the line *A*, *B*, at *c*, at right angles, would pass; then move the end of the tape-line to the point *d* and take 50 feet upon it—the point *e*, on the 40-foot circle where the 50-foot circle will cross, establishes a line *c-e*, exactly at right angles to the base line. This is easy and takes less time to do than to tell.

The field is properly marked in two directions by a variety of implements. Three cultivator teeth, set in a joist to which shafts and handles are attached, makes a good marker. Two wheels on a 4-foot axle-tree, with a pole having a chain dragging upon the ground to mark where the wheel must run the next time, is also a convenient implement. If the furrows are made fully and uniformly two inches deep, the corn may be dropped and then covered on smooth land, with a harrow on its back, or with one without teeth, and loaded.

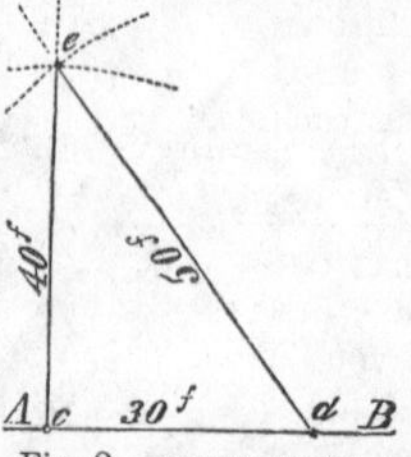

Fig. 2.—RIGHT ANGLE.

We describe a good corn-marker, which makes deep marks, in another article. There are some pretty good corn droppers, but none exactly adapted to planting in rows both ways. The dropping is a rather tedious hand process. It may be somewhat lightened by having a small tin measure, which will hold five or six kernels. A little skill is required to throw the kernels scattering, but it is done rapidly. James Corlis, of Long Branch, N. J., uses a home-made implement for covering corn, of which he sends us a drawing. The runners or scrapers (fig. 1) are of $1^1|_2$-inch oak plank 10 inches high, and 33 inches long. These are set 34 inches apart in front, and 9 inches at the rear. They are champfered on the outside three-quarters of an inch, and protected by a strip of iron on the inside. There are three cross-pieces of inch stuff. A smoother, 4 feet long, is attached to one runner behind to remove surplus soil, stones, and bits of sod. This smoother is attached by carriage bolts, so as to be set higher or lower. A 6-foot stick for a handle is bolted to the implement as shown. Two horses are used, and it may be loaded or run simply by its own weight, according to the character of the soil. We think it might also be used for covering potatoes.

Crows in the Corn Field.

Crows are not the only birds that trouble corn after it is planted, and before it is strong enough to defy them, but they are so active, knowing, and almost insatiable, that we regard it as a mistaken policy to spare the few which may be killed, in order to secure complete protection to our corn fields from their ravages. Mr. C. H. Rue traps crows in a way which we tried last year with complete success. He says he goes upon the principle that a crow will go almost anywhere for an egg. If he cannot find a secluded spot which will answer his purpose, he makes a sort of little bower of brush stuck into the ground, leaving a narrow entrance. In this he makes a rough nest and puts in some hens eggs. The sticks must stand close and meet at the top so a crow cannot get in, except by the entrance; here a steel trap is set, fastened by a cord, covered with tissue paper, and this sprinkled over with earth. We practiced placing an egg for bait just where the crow, when he eats it, will see the others. The first crow that flies over will almost surely be the victim, and as soon as caught, will make the welkin ring with his cries. This will draw a crowd of sympathizers. The bird may be taken out of the trap and fastened to a stake with his wings tied together and left so half a day. He will keep up his cries and not another crow will visit that field that year, if it be not more than 10 or 12 acres in extent. The crow, in case no bones are broken, may then be liberated, if the farmer is tender-hearted, or used upon another field.

CROW TRAP.

A Furrowing Corn Marker.

Most of the corn markers in general use, make too shallow marks, necessitating some other mode of deepening. A sled marker having two or three runners, 8 inches high and 2 inches wide, with their forward ends sawed out so that a cultivator tooth may be set in each, accomplishes all that is desired. The teeth may be set to run two or two and a half inches deep, and the runners following pack the earth a little, and prevent the furrows filling up. Such a corn marker may be made in half an hour.

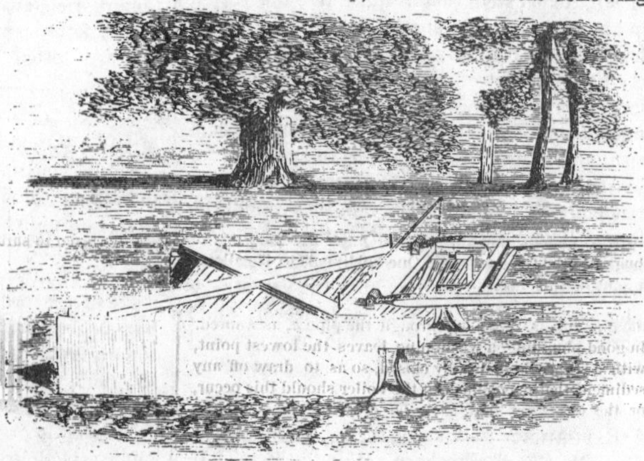

FURROW CORN MARKER.

Housekeeping Conveniences—Plumbing.

Our previous articles on convenient houses, have called out many inquiries, especially concerning Tanks, Boilers, Stationary Wash-*trays* (often called Stationary Wash-*tubs*), Water-Backs, etc. Many persons living remote from cities are desirous of knowing *how* they are arranged.

Fig. 1 gives a general view of a 2½-story house with the main apparatus in the first story. This is more frequently placed in the basement, in which case we can imagine the pipes carried up into another story. The tank, *t*, is filled from the roof, or by the pump. The cold water pipe, *c*, conveys the cold water down to near the bottom of the water-tight copper boiler, *b*, and forces the hot water out through the pipe *h*, entering a little way into the top, by which it is carried up to the sink *s*, in the next story, to the stopcocks in the wash-trays, *a*, *a*, *a*, and anywhere else throughout the house, to any point not higher than the tank. The force pump drives the

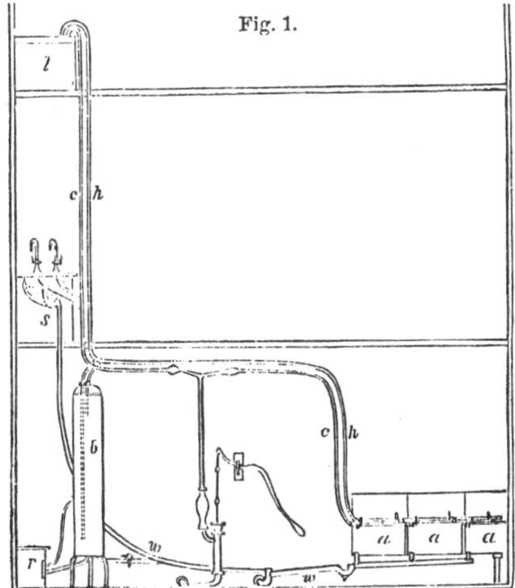

Fig. 1.

cold water into the trays, if the stopcocks be open; also, on closing these, it is forced into the boiler, and up into the tank when desired. The pressure from the tank keeps this cold water pipe always full for supplying wash-stands (*s*), bath-tubs, water-closets, etc. The waste-pipe, *w*, carries away filthy water from the trays, from the wash-stands, *s*, etc...... **Fig. 2** shows the lower part of the apparatus in a plainer form. The boiler, *b*, is of strong copper, riveted—its strength proportioned to the pressure. For a tank in 2d or 3d story we use what is called "Croton Pressure" strength. From near the bottom of the tank a pipe runs through the chimney back,

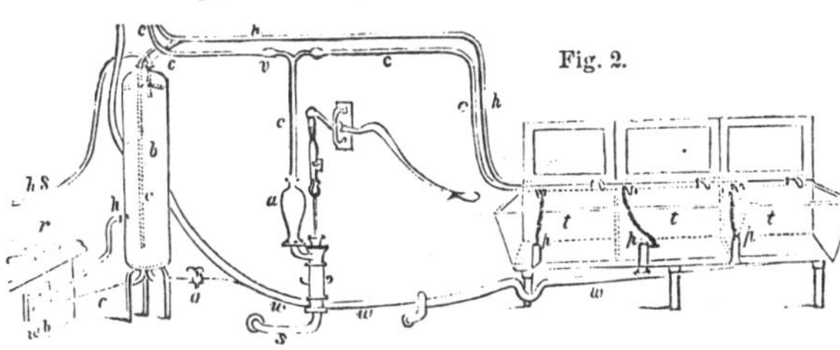

Fig. 2.

not shown here, to a *water-back*, *w b*, in the range, *r*, from which another pipe, *h*, runs back to the boiler at a higher point. This water-back is a strong iron box, say 10 to 14 inches long, 6 to 10 inches high, and about 2 inches thick; and it takes the place of the rear fire brick in the fire chamber of the range or cook-stove, so that it is always in direct contact with the fire, and being always full of water does not burn out. With the smallest amount of fire the water is warmed, rises through *h*, while cold water flows in through *c*. In this way the boiler is always kept hot as long as there is any fire. The boiler holds from 30 to 60 gallons, and the body of water will keep warm long after the fire goes out. One holding 35 to 45 gallons is sufficient for a medium house; 30 gallons will do for a small house. The hot-water of course always rises to the top, and is drawn off through the pipe *h*, as wanted. In good plumbing, a waste-pipe leaves the lowest point, with a stopcock, *o*, usually closed, so as to draw off any sediment that may collect in the boiler should this occur, or the water get foul, or it be desirable to empty the whole apparatus. From the hot-water pipe, branches run to any points desired, as *hs*, to a sink in the kitchen; one or more upwards, with side branches to the chambers. This also extends up over the top of the tank (as in fig. 3), and answers as a safety-valve to the boiler. In brisk boiling a little hot-water will sometimes rise over into the tank, and keep it from freezing. Another branch runs along the wall with a stopcock opening into each of

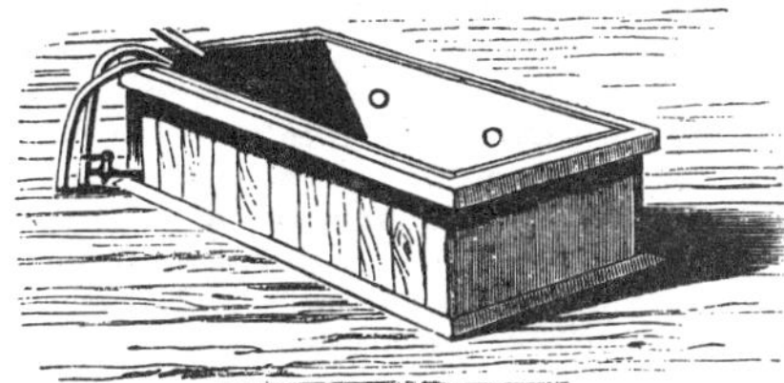

Fig. 3.

the wash-trays, *t*, *t*, *t*. The pump draws the water through the supply pipe *s*, from the cistern, reservoir, or well, and forces it up into the tank, or out at any point where a stopcock is opened; in the right branch a cold water stopcock opens into each of the trays. A valve *v*, in the horizontal pipe above the pump, prevents drawing water from the tank into the trays. It will be seen that by this arrangement, two pipes answer to carry hot and cold water *up* or *down*. The waste-pipes, *w*, *w*, *w*, collect and carry the water out through the wall into a drain. The bend in this, near the first tray, is called a *stench-trap*—the bend always keeping full of water so that no odors can pass back from the drain. These traps should always be provided with a broad screw in the lowest point of the bend, for removing any sand or other obstruction that may chance to settle there.—The wash-trays are made of strong clear pine plank, 1½ or 1¾ inches in thickness, let into each other at the joints, and these well smeared with white-lead ground in oil, before putting the pieces together. The case is divided into two or three compartments as desired. The back and end pieces are perpendicular and the front pieces inclined in at the bottom, so that while the tub is 19 or 20 inches wide at the top, inside measurement, it is only 13 to 14 inches wide at the bottom. Perpendicular depth inside, 14 inches. The top front edge should stand about 33 inches above the floor. The length of the trays may be about 2 feet each. The bottom board is a little inclined to the left, and in the lowest rear corner of each tray is a waste-pipe, *p*, with metallic plug attached to a chain from near the top, so that it can be opened readily. Two trays are convenient for different kinds of suds or clothes, and a third still more so, for rinsing. A wringer can stand upon the right end, or be changed to the division boards. The covers, if neatly fitted, form a good ironing table when shut down. It will be readily seen how great a convenience are these "stationary trays," and how much tugging and lifting they save. Turning a stopcock lets in hot-water as desired. Cold water is pumped directly in, or may be drawn from the tank if the supply be great enough, by omitting the valve *v*; and by lifting the plug, the wash water runs away. We can hardly wonder that when servants are asked to go to a house, the first question usually is "Have you stationary wash-tubs?" which implies boiler, etc., with hot-water for dishes and other purposes always ready for use. This apparatus is supposed to be in a separate wash-room or laundry. The whole may be in the Kitchen. The pump may stand at the edge of the sink with a cold water spout and cock near the air-chamber, *a*. This air-chamber is not necessary, but is desirable as a protection to the pipe and pump, and it makes the pumping easier, and gives a steady stream of water.——All pipes should be kept from the walls, at least by a board, and should be run up by the side of a chimney, or through a closet, and not against the outer wall. It is better also, when practicable, to have the tank, etc., on the warmest side of the building. Lead pipes, where passing through wood-work should fit closely, or be surrounded with cement and broken glass, otherwise rats in attempting to enlarge a passage will sometimes cut through the lead pipe. For size and weight of lead pipe, for apparatus of moderate size, see "Plumbing specifications," on page 29 (March No). For the great utility and profit of such an arrangement, see page 28. Size, arrangement, etc., may be varied to suit taste, means, location of rooms, etc.

Fig. 3 represents the Tank in some upper room, with its top a little below the eaves-trough, from which an inlet pipe is shown. The dark spot, just below it, indicates the over-flow pipe running down to the cistern. The hot and cold water-pipes are shown coming up from below. The short pipe (with stopcock for shutting off the water if ever desired for repairs or otherwise,) lets the water down or up. The extended curved part above, discharges water pumped up, and also answers as a "spring" in the pipe.—The tank is made with 1¼-inch matched plank, built inside a scantling or wall-strip frame, and is lined with sheet lead, well soldered. Two of the "plumber's tacks" are indicated on one side—that is, the lead is let into the wood at these points, the indenture being covered with solder to protect any weakness from the beating in of the lead. These "tacks" prevent the lead from sagging down at the sides, as it will usually do, after long exposure to alternate heat and cold. In building, stronger studding, and closer joists should be carried up under the tank to support the great weight of water. With this provision, it is better to make the tank large, and thus save pumping in dry weather. The cost is but a little greater for one of double size. Each cubic foot of contents holds nearly 7½ gallons, or about 4 feet to the barrel; 4×6 feet, 3½ feet deep, will hold about 650 gallons, or 21 barrels. In a house not constructed for the purpose, we made an aperture through the side from the attic floor out over the L which was half a story lower, and set the middle of a large tank over the heavy beam, supporting the outer end upon the L roof, and covered the outside portion over the roof with double boards, having 4 inches of shavings between, and surrounded the whole with a tin roof extending from the side of the main house over on to the roof of the L. This arrangement has worked well for several years.—It is on the south side, and has never frozen at all. It holds 60 barrels. The tank should always be closely covered, and have a trap-door in the cover to look into the tank. The water comes in from the upper roof only, and when full the surplus runs through a spout on to the L roof and thence to the cistern. The tank is cleaned of any sediment once a year or so. Pipes with branches extend from this to the boiler, to the bath-room, water-closet, wash-sinks, etc. A stopcock in the pipe where it leaves the bottom of the tank, is desirable, both to cut off the water from descending if leakage occur below, and also when cleaning out or washing the tank itself. A metallic strainer is put over the head of the discharge pipe, to stop any leaves or other material entering from the tank, to clog it.

Fig. 4.

Fig. 5.

Fig. 4 shows the "Butler's Sink," described last month, page 38, with the front door and casing below partly removed to show the waste-pipe, *w*, from which a branch extends to the strainer placed at the upper edge of the oval sink, to prevent its overflowing, if the stopcocks should by chance be left open. A chain plug in the bottom opens directly into the waste-pipe, *w*.

Fig. 5 shows sink, etc., *S—hcw*, in room *I*, fig. 4, page 28 (March number), with hot and cold water, and waste-pipes. The two supply pipes are seen in the corner. This is, of course, all cased in neatly, with door in front. The stopcocks are set high enough above the sink to admit a pail under them.

Fig. 6 shows the bath-tub and water-closet described last month, with the casing removed from the latter, in front, to show the internal arrangement... *p* is the earthen-ware basin, having a metal basin under it, which is turned down by raising the handle, and this movement also lets cold water into the rim of the porcelain, nearly horizontally, so as to wash it. The large 4-inch waste-pipe below has a stench-trap. The walnut cover, *a*, and seat, *b*, are both on hinges, so that by opening both, the wide porcelain bowl answers for receiving slops, and as a urinal. The hot and cold water pipes are seen in the corner, and the supply stopcocks at *h* and *c*; the bell-pull at *z*; the bathing-tub at *t*, neatly cased in with panels. The French pattern bathing-tub is now generally preferred—about 21 inches deep, 23 inches wide,

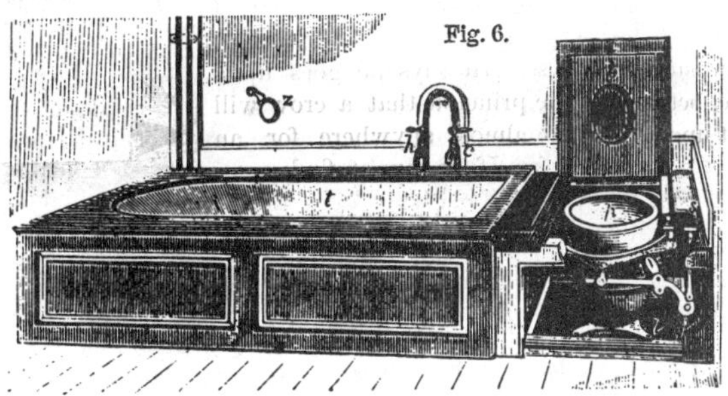

Fig. 6.

☞ For other Household Items, see "Basket"

An Aquarium.

A properly managed aquarium is not only a pleasing household ornament, but it is capable of affording no small amount of instruction. Some years ago, when the aquarium was a new thing, we gave considerable space to the subject. The excitement soon subsided, but all interest in the matter has not died out, as we have several letters asking how to start and manage an aquarium. The aquarium is a vessel of water containing plants and animals, and is in fact a miniature lake. A globe or other vessel containing fish, the water in which is daily changed, is not an aquarium proper. In the aquarium there is plant life as well as animal life, and the main condition of success consists in keeping the plants and animals properly balanced. The plants as they grow in the water, give off oxygen, a proper amount of which is necessary to the life of the fish or other animals that may be there. Reciprocally the fish, etc., give off carbonic acid, which is needed for the growth of plants. Too many plants are only objectionable, as they diminish the room needed by the fish; but an overstocking with fish will soon prove fatal. The first thing to be considered is the vessel, or tank, as it is generally called. A very pleasing aquarium may be made by using a large glass jar, holding a gallon or more. Such jars as confectionary is kept in (fig. 1) answer well, and they are to be preferred without bands or hoops. Large glass covers, such as are used by confectioners to cover up cake, make a good tank. They have to be supported in an inverted position by means of a base, which has a hole for the reception of the knob upon the glass as in fig. 2. Vessels with curved surfaces have the objection that they distort the objects within when viewed through the sides. Regular tanks (fig. 3) are made with metallic bottom and frame with the sides formed of glass; these are sold by the manufacturers at prices varying from $6 to $20, according to size and finish. The flat, glass sides allow the interior to be distinctly seen. An aquarium may be made with a wooden frame, black walnut being the wood usually preferred. The glass is fastened in with a cement of rosin and beeswax applied hot. The bottom of the tank and all the parts of the wood-work that come in contact with the water should be covered with the same cement, spread on while hot. Four parts of rosin and one of beeswax, with a small proportion of tallow, are used for the cement. Enough tallow is used to give toughness to the cement, so that it will not break readily when cold. The quantity is best found by experiment, probably a fourth as much as there is of wax will answer. Those who do not care to procure a tank of the regular style, can find sufficient to interest and amuse them in an aquarium made in a jar. An inch or so of clean gravel from which all fine particles have been washed, is to be put into the bottom of the jar or tank, and then the vessel is nearly filled with river or rain-water. The plants are next in order, and it is a little difficult to indicate which are most desirable, as there are few that are known by common names. Almost any plant which naturally grows quite under water in ponds or slow streams, will answer. One of the best is the Tape-grass, or Eel-grass, figured in August last, and those who will take the trouble to search the ponds, will find a number of others well suited to the purpose. The plants, being washed clean, are made into convenient bunches, to the lower end of each of which a small stone is tied to sink it, and as many plants as are desirable are anchored in the tank. It is best to allow the aquarium to remain thus for several days, before any animals are added, exposed to the light, at a window that has the sun for a part of the day. For animals, very small fish, water newts, snails, mussels, and tadpoles, are the principal ones. If the vessel is quite small, care must be taken in introducing fish, as they consume oxygen much more rapidly than newts, snails and less active animals. We shall speak of the management of the aquarium in another article.

Fig. 1.—JAR.

Fig. 2.—GLASS COVER.

Fig. 3.—SQUARE TANK.

How I brought Water into the House.

I live in one of those farm houses built about fifty years ago, which are only remarkable for having no modern conveniences. No gas, no bells, no soft water, no sinks, no drains, no furnace. It is true, there *is* a sink; but I turn it out of doors every summer, preferring to have the refuse water wheeled away to having a cholera bog near the house, which stands on a level space, and the water cannot be easily carried away by surface drains.

Men do not always know that they suffer from sink drains and neglected ventilation, for they are abroad in the breezy fields and acquire a stock of hardy health, which does not succumb to malarial exhalations; but women, who are obliged by their duties to keep in the house, often breathe an atmosphere of sickening odor from sink and cellar, and drift into consumption, fever, or paralysis. A farmer's wife usually rises at about half-past four o'clock in the summer, and six in the winter. Therefore, at half-past two she has worked ten hours and has a *possibility* of feeling weary, and before she goes to her chamber for an hour's rest, she remembers that the water for the chambers is to be carried up. It is work that she will not, if she can avoid it, allow her daughters to do, as she knows by experience the strain on back and arms; so she toils painfully up the stairs with a bucket of water, knowing it must be brought wearily down again, wishing, perhaps, as I have done for years, that she *could* have water brought into her second story. After wishing for it many years, I decided that I could and would bring it in. Perhaps others will do better than I have, if they will plan instead of regretting. Our house formerly had no gutters at at the eaves, but a few years since it was slated and a gutter added with a conductor, which delivers the water to a hogshead at the side of our south door. Above that south door is a large window, which lights a closet or bathing-room. As the small room adjoining was deficient in ventilation (for none of our windows can be opened at the top), several years ago I broke out the upper right-hand pane of glass in the closet, knowing it would be so difficult to re-set it that I should have constant ventilation. Why not bring a water pipe from the gutter above through that open sash, as well as air from without? My plans were soon laid. I could place a barrel with faucet upon a high stand by the window, curtaining off a space that no fastidious taste be offended. I would have a water pipe introduced into the gutter about 18 inches long with one elbow, bringing the water near the window, indeed quite close to the open sash, and at the lower end of this pipe there should be a rim sufficiently large to allow me to attach a section of hose which could either deliver water into my barrel or into the hogshead below, as I might choose.

At the left-hand corner outside the window I attached 20 feet of hose to deliver the refuse water into a small reservoir on wheels, which will be daily wheeled away. The house being covered on the south side by a luxuriant Virginia Creeper, the hose passes behind the net-work of branches, and is therefore not conspicuous, and will not be seen at all in summer. I have no doubt there are many country houses which might have similar or better arrangements, if women will once think for themselves, instead of waiting for others to plan for them. W.

A Comforting Use of Flowers.

Miss F. Hudson writes: "A friend lost a little child. When my mother heard of it, her sympathetic feeling urged immediate action. The universal desire to assist or relieve the mourning family, which is always felt when such tidings reach one, was always hers. So she went to our beautiful cemetery and gave directions for a pile of evergreen boughs to be placed in the yard where the little one was to be laid. While this order was being executed, she procured several baskets of exquisite flowers and returned to the yard where the grave was already excavated. Under her direction it was then completely lined with Spruce and Hemlock boughs, the heap of earth taken from the grave was also covered with them. Then, with the assistance of a friend, mother arranged flowers amid all the green, literally lining the grave with flowers. They were secured in their places either singly, or in tiny bunches by hair-pins. The effect of the arrangement was most beautiful, but its comparative effect still more so, when one saw and felt the difference between a bed of sweetest flowers, and the bare open grave. The earth used in the burial service being but a symbol, certainly the single lump softly dropped by our pastor fulfilled its purpose better than the ordinary unerring spadeful."

Hints on Cooking, Etc.

German Mustard.—Wm. Logler, Iowa, sends the following as his method of preparing mustard: "To half a pound of ground mustard add two ounces of sugar, and moisten with boiling vinegar; stir for half an hour with a wooden spoon, and set it aside, well covered, for an hour. Finally, add as much vinegar as may be necessary to thin it. Keep it well covered in a stone or glass jar." The mustard as sold in New York has some aromatic addition.—Who has another recipe?

An Excellent Rice Pudding.—By Mrs. W. Two qts. of milk, one cup each of rice and sugar, a teaspoonful of salt. Wash the rice and add it to the milk cold, and bake. The secret of having it nice consists in its being taken out of the oven before the milk is all dried away. It should be creamy in consistency, and when cool it is better than a pudding made with eggs, as there is no watery whey. Essence of lemon or raisins are an improvement.

Cream Pie.—By Mrs. R. J. R., Minneapolis, Minn. ½ cup of flour and 1 pint of milk boiled together, add the yolks of 2 eggs; ½ cup of sugar (white or coffee) and lemon to flavor the milk, and flour while boiling. Let all boil a few minutes. Make the crust and bake it, then put in the above mixture. Beat the whites of the eggs to a stiff froth, add enough white sugar to sweeten, and put this over the pie and bake to a light brown. This is enough for two pies.

Tea Stains on Table Linen.—Mrs. W. says: Japan teas stain table cloths more indelibly than other black teas, and for a long time it seemed impossible to take the stains out, but I find a weak solution of Chloride of lime will remove them. The solution must not be strong, and must be carefully strained; the cloth must not remain in the lime-water but a short time, and then must be thoroughly rinsed. The preparation sold as Javelle water would answer the same purpose.

Dominique Fowls.

This is an old-fashioned breed, so to speak, which is probably quite as widely known and has been as well liked as any of the now fashionable ones which are sought after at extraordinary prices, and bred with all the care and skill that man can apply. The Dominique is a breed adapted to all uses, and occupies a position in this country very similar to that which the Dorking does in England, being, however, hardy, and free from that deformity of the Dorking—the fifth toe. The characteristics of the Dominiques may be briefly stated as follows: Good size; aptitude for the table and for market; general excellence as layers, and as winter layers; hardiness, both as fowls and chicks; uniformity of style and plumage. The hens are good setters and mothers. The breed is, besides, remarkable for the degree to which it impresses its character upon the common fowls. There are several breeders in the country who take great pains to maintain it in its perfection, and we are gratified to learn that in Illinois, and perhaps elsewhere at the West, it is held in such esteem that choice breeding birds are held at $40 to $50 a trio.

DOMINIQUE FOWLS—PROPERTY OF COL. HENRY HOWLAND, OF CHICAGO.

Both single and rose combs are admissible, but should not be found in the same flocks. The legs are yellow and clean, short and strong. At present the rose-combed variety is the favorite, although in no way superior, unless better bred. The color of the plumage distinguishes the Dominiques from all other fowls with similar characteristics of form. It may be described as bluish-gray — each feather having a light gray ground, barred crosswise with a dark slaty-blue penciling. The cocks have a proud carriage, full neck, and saddle hackle, and full tails, with well-curved sickle feathers. The hens are domestic and active. The cocks should attain a weight of 5 to 7 pounds at eight or nine months old, and at full maturity of 6 to 8 pounds, and hens should weigh 4 to 6 pounds. If bred without regard to size, they run down to 4 pounds for the cocks, and 3 for the hens. This, some otherwise good breeders have allowed, by breeding too close, or too much for brilliancy of color. It deprives the breed of one of its chief recommendations—size and quality as a market fowl—and should not be practiced. We should add that a broad back, and full, deep breast, with short thighs, set wide apart, and well-tucked wings, are essential points. The face should be coral-red, the ear-lobes red, and wattles of medium size, and not too meaty.

The fowls, of which we present portraits, were bred by Col. Henry Howland, of Chicago, and received the first prize at the N.W. Poultry Association last February. Col. H. has none to sell, and this article is not written to advertise anybody's fowls, but to call attention to this excellent, though somewhat neglected breed.

The Ferret.—*Putorius furor*

This little animal, which will be at once recognized as belonging to the family of the Minks and Weasels, is much better known in Europe than in this country. It is a native of the Barbary States in Africa, but we believe sustains itself in Spain and Southern Europe in freedom. In England, however, and wherever the winters are cold, if one escapes and lives a wild life during the summer, it either returns to captivity, which is often voluntarily done, or it perishes. Hence there is no danger, as sometimes apprehended, that ferrets will run wild and become a great nuisance in every cold country. As is well known, they are employed in hunting rats and rabbits, in which exercise they enter their burrows or holes and drive them out. The hunting ferret is invariably muzzled, so that he shall not kill the rabbit, suck its blood, and leave it dead in its burrow. The training of ferrets for the different uses is a task requiring no little skill. A ferret must be a good fighter, or it will be worsted by a rat, and one accustomed to driving rabbits is, it is said, useless for catching rats. Ferrets are kept in close boxes, well supplied with wool to keep them warm, and are fed on bread and milk and meat. They breed readily and acknowledge the care of their breeder by seldom biting him, if properly handled. Sometimes, however, they inflict very severe wounds. It is a practice of European breeders to cross the ferret with the common European Polecat (*Putorius fœtidus*). The progeny is larger, darker colored, and hardier. It is said also that similar crosses may be made with the mink and some other members of the genus. The color of the ferret is very light yellowish, or white with red eyes. Our engraving represents two ferrets of very dis-similar form and size. The smaller one is very highly bred and well trained. The other, a Polecat Ferret, a famous rat-killer. Ferrets are sold by dealers in birds, dogs, and other pets, at prices varying with the training which the animals have received.

THE FERRET.—(*Putorius furor.*)

containing a great variety of Items, including many good Hints and Suggestions which we throw into smaller type and condensed form, for want of space elsewhere.

Postage 12 Cents a Year in Advance.—The postage on the *American Agriculturist* anywhere in the United States and Territories, *paid in advance*, is 3 cents a quarter, 12 cents a year. If not paid in advance, twice these rates may be charged.

How to Remit:—Checks on New York Banks or Bankers are best for large sums; made payable to the order of **Orange Judd & Co.**

Post-Office Money Orders may be obtained at nearly every county seat, in all the cities, and in many of the large towns. We consider them perfectly safe, and the best means of remitting fifty dollars or less, as thousands have been sent to us *without any loss.*

Registered Letters, under the new system, which went into effect Oct. 1, 1868, are a very safe means of sending small sums of money where P. O. Money Orders cannot be easily obtained. *Observe*, the *Registry fee*, as well as postage, *must be paid in stamps* at the office where the letter is mailed, or it will be liable to be sent to the Dead Letter Office. *Buy and affix the stamps both for postage and registry, put in the money, and seal the letter in the presence of the postmaster, and take his receipt for it.* Letters thus sent to us are at our risk.

Clubs can at any time be increased by remitting for each addition the price paid by the original members, if the subscriptions all date at the same starting point. The back numbers will, of course, be sent to added names.

Bound Copies of Volume XXVIII (1869) are now ready. Price, $2, at our office; or $2.50 each, if sent by mail. Any of the previous twelve volumes (16 to 28) will be forwarded at the same price. Sets of numbers sent to our office will be neatly bound in our regular style at 75 cents per vol., (50 cents extra, if returned by mail.) Missing numbers supplied at 12 cents each.

1,000 Eumelan Vines.—Any subscriber to the *American Agriculturist* may order of us one or more of these remarkable grape-vines before the 15th inst. at the 1,000 price, viz.: No. 1 Vines, $1 each; Extra, $1.75 each. Post-paid by mail, or by express, the receiver paying express charges.

Letters Without Names Signed.—It is useless to send these; we can not give them attention. Here are a lot of *anonymous* inquiries about Kansas lands, White Wire, sundry Doctors, various implements, Dealers, etc., several of which would be answered by letter if we knew just where to direct. If we answered all such questions *in* the paper, it would contain little else; and if we answered all by letter, we could do little else. Better save your time, paper, and postage, than to write anything whatever without giving your *true* name and address. These are not published when the writer desires them withheld.

VALUABLE BOOK—FREE!—The Publishers of this Journal issue, every year, two Volumes, prepared with great labor and care, containing a large amount of valuable information, finely illustrated, in neat illuminated covers. These volumes should be in every household. They also contain very complete Almanacs, with calendar of work to be done each month, on the Farm, in the Garden, etc. One is the **"Agricultural Annual,"** and the other the **"Horticultural Annual,"**—entirely distinct in matter, illustrations, etc. Each volume contains **152** 12mo pages. They are of permanent value, and every one should have each year's issue. There are now Nos. **1, 2, 3,** and **4,** (for 1867, 1868, 1869, and 1870,) of each work. ....To make these more widely known, and as a premium, also, the Publishers offer to send any one of these eight volumes, post-paid, to any person who will, during MAY or JUNE, forward a subscriber for the current volume of the *American Agriculturist* (that is, for 1870), at the regular subscription price of $1.50. One copy of any volume of the Annuals desired, will be presented for *each* subscriber thus sent between May 1st and June 30. Such names will not, of course, be counted in lists for the general premiums

Out-of-door Whitewash.—Chas. Stierlin, of Washington, D. C., asks us "to publish in the May issue of the *American Agriculturist* a recipe for a good whitewash for fences and out-buildings, that will not be washed off by a few rain showers."—Here it is. Take good quick-lime, in lumps, slake it with hot water, and while slaking add, to what will make a pailful, a pound of tallow, or any grease free from dirt. It may be rancid, smoked, or otherwise unfit for kitchen use. As soon as the violent slaking is over, stir thoroughly. All the water should be added before the slaking ceases, and the mixture should be very complete. This forms in the whitewash an insoluble lime soap, which, if the whitewash is diluted with cold water, often separates in minute clots. If the mixture be well made, it will be very smooth, and is little affected by rains.

Harris on the Pig.—A slight delay occurred in the issuing of this work, owing to the inability of the author to furnish the final pages, on account of illness. The work is now ready, and will commend itself to all who take an interest in the improvement of swine, as well as in breeding and rearing them in the best and most profitable manner. The advertised table of contents will show how full of information it is.

Every Saturday.—Fields, Osgood & Co. have made a hit with this journal in its new form. It gives the best foreign engravings, and excellent selections from English magazines, at a very popular price. Admirers of Dickens—and that includes almost everybody—will be glad to know that he has begun a new story, which is published in Every Saturday from the author's advanced sheets.

Mark Twain has been engaged to edit an agricultural department in the Galaxy Magazine. Mark is a humorist by profession, but we doubt if he will succeed in doing anything as funny as the reports of what is done at the Farmers' Club in earnest.

The Handy-Book of Husbandry, is the title of a work by Col. George E. Waring, Jr., which E. B. Treat & Co. announce as in preparation, and of which they send us specimen sheets. In accordance with our custom, we defer any extended comments until the complete work is received.

Haying and Harvesting Machines. Mowers and Reapers, Tedders, and Horse-rakes, of the best kinds, must be ordered early, in order to be reasonably sure of getting much use out of them this year. Thousands of farmers are disappointed every year, and some, strange to say, year after year, because they want the best, and order too late.

Pleasant Valley Grape Growers' Association.—At a recent election, C. D. Champlin was chosen President, and H. Gardener, Secretary of this Association. The 11th Annual Fair will be held at Hammondsport, Steuben Co., N. Y., Sept. 25th, 26th and 27th.

Cement for a Manure Pit.—H. F. Goodban, Erie Co., Pa., asks—"Will water-lime answer for laying the wall of a manure pit?" and adds, "I want to make a cheap, durable cement, that will not spoil by freezing."—Hydraulic lime occurs in market in two principal forms—"Water-lime" and "Cement." The cement is ground, and is the superior article. The water-lime, however, is of fair quality, answers for all coarse purposes, and would, we think, make an excellent mortar for such a wall, and for the grouting and cementing of the bottom. If, on trial, the sample is found to make a close, firm mortar, none other need be used, even for the top, which will be exposed to freezing. This portion might be laid in cement. Walls laid early in the season will not be injured by frost if the surfaces be smooth.

Pear Duchesse de Bordeaux.—We have before spoken of the excellence of this winter pear. Messrs. Smith, Clark & Powell, of Syracuse, N. Y., send us a colored plate, which is a very good representation of this large and fine variety.

Plow Deep, but not too deep at one time. F. B. Minch is right. He writes: "I sometimes feel like writing after reading the New York Farmers' Club reports about shallow culture. I have a little over 100 acres under cultivation, and my motto is, 'Plow deep, and manure as you go.' It does best for me. I think that the teaching of the shallow culture theory is productive of much error in the community."

Death of Seth Boyden.—Mr. Seth Boyden died on March 31st, at the advanced age of 82, in the City of Newark. Mr. B. was the pioneer in many of those branches of business to which Newark owes a large share of its prosperity, and by his inventions, contributed much to the wealth of others, though he acquired little himself. It is on account of his labors in horticulture that we notice his demise. He was particularly successful in raising seedling strawberries, several of which maintain a rank among the valuable varieties. The Agriculturist, Green Prolific, and Boyden's No. 30, are among Mr. Boyden's contributions to our list of varieties.

Croquet.—B. Brown. This game is not patented. A particular style may be.

A Suggestion to Farmers' Clubs.—"D. R.," Beverly, Mass., suggests that Farmers' Clubs make their President a member of their State Agricultural Society. This would connect the State with the local organizations, and be a proper acknowledgment of services which are not often requited.

Cement for a Cellar Bottom.—We are asked what kind of cement should be used for a cellar floor, and how much surface will be covered by one barrel....There are several good kinds of water cement. In the vicinity of New York, builders give preference to the "Rosedale." If the cellar bottom is firm and level, especially if it be stony or gravelly, the cement need not be more than an inch or an inch and a half thick, if mixed with sharp sand or screened gravel; a barrel will cover about one hundred square feet—what the builders call a "square." If the ground is soft, and a grouting of broken stones is necessary, it takes more. In this case, the floor should be covered with stones of the size of an egg, to the depth of three inches, at least, well pounded down to a level surface. The mortar is spread upon this, and worked down into the grouting, and when set, but still moist, the top coat is applied.

Slugs.—B. Thornton, Tioga Co., Pa. The specimens are common garden slugs. Ducks will destroy them in large numbers. Lime-water, or dusting with dry lime, may be used. Another plan is to lay cabbage or lettuce leaves on the ground; early in the morning the slugs will be found gathered under them.

Osiers.—"Subscriber." Osiers are cut when vegetation is at rest. Machines have been invented for peeling them, which the makers should advertise. For home use upon the farm, there is no necessity for peeling.

To Get Rid of the Ox-eye Daisy.—"T. B. R.," Alexandria, Va., asks: "Do you know any method (other than repeated cultivation) by which the 'Ox-eye Daisy' may be destroyed? It spreads very rapidly, threatening serious injury to grass fields that many of our people wish to continue in grass. This weed was much introduced during the late war in the hay used by the troops of the general government."—One of the minor results of war is the weeds that it carries in its train. There is probably no way to get rid of this weed except clean culture, and early cutting before it blossoms, so that no mature seeds shall get into the manure.

"Black-leg" in Cattle.—"E. B. C.," La Porte, Col. With reference to black-leg, Dr. Liautard refers us to Professor Gamgee's statement that Black-quarter, which is probably the same thing, has been successfully prevented by proper drainage, by keeping up the condition with oil-cake, and the use of purgatives and setons. The administration of half an ounce to an ounce of nitre in food is of great benefit. Purgatives and stimulants also should be freely administered.

Brewers' Grains Bad for Fowls.—The experience of "J. F. C.," Wrentham, Mass., indicates that brewers' grains fed freely to fowls, induces disease. Are not the fowls which run at large about breweries, healthy?

Hen-plucked Roosters.—"B. F. H.," Natick, Mass., found that his hens plucked off the feathers from the neck of a favorite Black Cochin cock. He applied castor oil to the bare neck, and to the feathers, and it proved a complete remedy.

The Hatching of Eggs can not be expedited by artificial means. It can only be accomplished in the regular time, or a little longer, if at all.

Keeping Eggs.—"H. R. D.," of Belmont, N. Y., has, for forty years, been in the habit of keeping eggs by taking them fresh from the nest and greasing them with good lard or butter, and setting them in a cool, dry place, handling them over once in a while, to keep the yolks from settling down to the shell. With him they keep six or eight months. We have had best, in fact, perfect success, by plunging the fresh eggs, a few at a time, held in a wire ladle, into a kettle of boiling water, and keeping them there long enough to count 10—say five seconds. The water must not stop boiling.

AMERICAN AGRICULTURIST

FOR THE

Farm, Garden, and Household.

"AGRICULTURE IS THE MOST HEALTHFUL, MOST USEFUL, AND MOST NOBLE EMPLOYMENT OF MAN."—WASHINGTON.

VOLUME XXIX.—No. 6. NEW YORK, JUNE, 1870. NEW SERIES—No. 281.

THOROUGH-BREDS OF CLIFTON STUD.—*Drawn and Engraved for the American Agriculturist.*

The Thorough-bred compares with other horses, as oak compares with pine, or steel with iron. Blood tells for generations; and the best blood among the thorough-breds is of course very valuable. We give above, striking pictures of some of the best of Mr. R. W. Cameron's horses. Clifton Stud Farm, Staten Island, has been for several years famous for thorough-bred stock of several kinds—Shorthorns and Jerseys, Berkshires and Dorkings—besides its stud of mares, stallions, and young horses.—*Leamington*, the horse on the lower left-hand side, is regarded as the handsomest model for a race-horse in this country if not in the world. He is so well known, both here and in Europe, that it is only necessary to state that he is a very powerful, dark brown horse, 16½ hands high. He was got by Foig-a-Ballagh, out of a Pantaloon mare. Foig-a-Ballagh was full brother of Irish Birdcatcher, and all the family are famed race-horses. Leamington was imported by Mr. Cameron in 1866, and was sold last autumn to Mr. A. Welch, of Chestnut Hill, Philadelphia, for $15,000. He stands at Patterson, N. J., this season, at $200.

This horse (in England) twice won the Chester Cup, 2¼ miles, carrying the heaviest weights, and beating in one race 37, and in the second, 41 horses. He also won the Goodwood Stakes, and was considered the best horse of his day. His get in this country are already famous: Lynchburg and Enquirer, in Kentucky; Miss Alice and Anna Mase, here, established his superiority. Lynchburg was sold for $5,000, and this price was refused for Enquirer. $7,000 was offered and refused for Miss Alice after her famous match last year, but both she and Anna Mase were unfortunately killed by accidents.

BULLARD'S IMPROVED HAY TEDDER.

Important Improvements! See Pamphlet!

Spreads from behind the wheels and does not run over the grass after it is spread. Wheels 4 feet high. Light draft for one horse. Its use enables the Farmer to *cut*, cure, and *store away* hay in *one day*, and adds 20 per cent to value of crop. A large farmer says: "*Its use in a single season will more than pay its cost.* The enterprising Farmer *will not, cannot* do *without it*; the longer he puts off buying, the poorer he will be." It is the most popular Harvesting Machine ever offered.

Union Mower and Reaper

Is the height of perfection. Light, easy draft, simple in construction, neat and accurate in workmanship, convenient to operate, perfectly adapted to cut on salt marsh, uneven hill-side or lawn. Those who have used it pronounce it the **Most Durable Machine Made.**

Eagle Sulky Hay Rake.

Light, simple, durable, and cheap.

Agents wanted to sell any or all of the above. Liberal inducements offered to enterprising men. Illustrated pamphlets and terms to Agents mailed upon application.

DUANE H. NASH, Gen. Agt.,
29 Cortlandt St., New York City.

H. KILLAM & CO.,

Chestnut St., New Haven, Conn.

We manufacture the finest class of carriages for city use, consisting of Landaus, Landaulettes, Clarences, Coaches, Coupes, Coupelettes, Barouches, Bretts and Phætons. Which we warrant equal in point of style, finish and durability to any built in this country.

Messrs. DEMAREST & WOODRUFF, 628 Broadway, are our Agents in New York City.

NEW YORK STATE AGRICULTURAL WORKS,

PATENTEES AND MANUFACTURERS OF

Railway Chain and Lever HORSE POWERS,

Combined Threshers and Winnowers, Overshot Threshers, Clover Hullers, Feed Cutters, Saw-Mills, Horse Rakes, Horse Pitchforks, Shingle Machines, &c., &c.

ALBANY, N. Y.

Buckeye Thresher and Cleaner

FOR 4 OR 6 HORSES. THE BEST, MOST COMPACT and Cheapest. Send for Illustrated Circular.

BLYMYER, DAY & CO., Mansfield, O.

Victor Cane Mills, Cook's Evaporator, Victor Grain Drill.

BUY THE BEST

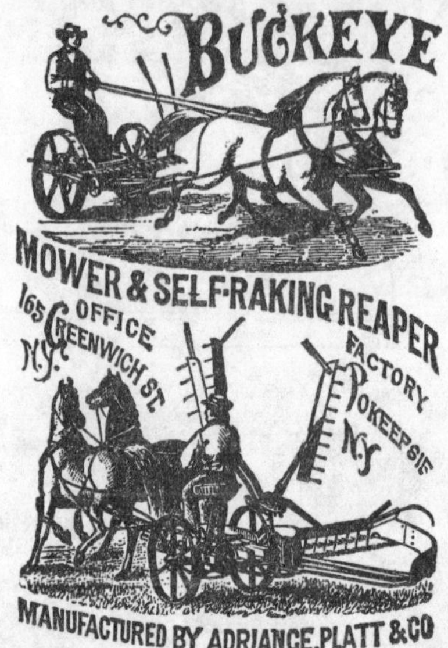

STYLES, SIZES, and PRICES, to Suit All Classes of Farmers.

Descriptive Circulars forwarded by mail.

THE CLIPPER MOWERS AND REAPERS.

The great field trials at Sedalia, Mo., and at Xenia, Ohio, in the summer of 1869, where the Clipper met some forty competitors, including all the leading machines built, and was awarded the First Premiums and Gold Medal—prove that the proprietors have been able to *sustain, in the field, their claims of superiority for the Clipper Machine* over all others.

Its EXTRAORDINARY LIGHTNESS of draft takes one-third of the labor off the team. The adjustability of the finger-bar to cut high or low, without stopping the team, is worth $25 on any machine. It is the only MOWER that has the true center draft, in accordance with correct mechanical principles. In short, it is the lightest, the most durable, the least complicated, the most perfectly constructed, the most symmetrical in design of any implement of its class. Farmers, the Clipper is the **Horses' Friend**, and you cannot afford to ignore their welfare. Send for pamphlet. Investigate, see or write to some one who has used the Clipper. See the machine yourself before you buy any other.

THE CLIPPER MOWER & REAPER CO.,
P. O. Box 6173. 154 Chambers St., New York.

THE JOHNSTON *Self-raking Reaper*.

is the MOST POPULAR MACHINE of the kind now before the public; it has more practical points that universally adapt it to the work of reaping grain than all others put together. It cuts

ANY GRAIN THAT GROWS,

whether high or low, thin or thick, lodged or standing, in the MOST SATISFACTORY MANNER; it is easily handled by an ordinary pair of horses, to cut from 15 to 20 acres between sunrise and sunset. It swaths, or lays the grain in gavels out of the way of returning team; is worked by almost any boy or ordinary field hand, or automatically, as you prefer. For further particulars send for Descriptive Pamphlets to the manufacturers,

C. C. BRADLEY & SON, Syracuse, N. Y.

Holbrook's Regulator Seed Drill.

A perfect hand-machine for sowing with *regularity and in proper quantity*, Beet, Carrot, Onion, Parsnip, Turnips, Sage, Spinach, Sorghum, Broom-Corn, Peas, Beans, etc. It is very simple, compact, durable, easy to operate, and *shows the seed as it drops* to prevent any mistake or failure, ridges the earth over the seed, and marks the rows straight any width desired. All springs, slides, reeds, and brushes dispensed with. Price $12.00.

A Hand Cultivator.

An improved implement to weed between the rows and mellow the soil. It expands from 8 to 14 inches in width, is simple and efficient. Price $6.00. Manufactured and sold by F. F. HOLBROOK & CO., Boston, Mass.

☞ Sent on receipt of price by P. O. money order or otherwise.

E. FRANK COE'S AMMONIATED BONE SUPERPHOSPHATE, price $56 per ton.

FISH GUANO.

Warranted pure, $65 per ton.

For sale by ENOCH COE, Williamsburg, N. Y.

SHEEP SHEARS, WITH THUMB-PIECE,

MANUFACTURED BY

HENRY SEYMOUR & CO.,

52 BEEKMAN ST., NEW YORK.

Two first Premiums awarded at American Institute Fair, October, 1867 and 1869.

"VALUABLE AND WELL-MADE IMPLEMENTS."
HORACE GREELEY, Pres't.

"The Sheep Shears work splendidly, and I would not ask for any better, so says Mr. Jepson, who shears our sheep and he has sheared thousands." Resp'y yours,
(Signed) L. A. CHASE,
AMERICAN AGRICULTURIST.

Twelve First Premiums at twelve different State Fairs, Oct. and Nov., 1868. Highest recommendations from Hon. H. S. RANDALL, Pres't Wool Growers' Association and Author of the Practical Shepherd.

☞ Every pair warranted. Sent free by mail on receipt of

$2.25,	$2.50,	$2.50,	$2.75.
4½ in.,	5 in.,	5½ in.,	6 in.

Length of Blade—Thumb-piece.

SHEEP TOE SHEARS, $2 per pair.

Send Post-office Order, Registered Letter, or Draft, if possible.

HOLBROOK'S PREMIUM HORSE HOE,

FOR THE CULTIVATION OF

Corn, Potatoes, and the various kinds of Root Crops. Is extensively used. Price $12.00.

It cannot be clogged or choked with weeds, witch grass, stubble, sods, or manure, running free and clear in all kinds of soil, and is the most valuable implement ever invented for destroying witch grass.

The rear plows can be expanded from 15 to 36 inches in width; they are also reversible, so that the soil may be turned towards or from the row at pleasure, the wheel gauging the depth from 3 to 7 inches.

A pair of extra large rear plows, made for hilling up, $1.50. An extra large middle plow to furrow, and use among potatoes, $1.50.

It is a complete pulverizer of the soil, and is generally preferred to the crossing plow, or to the harrow, in mellowing the surface of plowed land preparatory to putting in the crop, and for covering manure near the surface without disturbing the inverted sod or other vegetation previously buried by the plow.

☞ A farmer cannot afford to do without one.

Manufactured and sold by
F. F. HOLBROOK & CO., Boston, Mass.

FRUIT BASKETS.

QUART. PINT. ½ PINT.

GET THE VERY BEST. In nothing does this injunction better apply than in the selection of a basket for marketing small fruits. It is an undeniable fact that no article yet offered for this purpose comes so near perfection as **THE BEECHER VENEER FRUIT BASKET.** These baskets are not only cheap, but combine greater strength, beauty, durability, and capability for ventilation than any other known style, hence for use and convenience are by far the most economical made. *Full measure guaranteed.*—no "short quarts" when the *Beecher Basket* is used. We have the voluntary testimony of many of the largest growers and commission fruit-dealers sustaining our claims for these baskets. In short, they are *the* basket,—try them and prove the correctness of above statements. Light, neat, and strong, crates of various sizes to accompany baskets, on hand, and made to order.

Send for circular and price list to

THE BEECHER BASKET COMPANY,
Westville, Conn.

Improved Foot Lathes,

With Slide, Rest, and Fittings. Elegant, durable, cheap and portable. Just the thing for the Artisan or Amateur Turner.

Send for descriptive circular.

N. H. BALDWIN, Laconia, N. H.

C.W.Co.

Motive Power for Nothing.

Our Patent, Self-regulating, Storm-defying Wind-mill is superior for pumping water for Railroads, Country Residences, Hotels, Farms, Stock-Fields, Drainage, Irrigation, etc. For circulars address

"CONTINENTAL WIND-MILL CO.,"
5 College Place, corner Park Place, New York.

Holbrook's Swivel Plows

Leave no "dead furrows," "nor ridges," turn furrow-slices flat on level land, lay the fields down level and smooth for the Mowing Machine, Hay Tedder, etc., work equally well on side-hill, save considerable time, pulverize thoroughly, will not clog, and equalize the work of team. All sizes, from a Garden Plow one horse, to a Plow for four cattle. Manufactured and Sold by

F. F. HOLBROOK & CO., Boston, Mass.

AMERICAN AGRICULTURIST.

NEW YORK, JUNE, 1870.

Farmers who have their work properly planned and laid out, will find little time for reading. We would remind those who look to us for hints, that the moist and fertile soil, as it warms under the summer sun, will give life to thousands of weeds, which must be despatched *now*, in their seed-leaf and tender beginning, or day by day the labor of subduing them will be increased. Procure early, suitable implements for all farm work, and keep tools sharp and clean. Sets of duplicate tools to supply the places of those particularly liable to be broken are of great service.

Hints About Work.

Clean out the Barns.—This is the only time in the whole year when the barns may be emptied of every thing. It is barn-cleaning time as much as April is house-cleaning time. If you have several tons of old hay, let it alone—but if the mows and bays are pretty low, make a clean sweep, and see the poles and timbers everywhere. Most of the stock are then in the pastures, and stable floors can be lifted and examined. Timbers may be decaying, and should be renewed. Accumulations will almost surely be found under the floors and near the timbers which may soon cause decay unless removed, and which will be valuable additions to the compost heap. Soon the hay will be coming in; and so this work must be done betimes. Store old hay and straw where they can be conveniently got at after the barns are full.

Pasture-land.—Top-dress with plaster, leached or quick ashes, bone-dust, or any hand manure. Bone-dust 2 bushels, wood-ashes (quick) 4 to 6 bushels, plaster 1 bushel, makes an admirable top-dressing for any kind of grass land. It is excellent also for potatoes and for corn. If it stand three weeks, and is well shoveled over two or three times, it is much improved. Take the stock off top-dressed land until after a soaking rain.

Orchard and Nursery.

With established trees, the chief things to do are to thin the fruit and keep the insects in check.

Insects, do not wait until some of the legion have made sad havoc, and then write to us for a remedy which shall despatch them at one application. We know of no sovereign cure for insects or weeds except work. One should be among his trees frequently, and note the first inroads. The eye should be trained to notice the beginning of trouble. The caterpillars that are to-day feeding upon the leaves of one twig only, in a week will have scattered over the whole tree. Neglect the little web that is just observable, and in a short time the tree will look as if a blight had fallen upon it. The nests of the tent-caterpillar are easily pulled out by a gloved or bare hand. Take them early in the morning when the inhabitants are at home.

The Curculio is only to be caught by jarring the tree. It will fall and may be caught on a sheet. Let all secret insect remedies alone.

Borers.—We are accustomed to speak of "the borer," when there are several, but they are all alike in being hatched from eggs laid upon the bark of the tree by the mother insect. The eggs are usually deposited this month; various obstacles to the insect have been proposed; about the simplest of them is to wrap the lower part of the trunk of the tree with coarse paper, the lower edge of the paper going just below the surface of the soil, and the other extending about a foot above it.

Slugs which attack the leaves of pear and cherry trees are destroyed by a dusting of lime.

Grafts must be looked after. If two were put in where but one is needed, cut one out. If the growth of the graft is one-sided, or one of its several shoots gets the better of the others, pinch the end of the strong grower.

Pruning is done this month by many, as the wounds heal most readily. Others object to any other than winter pruning, on account of the check to the tree by removing so much foliage.

Young Trees.—Attention to young trees will avoid the necessity of pruning large limbs hereafter. Shoots will become branches; if they grow where branches are not needed, remove them while young; better to have done it when the bud was first pushing, but if done while the wood is still tender, it will save much cutting. It will be a great help to young trees to mulch them with litter of some kind.

Pinching.—The nipping of the growing point of a tender shoot, of course stops it from prolonging. One of its upper buds will probably push again; but the growth will be checked, neighboring and weaker shoots will have a chance to grow, and thus control the growth and shape of the young tree.

Seedlings.—Keep young seedling trees well weeded and cultivated. Shade young evergreens and deciduous forest trees.

Fruit Garden.

Grape-Vines.—Vines planted this spring should have but one shoot allowed to grow. Tie up the young shoots, handling carefully to avoid breaking. If too much fruit is left upon the vines, thin early. As the laterals push, pinch back to one leaf. Layers may be made of the present season's growth.—Use the hoe around the vines, and hand-pick beetles and caterpillars. Use sulphur, applied by a bellows, on the first appearance of mildew.

Strawberries.—Upon beds not covered, put on a mulch to keep the fruit clean, unless the vines are grown in matted beds, when it is not necessary; straw, hay, or corn-stalks are used. In marketing berries give full measure, and let the baskets run of even quality throughout. Reject all sandy or over-ripe berries. Make pot-layers which may be planted as soon as well rooted; they will bear next year.

Currants.—Use powdered white hellebore for insects which attack the leaves.

Gooseberries are usually marketed green. They are most expeditiously cleaned from leaves and sticks by allowing them to roll down an inclined trough.

Raspberries.—Tie up the new growth, three or four canes to a stool, and remove all others.

Blackberries.—Stop the growth of new canes when 4 or 5 feet high. Tie up fruiting canes. Hoe up all suckers not needed for planting.

Dwarf Fruit Trees will need the attention in thinning, pinching, and keeping free of insects as noted for trees in the orchard.

Kitchen Garden.

The contest with weeds must be kept up. It makes a great difference which gets the advantage of the start. The ease with which weeds can be destroyed in their young state, is in marked contrast with their persistence when of larger size.—Much weed killing can be done with a rake. Have one with sharp, steel teeth, and use it frequently. We use Comstock's hand cultivator with much satisfaction.

Asparagus.—Many forget that next season's crop depends upon a good growth of tops after cutting stops. Do not cut too late. Hoe over the bed and it will be all the better if a dressing of bone or phosphate can be given.

Beans.—Bush and running varieties may still be planted. For Limas, set poles 6 or 8 feet high, 4 feet each way, and make a rich spot to receive the beans. Plant 5 or 6 beans, by pushing them into the ground eye downward, and covering about an inch. Leave 4 plants to a pole, and twine them about it if they do not climb of their own accord.

Beets.—Weed and thin. We always sow rather thickly in order to have a plenty of young beets to use as spinach.

Cabbages and *Cauliflowers.*—Early sorts will now be ready for the table or market. Transplant second early, and sow late varieties.

Carrots.—Keep well cultivated. Sow seed.

Corn.—Sow every week or two for succession.—The late sowings should be of early kinds.

Capsicums or *Peppers.*—Give a warm rich place, and cultivate frequently.

Cucumbers.—Sow seed in rich hills 4 feet apart, using plenty of seed to guard against loss by bugs. When past danger take out all the plants but two.

Egg-Plants.—These need all the encouragement that frequent hoeing and liquid manure can give them. The Tomato-worm is fond of them.

Lettuce.—Sow for late supply, selecting a somewhat shaded and moist spot. See article on culture, last month.

Melons need the same care as cucumbers.

Onions.—Thorough weeding and frequent cultivation between the rows are essential to success.—Salt sown broadcast at the rate of 2 or 3 bushels to the acre is beneficial, as are dressings of wood-ashes.

Parsnips need working until the leaves prevent.

Peas.—Plant for late crop if you choose to run the risk of mildew.

Radishes.—Sow now and then for succession.

Rhubarb.—Keep the flower-stalks cut off, as they needlessly exhaust the plants.

Ruta-bagas.—The variety known as the Long White French cannot be too highly commended for family use. Sow the latter part of the month; and if insects trouble the young plants, dust with ashes or plaster.

Spinach.—Sow the New-Zealand for summer use, three or four plants to a hill, which should be rich, and about 6 feet apart, as the plant spreads.

Salsify.—Sow if not already done, and cultivate the same as parsnips.

Tomatoes.—As the vines grow, keep tied up to a trellis, or place brush to keep them from the ground.

Flower-Garden and Lawn.

Lawns.—Mow often and leave the clippings on the grass. A few hours of sun make them invisible, and they serve both as mulch and manure.

Bedding Plants will now need attention. Keep the weeds down until the plants cover the beds.

Annuals may be transplanted and seed be sown.

Tuberoses need a warm and rich spot. Plants that have been started under glass may be obtained and will grow rapidly.

Bulbs.—Do not remove the leaves from those that have passed out of flower until they begin to turn yellow. Take up Hyacinths, Tulips, Narcissuses as soon as the failure of the leaves shows that the growth is over. Spread under cover until the leaves are dry, and then store in a cool and dry place until time to plant in the fall.

Roses.—When the Remontants pass out of flower, cut them back. Remove the remains of the flowers of the Ever-blooming varieties. Shake off rose-bugs.

Neatness is secured by constant attention to little things. Keep straggling plants tied up. Remove unsightly flower-stalks unless seeds are wanted.—Rake off remains of flowers and dead leaves. Clip or pinch shrubs disposed to grow out of shape.

Green-house and Window Plants.

These will likely all be out of doors, and they must be disposed of according to their needs.—Some may be set in full sun, while Fuchsias, Camellias and other broad-leaved evergreens need a partial shade. Some may be disposed of as ornaments to the veranda, while others may be plunged in the borders. Place a little coal-ashes under plunged plants to keep worms from entering. The earth in the pots dries rapidly, and care must be taken that they do not suffer for want of water. The pots containing tall plants should be protected from strong winds. Procure sods for potting soil, and look after supplies of manure.

Fire Hot-beds.

BY PETER HENDERSON.

In the St. Louis Journal of Agriculture for February 24th, there is an article upon "Fire Hot-beds," by E. A. Rheil, and in the Prairie Farmer Annual for 1870, Dr. E. S. Hull, of Alton, Ill., in an article on "Propagating the Grape," treats on the same subject.

The plan, if I rightly understand the gentlemen, is, in substance, to run under the soil smoke or heated air flues from the furnace, at such distances apart as will heat the space of hot-bed wanted for planting or sowing upon, or to form the base on which to rear the superstructure of green-house or frame. Both writers fail to tell in what way the superstructure of glass is to be made; they do not give the hight, and we are left to conjecture, whether they recommend a *green-house* or a *hot-bed*. If they intend a green-house, then if walking room is to be obtained inside, the bed must be too far from the glass for the health of the plants; if simply a hot-bed is intended, where access is had only by lifting the sashes outside, then all who have had experience know that it cannot be worked to advantage in the winter months.

Inarching the Grape-vine.

An enthusiastic amateur gives in the Horticultural Annual for 1870, an article on inarching the grape. He finds that the Clinton and others of our native species, have much more vigorous and hardy roots than the exotic grapes; and he has in his extensive vinery a collection of the choicest foreign varieties inarched upon natives. This method may be practised on out-door vines, in cases where we wish to get a feeble grower upon a strong root, or where we wish to fruit a new variety as soon as possible. In inarching, the plant that is to serve as a cion must be planted in a pot or box

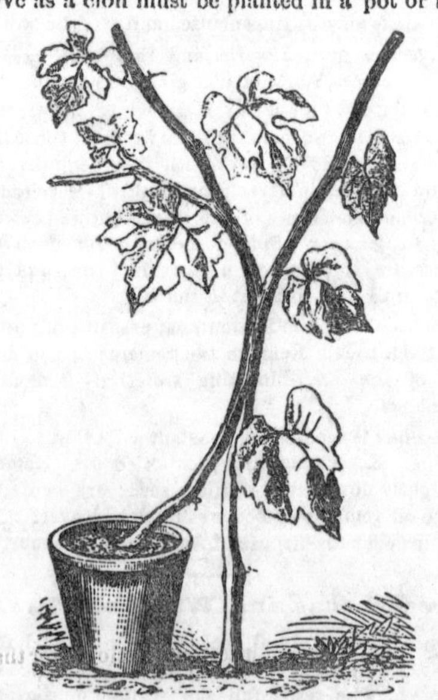

Fig. 1.—INARCHING.

so as to be movable. The operation is performed in the open air in the early part of July. The vines to be operated upon being brought into convenient position, a slice from 2 to 4 inches in length is cut from stock and cion, cutting through nearly half their diameter. The slice is thickest in the center, and tapers to each end. If the branches operated on are small, no tongues are made; but if a quarter of an inch or more in diameter, the cion may be tongued from below *upwards*, and the stock from above *downwards*, and the two put together as shown in figure 2. A tie is placed above and one below the wound, which is then rather firmly bound with a strip of oil-silk, and over this is wrapped a strip of muslin. A few vine leaves are bound on to keep off the sun. The green shoots of the vine are very brittle at the nodes, or joints, and in operating, care should be taken to avoid breaking them at these points. In about a month the union will have so far perfected that the muslin may be removed, and the oil-silk be cut by running a knife along the stock lengthwise, to allow the vine to increase in size. The oil-cloth is not removed, but the muslin is loosely replaced. At the end of six weeks all ties and bandages may be removed. Three weeks or a month after operating, the growing ends upon the stock are pinched to direct the sap into the cion, and the growth of the stock kept checked by frequent pinching, but it is not cut away above the union until the wood has ripened. At the end of six weeks the cion is cut half through below the point of junction, and before the leaves commence falling it may be severed completely. Figure 1 shows the union between the two vines perfected, and before the cion has been cut away from its own root.

Fig. 2.

PRESIDENT WILDER AND CHARLES DOWNING STRAWBERRIES.—There are two varieties of strawberry called President Wilder, and Charles Downing. The European President Wilder, was raised by De Jonghe, and the American one originated with Col. Wilder himself. The European Charles Downing is a seedling by De Jonghe and the American variety of that name was obtained by J. S. Downer of Kentucky. It is very unfortunate that this doubling of names should occur. According to pomological rules, the name first published with a description in a pomological work, takes the priority. This is a matter that will properly come before the committee on synonyms of the Am. Pomological Society. Those purchasing varieties with the above name, should ascertain whether they wish the American or European, and order accordingly.

Fencing across Streams.

We give herewith an ingenious and simple flood-gate or bars for fencing across streams. A model and description of it was sent to the *Agriculturist* by Mr. Asa Hartshorn, of Ashtabula Co., Ohio. The engraving, prepared from the neat model, precludes the necessity of a particular description. Mr. H. writes as follows: "I assisted my father 50 years ago in making a set of bars in this way, in New London Co., Conn., and they remained good for 30 years or more. The posts or crotches should be fastened firmly by stones, or be set well into the bank of the stream. Two staples and one link are required to each bar. These should be made of $1\frac{1}{2}$-inch iron. The other ends of the bars rest on $\frac{3}{4}$-inch pins. In time of a flood, one or more of the bars may float up and swing around down stream, and when the water subsides, they may be returned to their places. These floating bars have the advantage over any other gate, as they open and let all flood-wood and trash pass. I am an old man and have whittled it out, so you can see how it should be on the stream. The round top pole rests in the crotch of the posts above high water, and in this way any stream 25 feet wide can be fenced with safety."

Fig. 1.—FENCE.

Fence Posts for "Heavy Land."

There are many sections in which limestone or sandstone rocks or boulders occur abundantly, which may be split into slabs of convenient size for the purpose indicated in the following suggestive communication. They are not unfrequently used for gate-posts and fence-posts, and of course are durable and excellent. "L. W. S.," of New Haven, Vt., writes, describing a plan which he has had in use and highly approves. There are many farms on which it may be imitated with profit. Wooden posts are expensive on account of the constant care attending their use from their decay and heaving by the frost. A post of this kind is valuable on account of the facility with which the fence may be shifted, as it is altogether upon the surface.

"Take flat stones (ledge limestone is best) from $2\frac{1}{2}$ to 5 feet long, 4 to 6 inches thick, and any convenient width, and drill inch holes 3 inches deep in one side of each stone, about 18 inches apart, or 9 inches from the center each way. Take $\frac{5}{8}$-inch round iron, cut 10 inches long; flatten one end about 4 inches, and punch two holes for screws. This is fastened on one side of each post with two screws, and inserted in one of the holes as in fig. 2. Then, for the brace, take pieces of the same-sized iron 25 inches long, flatten and punch as above stated, and insert the round end in the other hole, bend the iron to the post, and fasten; then fill the holes with melted brimstone, and you have a fence that will keep its place. I built some 25 rods in this way, three years ago, and it stands as true as when first finished. If it heaves in the winter, it settles back to its former place."

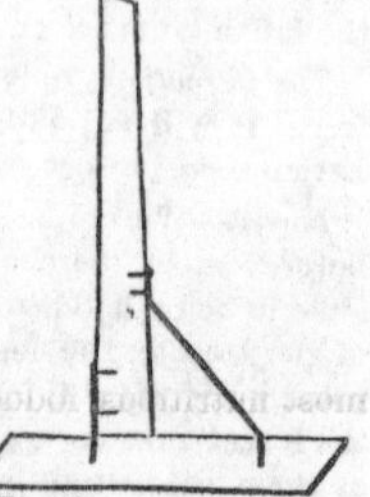

Fig. 2.—FENCE-POST.

Another suggestion comes from Mr. A. E. Smith, of New Haven Co., Conn. He recommends the use of a ³|₄-inch rod of iron, bent in the middle over a mandrel of proper diameter, having its ends inserted in holes drilled about 3 inches deep, in flat stones about 18 inches square, as shown in figure 3. The fence is made by inserting strips or rails either lapping or cut beveling, and the beveled portions lapping between the parts of the post. Blocks of wood or pieces of board placed between the different ranges of strips would hold them at the required hights, as shown in figure 4. Several modifications of this fence are suggested. Mr. S. says: "Ordinary split rails trimmed a little at the ends, can be placed with the ends a little through the posts resting on each other;

Fig. 3.—IRON FENCE POST. Fig. 4.

making a fence from which a length may be taken with little trouble, and which may be moved by taking apart, without injury, and with little expense. Two-inch plank, 5 inches wide, and 12 feet long, ship-lapped together at the ends with blocks 6 inches long, and any width desired, make a neat and durable fence. The stones should be left a few inches above the surface, and should be heavy enough to keep the fence in its place in a gale. Cement blocks might be used in the absence of stone—but in this State we can hardly imagine such a contingency."

We have seen fences made by boring holes through rails, or bars, and slipping the bars over single rods of iron set in stones in the way described; the bars forming adjoining lengths, alternately supporting one another. Mr. Smith's plan is a better one, as it admits of the easy removal of the rails, or the shifting of the fence.

Early Made Hay.

The practice of beginning to cut the grasses as soon as they are fairly in blossom is rapidly gaining favor. This favored time comes to many fields in this latitude, from the middle to the last of June. With the improved implements of hay-making—the horse-mower, rake and tedder, it is not difficult now for the farmer to gather all his hay crop at the time when it makes the most nutritious fodder. The early cured grass does not give so heavy a yield of hay to the acre at one cutting, and the same bulk will not weigh so much as the grass cured two or three weeks later. But very careful experiments show that the hay thus treated is much more nutritious; it is relished better by all kinds of of neat stock, and nothing is left in the manger. This tender, sweet hay is particularly important for sheep and young stock in the winter. If it do not bring so much in market or at the stables, at least, the hay, that is retained for home use, should be early cut. It does not exhaust the soil so much as where the seeds are allowed to mature. If the ground is very rich it allows of a second cutting in August. If not, it makes a heavy after-math and shelters the roots of grasses in the winter. The practice is increasing among our reflecting farmers, which is pretty good evidence that it is safe to follow.

Marketing Butter.

Having had several inquiries as to the best manner of preparing butter for retail marketing, we give the following account of the practice at Ogden Farm, which is similar to that of the producers of the celebrated Philadelphia butter, well known for both its excellence and high price.

The butter is all sold in half-pound prints, each of which bears the monogram or trade mark O F, as shown in fig. 1. These pats are 3¹|₂ inches in diameter, rounded up in the center and about an inch thick at the edges. As soon as made, they are wrapped in damp cloths, about 8 inches square. The printing is done in a wooden cylinder 3¹|₂ inches in diameter, of which the printing die forms the bottom. Fig. 2 shows a section of the cylinder with the parts in place. The walls of the cylinder are ³|₄ of an inch thick; *a a*, are brass screws, 2 inches long. There are three screws, which serve as legs to the mould; they may be turned in or out to regulate the quantity of butter; *b*, is the printing block, and *c*, the block on which it is supported. By substituting a thinner block in the place of *c*, heavier lumps may be printed. The butter is pressed closely into the mould and "struck off" even with the top of the cylinder. The pat is thrown out by pushing the block *c*, from below, mould and all going into cold water for a moment. It is necessary that all parts of the apparatus be kept soaking wet. The printing is laborious but it pays.

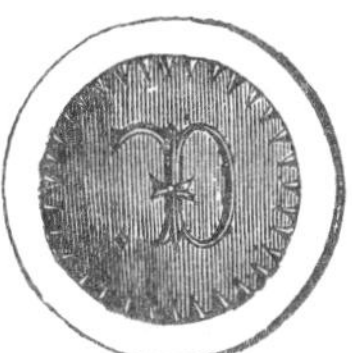

Fig. 1.—STAMP.

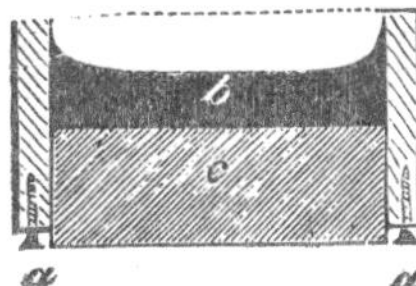

Fig. 2.—MOULD.

As fast as the pats are made and wrapped in cloths, they are packed in the market tub, fig. 3, which is 2 feet high, 2¹|₂ feet long on top and 1 foot 7 inches wide. It is made of cedar, and is brass hooped and painted; the lid is hinged in the middle so that either side may be raised, or the whole may be lifted entirely off. One of the staves at each end rises about 6 inches above the top and has holes cut through for handles. An oak bar, with a head on one end and a hinged hasp (brass is preferable) on the other, is passed through these handles to hold the lid down.

Fig. 3.—BUTTER-TUB.

When it is pushed home, the hasp can be turned down over a staple at the end of the tub and secured with a lock or pin. This tub is only a case for the tin, ice, and butter tray which lines it. The ends of this tin tray are partitioned off at a width of about 6 inches, as shown in fig. 4, and the compartments so formed at the ends, which are for broken ice, are closed with hinged tin covers. The center compartment is fitted with tin studs supporting five tiers of movable wooden shelves; and each shelf consists of two pieces for facility of handling. The shelves allow, with the bottom, six tiers of prints. A section of this tray, packed with butter and ice, is shown in fig. 5. Packed in this way, a blanket being thrown over the tub in the wagon, butter may be carried around for a whole day in the hottest summer weather, without losing its hardness.

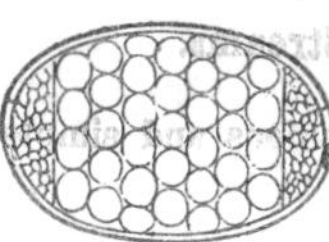

Fig. 4.—TOP VIEW.

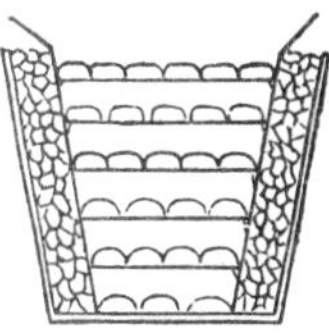

Fig. 5.—SECTION.

More Stvles of Tail-board Fastenings.

During the year past we have exhibited several forms of tail-board fastenings, which to us, and most of our readers, have, at least, had the merit of novelty. Some have been great improvements upon the old rod and tail-nut, and all have had some feature to recommend them either for dumping carts or wagons. Now we have two more to show. The first, fig. 1, is from Isaac S. Bachtel, of Stark Co., O., who writes: "I first saw this tail-board fifteen years ago. You will observe it may be loosened from either side, and swung open like a gate, or it may be taken out altogether. A chief merit is that there are no strips in the corners of the box. I use it behind and before on that account." The fastening of each end consists of two semi-circular straps of ¹|₄-inch iron, one

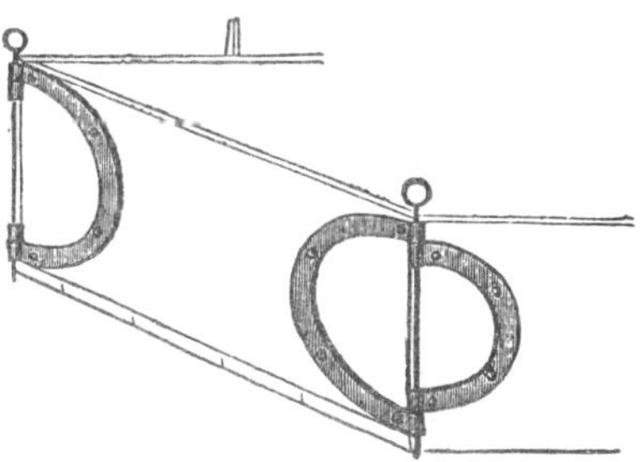

Fig. 1.—TAIL-BOARD FASTENING.

attached to the tail-board, the other to the outside of the box, and forming a hinge when united by a ¹|₄-inch rod passing through their bent ends. This rod when withdrawn to disengage the board, should be immediately replaced.

The other plan is sent us by Mr. C. H. Rue, and consists of a slight modification of the common rod. A rod of the usual size, say ¹|₄-inch iron, is fixed upon the tail-board, as shown in fig. 2. This rod terminates in a strong hook at each end. These hooks go into eyes in bolts, one or both of which are fastened with tail-nuts, and having some play, so that after the hooks are in they may be screwed up tight. A few turns of the nut are sufficient to relieve the strain upon the hooks, and the rod having a little play when one hook is loose, the other is easily unfastened. The bolts are inserted a little lower than

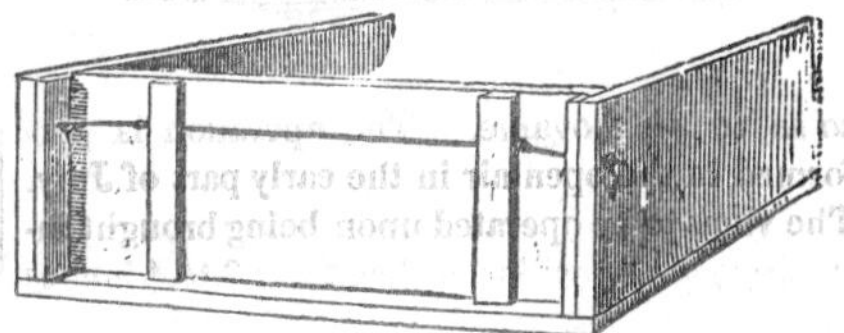

Fig. 2.—TAIL-BOARD FASTENING.

the line of the rod, so that when screwed up the board will be held down and kept from rattling. The cheapness and convenience of the arrangement are apparent and well worth imitating.

The "Prairie Apple"—(*Pomme Blanche.*)

The species of our native plants are very numerous, but among these there are but few which furnish articles of food. Berries and perishable fruits are more or less abundant in their season, but those native products which can be stored up are limited in number, and as articles of food are at best indifferent. Neither in the variety nor in the quality of his food does the savage equal the poorest among the civilized. Acorns and grass seeds are poor substitutes for corn and wheat; and among the several more or less edible roots used by the Indians there is none which approaches the potato in excellence and nutritious quality. A large share of the vegetable food of some of the Western tribes of Indians is the Prairie Apple, or Pomme Blanche, as it was named by the French *voyageurs.* It is the root of a Psoralea (*P. esculenta*), which is found from Wisconsin westward to the Rocky Mountains. The plant grows about a foot high, has leaves with five divisions, and its flowers are clustered in a dense head, much resembling a large clover; the flowers are purplish-blue. The root is turnip-shaped and somewhat farinaceous; and though it would be considered scarcely edible by us, is gathered in large quantities by the Indians and stored for the winter. Mr. Cary, who has given us many illustrations from the far West, has sketched a party of squaws—it being one of the Indian "women's rights"—engaged in collecting their supplies of this much valued Pomme Blanche.

SQUAWS COLLECTING THE PRAIRIE APPLE.

Our Native Frogs.

There is a strong resemblance between the Frogs proper, the Tree-toads, the Cricket-toads, and the true Toads. They are all remarkable in having their young strikingly unlike the adult animals. The young are hatched in a very incomplete state, and go through their developments in the water. At first they breathe by external gills, and have no organs of locomotion, except a tail. As the animal grows, limbs appear, the tail disappears, and the office at first performed by the gills is afterward assumed by the lungs. Some, such as the toad, when fully developed, leave the water, and only return to it at the breeding season; while the frogs are more or less aquatic. While in the young state these animals all feed upon vegetable food; but after they have reached their perfect development they are carnivorous. The frogs proper have a visible tympanum, or ear-drum; the upper jaw and palate are furnished with teeth, and in the male the throat has what are called "vocal vesicles," membranous appendages communicating internally with the mouth. These are capable of being enormously distended, and by the aid of them, the well-known piping and croaking are produced. The best known, as well as the largest of our species, is the Bull-frog, (*Rana Catesbiana.*) This sometimes reaches over 20 inches in length; and its deep bellowing may be heard for a long distance. The color of the upper portions varies from green to dirty olive, upon which are irregular black blotches; the body is yellowish-white below. The Bull-frog lives more generally in the water than our other species. It bites readily at a bait, if held near to its mouth, and even may be caught by a hook baited with a bit of cloth. This frog has the reputation of being destructive to young ducklings, seizing them from beneath the water. The hinder legs of the Bull-frog afford a highly esteemed article of food, occupying the place in the estimation of our gastronomers that the Edible-frog (*Rana esculenta*) does with those of Continental Europe. In the times when the English were strongly prejudiced against everything French, the French were looked upon as a nation of frog-eaters. The fact is that frogs are not, as is popularly supposed, a common article of food with the French, but are only used as a luxury. It may be that if the Edible-frog were found in England, the prejudice against frogs and those who ate them would not have been so strong. In this country much of this prejudice exists, and many who live where frogs are abundant deprive themselves of a most delicate article of food. A few years ago we met among the lakes of the New York wilderness, a person who annually caught barrels of frogs for the New York market, and who had never been induced to taste one. The hind-legs are the only portions used; these are skinned and fried in crumbs or made into a fricassee, in the same way as are spring chickens, which in delicacy and flavor they much resemble. The Green, or Bawling-frog, (*Rana clamitans*), is about 3½ inches long; it is green above and yellowish below. The Leopard-frog, (*R. halecina*), in some localities called the Shad-frog, is bright green above, with spots of dark brown, margined with yellow, and a yellow line extending from the eyes along the sides of the body. It displays great activity in leaping. The Pickerel-frog, (*R. palustris*), so called because it is often used as a bait for pickerel, is 2¾ inches long. It is yellowish-white below, and above it is marked with two longitudinal rows of square spots, which are dark brown upon a ground of pale brown. This species is shown in the engraving. The Wood-frog, (*R. sylvatica*), only visits the water at breeding time. It is found in thick woods, and leaps with great activity when disturbed. It is about 2 inches long, pale reddish-brown above, yellowish-white beneath, with a dark brown stripe upon each side of the head. The smallest specimen shown in the engraving, though it usually passes for a frog, is properly a Cricket-frog, and belongs to the genus *Hylodes* (*H. Pickeringii.*) It is found in woods and upon plants near the water. Its shrill note is familiar to all who live in the country. A related species is called the Savannah Cricket, which is often tamed and sings in confinement. It is said to soon learn to know its owner, and to accept flies from his hand. The related Tree-frogs, or Tree-toads, belong to the genus *Hyla.*

CRICKET-FROG. PICKEREL-FROG. BULL-FROG.

What Flowers will Grow in the Shade?

The question "What flowers will grow in the shade?" is put to me every spring by scores of city people, whose little patch which they wish to devote to flowers is so walled up by neighboring houses, that the direct rays of the sun, never touch it. But few plants will develope their flowers there, and none will do it as well as if it were lighted up by sunshine a part of the day. Fuchsias, Pansies, Forget-me-nots, Violets, Lobelias, Lily of the Valley, Phloxes, and other herbaceous plants whose native habitat is shady wood, will do best, but even these languish if denied all direct sunlight. The best effect in such situations is produced by ornamental-leaved plants, the beauty of which, is not dependent upon their flowers. Among these may be ranked the Gold and Silver Variegated leaved Geraniums, Achyranthes, Alternantheras, Begonias, Caladiums, Centaureas, Coleuses, etc., which, if planted so as to bring the various shades in contrast, produce a pleasing effect, which continues during the entire summer months, and is not surpassed by any display of flowers.

The cultivator of flowers in rooms should understand the necessity of sunlight to plants that are to flower, and endeavor to get these as close as possible to a window having an eastern or southern aspect. The higher the temperature, the more plants suffer for want of light. Many plants might remain in a temperature of 40 degrees, in a cellar for example, away from direct light, for months without material injury, while if the cellar contained a furnace keeping a temperature of 70 degrees, they would all die; such would particularly be the case with plants of a half hardy nature, such as monthly Roses, Carnations, Fuchsias, etc.

The Scarlet Crassula.

While we take a pleasure in welcoming all such new plants as the Bouvardia, described in another article, we have an equal satisfaction in calling attention to old and meritorious ones—excellent things which are thrust aside to make place for new comers, and with few to say a word in their favor. One of the first green-house plants that the writer learned the name of was the Scarlet Crassula (*Crassula coccinea*)—though the old gardener did call it with a very broad Scotch accent "*Cradjúly Cockseény*." Though for a long time a popular plant, and even now a leading one with the London flower dealers, we cannot learn that it has ever received a popular name. It is now placed by botanists in the genus *Rochea*, and its former botanical name, Crassula, may as well be adopted as the common one. It belongs to the same family with the Stone Crops (*Sedums*) and House-leeks (*Sempervivums*), and like them has fleshy leaves and a great amount of vitality. Its leaves are arranged in four rows upon the stem, which bears upon its summit a cluster of scarlet flowers of about the size of those in the engraving. It is propagated from cuttings which are laid aside to dry for a few days before they are potted—a precaution necessary with succulent plants to prevent them from decaying. The plants, when growing, need an abundant supply of water; but the pot should be so drained that none will remain stagnant about the roots. After flowering, the water should be gradually

THE SCARLET CRASSULA.—(*Rochea coccinea.*)

withheld, and the plants should have a season of rest and dryness. The Crassula is well suited to house culture, as it endures a dry atmosphere; but it is easily injured by frost. Two others belonging to the same genus, *Rochea falcata* and *R. perfoliata* are interesting green-house plants.

The Maze at Central Park.

In the gardening of a century or two ago, the Maze or Labyrinth was considered an essential appendage to grounds laid out in the then prevailing style. The Maze is a tortuous, intricate path, bordered on each side by a hedge. If ingeniously arranged it affords an amusing puzzle to reach the center, where there is usually a shaded seat or a fine view to reward the successful visitor. In the present style of landscape gardening the Maze is considered as too artificial, and it is mainly a thing of the past. At the New York Central Park one has been constructed by the efficient gardener, Mr. I. A. Pilat, and here, where the object is to furnish as much variety and amusement as possible, it comes in appropriately enough. The hedges may be of Norway Spruce; in England the Yew is a favorite plant for the purpose. The hedges are kept about five feet high, or they may be so tall that they cannot be seen over. In the labyrinth at the Park there is a commodious rustic shelter, A, at the center; at B, are rustic seats, and at C, circular seats around shade trees. The Maze is situated east of the old Croton reservoir, not far from 79th Street.

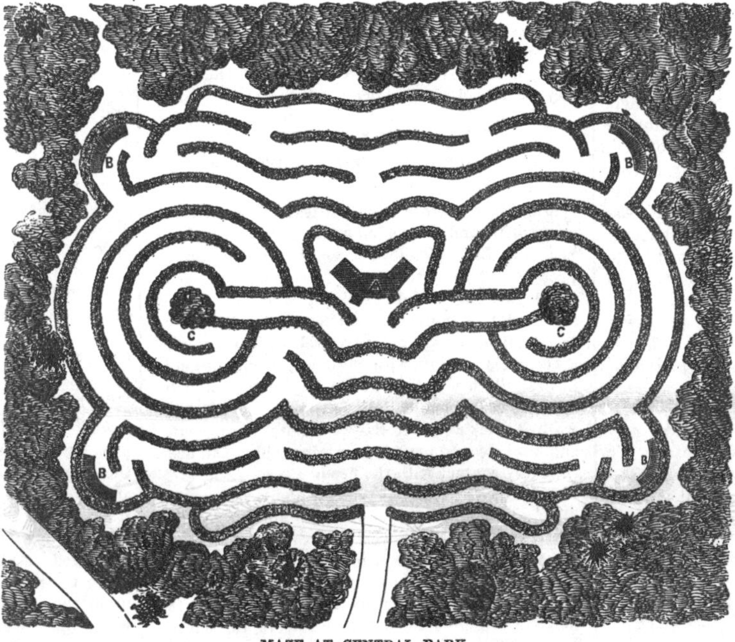

MAZE AT CENTRAL PARK.

Pruning Evergreens.—We are frequently in the receipt of letters asking how and when to prune evergreens; but each writer forgets the important point,—what he wishes to prune for, or whether he is growing a forest for spars, or a low hemlock hedge. Many evergreens, if they grow as single specimens, will take on a form which no art can improve. Specimen trees will need but little pruning; if the upper limbs are disposed to overhang the lower ones they should be shortened, to give those below them a fair share of light and air. This may be done while the young growth is yet tender, if a mere cutting back is all that is required. If limbs need to be taken out, it may be done late in autumn or early in spring. Hedges in their early growth are cut in autumn after the season's wood has ripened; but when the plants have become well established, and it is desired to check the growth, they are cut back in June while the young shoots are still immature. Pruning at this time has the same dwarfing effect that pinching has upon a deciduous tree. A very little trimming may afterwards be needed in early autumn to bring the hedge into the desired shape.

Seeds of Perennials.—One great obstacle to the cultivation of herbaceous perennial plants, is the difficulty with which the seeds of many germinate. A large proportion of them require to be sown as soon as ripe; they come up soon and make strong plants, which will bloom next year; while if kept until the following spring, they would remain a full year before they came up. Several of the Columbines behave in this manner. We notice in the seed catalogue of Mr. Ortgies, of Zurich, Switzerland, that those seeds requiring to be sown in summer are indicated. This is very useful information, which our dealers would do well to give in their catalogues also.

For other Household Items, see "Basket"

An Aquarium.

Having said something about the construction of an aquarium as a household ornament and a means of home instruction, a few hints as to its management are necessary. It was stated last month that most plants that grow partially or wholly submerged would do. Every pond or deep, quiet stream will furnish several. The one shown in figure 1 is the Mermaid-weed, (*Froserpinaca,*) which is frequently used, though it is not a true aquatic. It grows in shallow water, and its submerged leaves are divided, as shown in the engraving; but later in the season the upper portion grows above the surface, and the leaves then developed are so different in shape that they would not be recognized as belonging to the same plant. Another very common plant is the Water Starwort, (*Callitriche,*) figure 2. Its floating tuft of green leaves looks very pretty, but the portion below the surface is not so pleasing. One will soon find what plants flourish best and look best. When the aquarium has been in operation for some weeks, the glass will begin to look dim, and it will be found that a green film has covered it. This green film is a growth of microscopic plants, and while it in no wise interferes with the healthy condition of the tank, it is annoying, as it obstructs the view of what is going on within. Snails are useful in keeping this growth in check, and they also consume decaying vegetable matter. It is well to have a good supply of snails, as they require very little oxygen, and are excellent house-cleaners and scavengers. They crawl along the surface of the glass with their broad foot expanded, as shown in figure 3, and keeping in motion they usually destroy or prevent the troublesome green film. If the snails do not prove efficient, the glass may be occasionally cleaned by rubbing it with a swab. There are two or three species of snails to be found in streams and ponds. They will all answer in the tank; but if either is found to destroy particular plants, it should be rejected in favor of the others. Mussels, or fresh-water clams, are useful, as they strain out much impurity from the water. We have already cautioned against introducing too many fishes. Those that are kept, as well as other animals, will require feeding about twice a week. Small shreds of raw beef are generally used, and when gold-fish are kept, small pellets of bread may be given. Whatever food is used, all unconsumed fragments should be removed, otherwise they may produce trouble by their decay. A small net attached to a wire will be found convenient to remove objects from the tank; and a glass tube long enough to reach the bottom will be useful in removing unconsumed food, etc. Close one end of the tube tightly with the finger, immerse the other end of the tube and bring it directly over the object to be taken out; remove the finger from the upper end and the water will rush into the tube, carrying with it the piece of meat or bread; close the tube again with the finger and it may be lifted out with its contents. In a properly managed aquarium the water will remain for an indefinite time without the necessity for changing; but sometimes from neglect or accident it will become impure, and must be exchanged for a fresh supply.

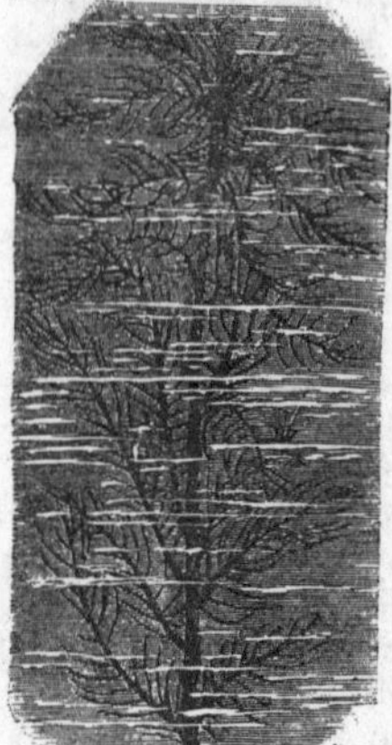
Fig. 1.—MERMAID-WEED.

Fig. 2. WATER STARWORT.

Fig. 3.—SNAIL.

Fig. 4.—SNAIL.

An Unpatented Clothes-Horse.

Many of the contrivances to hold clothes while they are drying or airing, are patented. We give one invented by Mr. Chas. F. Diebert, Schuylkill Co., Pa., that is not patented, and which is so simple that it can be readily made. The engraving shows a perspective view of the horse, which is 5 feet long and 4½ feet high. It requires in its construction neither nails nor screws, the whole being fastened together by wooden pins through the ends of the bars, which project through holes in the uprights. The bars are 1 inch square, but rounded at the ends which go through the uprights, which are ¾ inch by 2 inches. The end view is given in fig. 2; perpendicular dotted lines show the manner in which the clothes hang, and it will be seen that those upon one bar will not come in contact with those upon another. The cross-pieces at the ends are attached by one of their ends to one of the horizontal bars, and hook upon another bar by means of a notch near the opposite end. By unhooking the cross-pieces, as shown by the dotted lines, and removing the lower center bar, the horse can be folded up and occupy but a small space.

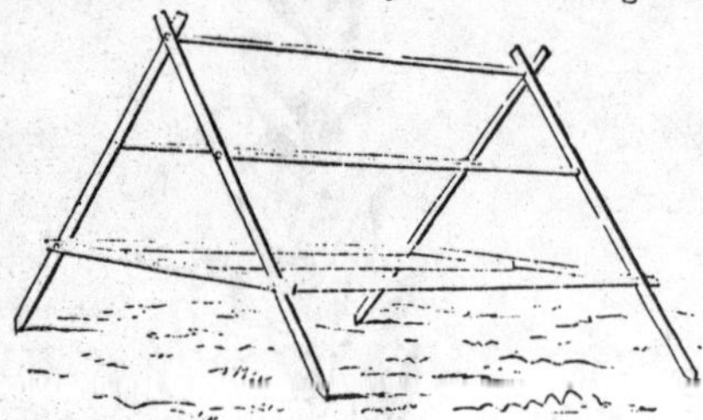
Fig. 1.—CLOTHES-HORSE EXTENDED.

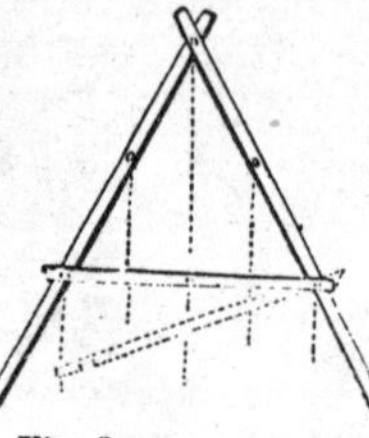
Fig. 2.—END VIEW.

Cleaning Coat Collars.

Mrs. C., Montrose, Md., writes: "For cleaning coat collars and all woolen goods I recommend the Soap-tree Bark (*Quillaya saponaria*) which can be procured at the drug stores. Break a piece about two inches square, into small bits, and pour over it a half pint of boiling water; let it stand an hour or two, then sponge the collar well with the liquor; a second sponging with clear water will clean it nicely. Both washing and rinsing water should be as warm as for flannel. We have by using this bark washed black and blue Empress cloths successfully and have cleaned hair cloth chairs, which had been soiled by contact with the head."

There are several vegetables which are in use in different countries as substitutes for soap. The natives on the North-west coast use a soap root; the Mexicans use one or more vegetables as soap, and the one refered to by Mrs. C., the Soap-tree bark, is largely employed by the Chilians. All these make a lather with water and serve to remove grease without injury to the fabric. The Soap-tree bark has been used to some extent in tooth washes and in preparation for cleansing the hair.

How to Cook Green Peas.

Pick the peas in the afternoon, let them stand until time to prepare next day's dinner, or, if not convenient to shell them, until the day after. Put the shelled peas into a large kettle of water with a piece of pork, and let them cook until they are boiled out of their skins.—That is how not to treat peas, and yet it is the style in which many treat one of the most delicate of vegetables. Every hour that passes between the picking and the cooking of peas is attended with a deterioration in their quality. Those who buy peas must submit to having them stale, but those who raise them have only themselves to blame if they do not enjoy them at their best. Peas should be boiled in just water enough to cover them. It should be salted and boiling when the peas are put in. Allow them to boil uncovered. The time required varies with the age of the peas, from 15 to 25 minutes. Peas that require longer boiling, or need the aid of a lump of soda to make them tender, are not to be considered as *green* peas. When the peas are done, skim them from the water or drain on a colander, and place them in a vegetable dish with a good lump of butter. The English frequently cook mint with their peas, and the French stew them with onions, butter, and parsley. These additions disguise the delicate flavor. Peas need only salt and butter; sweet cream is an acceptable substitute for butter.

Strawberry Short-cake. We believe that the Strawberry Short-cake is an American institution. It is in its season the popular dish at the restaurants, most of them announcing it by placards, and some of them advertising the "Original Strawberry Short-cake." We have always looked upon it as an ingenious device for spoiling strawberries; but as we are probably in the minority, we consult the wishes of the majority by giving the following recipe, which comes well recommended: "To 2 teacupfuls of sour milk add 1 teaspoonful of soda; when this is dissolved, add 1 cup of butter or lard, and flour enough to make a soft dough. Roll into thin cakes large enough to fill the pan in which they are to be baked. Dust a frying-pan with flour, place in the cake and bake over the fire, turning as soon as the underside is done. Split the cakes while hot, and butter well. Lay on a plate a half of the cake, put on it a layer of well-sugared strawberries, put another half of a cake, more strawberries, and so on until there are five or six layers, and serve."....Another says: "Mix the dough with buttermilk as for short-biscuit, roll so that it will be from 1 to 2 inches thick when baked, and bake in an oven. Mash the strawberries slightly and add sugar to your taste, and water, to make juice sufficient to moisten well the cake; split the cake while hot, butter each part well, and pour the strawberries, etc., upon the under half and cover with the upper."

Lyonnaise Potatoes.—Prof. Blot gives the following: "If you have not any cold potatoes, steam or boil some, let them cool, and peel and slice them. For about a quart of potatoes, put 2 ounces of butter in a frying-pan on the fire, and when melted put as much onion as you please, either sliced or chopped, into the pan, and fry it until about half done, when add the potatoes and again 2 ounces of butter; salt, pepper, and stir and toss gently until the potatoes are all fried of a fine, light-brown color."

Canned Rhubarb.—Rhubarb in tarts and pies, or eaten as a preserve, is highly esteemed in the spring. It fills, acceptably, the gap which occurs after apples are gone, and before fresh fruits come. Rhubarb can be enjoyed in winter by putting it up in cans, the same as peaches and other fruits are preserved. As soon as strawberries and other fruits make their appearance, there are few calls made upon the rhubarb patch, and much is allowed to go to waste which might be preserved for winter use. The stalks are prepared as for making pies, —stewed tender, sufficiently sweetened, and put in jars or bottles while hot, observing the same precautions as are necessary in canning fruit.

THE GREAT NORTHERN SHRIKE OR BUTCHER-BIRD—(*Collyrio borealis.*)—*Drawn by H. W. Herrick, and Engraved for the American Agriculturist.*

The Shrikes and the Fly-catchers form a small family, of birds, the members of which have a strong and compressed bill, which is abruptly hooked at the tip, and the upper and lower half both strongly notched. They are all carnivorous, feeding on insects, reptiles, and smaller birds. We have but four Shrikes in North America, and only one of them common in the Northern Atlantic States, where it is generally known as the Butcher-Bird. Its name of Great Northern Shrike would convey the idea of a large bird, but it is only a little larger than the other species, and at most only about 10 inches long, and of the form shown in the spirited drawing by Mr. Herrick. The upper parts of the head and body are of a bluish-ash color, the under parts white; the wings and tail are black. The bird is found from the Eastern United States to Vancouver Territory, building its nest in the smaller trees or tall shrubs. It is especially noticeable for its disagreeable qualities. It possesses a remarkable power of imitating the sounds of other birds in distress, and thus attracts the attention of smaller birds, which come to the aid of their supposed fellows in trouble and fall victims to the strong bill and sharp claws of the Butcher-Bird. It is said that it will even attack birds confined in cages, coming fearlessly near dwellings for the purpose. It is altogether a very quarrelsome and unamiable character. Not only does it destroy defenceless birds, but it seems to be proud of the achievement, and hangs them up as trophies. Instead of devouring its prey at once, it impales it upon a thorn or sharp twig. The slender thorns of the Three-thorned Acacia are favorites with the bird for this purpose, and it is not rare to find this tree decorated with the dead bodies of small birds, as well as those of different insects which have been suspended there by the Butcher-Bird. What object the bird can have in thus suspending its dead prey, seems to be a puzzle to naturalists. It probably has similar tastes to those gourmands, who prefer to have their game hang until it is on the verge of decay. A related species is common in Continental Europe, and has similar habits.

To prevent Birds pulling Corn.

There are numerous devices for keeping the birds away from corn, among which are scare-crows, wind-mills, bits of tin hung on strings, suspended twine, and powder and shot. We had faith in the stuffed man, until we saw the birds use him as a roost; in the twine, until we found corn pulled up right under the strings; and in shooting, until we found, we had to keep a man in the field all the while there was danger, in which case, the man would answer without the powder. Where the services of children are available, continual watching for two weeks, will save the corn. But this is not always the case, and the cornfield is often at a distance from the dwelling. Coating the seed in tar, is one of the best safeguards we have ever tried. The strong odor is offensive to the birds, and after a few trials of the article they quit in disgust. Caution however should be used in putting on the tar. The thinnest coating is just as effectual as one that will prevent germination. Stir in with the seed the smallest quantity of tar that will give a coating, and to prevent the adhesion of the kernels, stir in plaster, ashes, or dry earth. If the tarring has been neglected, and the corn is already up, it is a great safeguard, to sow corn broadcast at the rate of two quarts or more to the acre. The birds will take the corn that is most accessible, and leave that which is below the surface, or just sprouting. If one is not a disciple of Bergh, and wishes to multiply insects, let him poison the sowed corn. He can bag his game at leisure, but he would also be likely to bag less fruit and grain.

Saving labor, or what is equivalent, saving "steps," is an important point to be always kept in mind, in all plans for the arrangement of rooms, pantries, etc. A house-keeper writes us, that, from the suggestions thrown out in our articles on dwellings, she has been led to re-arrange the position of the sink, pump, table and pantries, in her kitchen and dining-room; and that by a simple arithmetical calculation she finds that during the last 12 years she has traveled 2,190 miles more than she will have to, during the next dozen years, with the new arrangements. It cost her about $50 to put things just where they ought to have been placed at first without any extra cost.

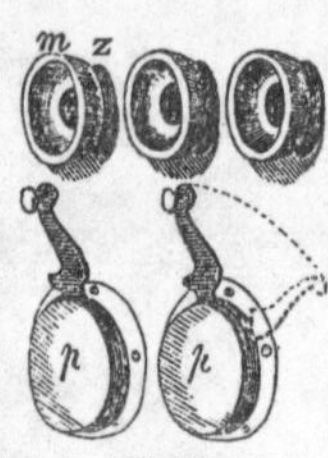

Fig. 1.

Bells.—These are very simple contrivances, easily provided at trifling cost, and they not only save steps but promote quietude. Very good bells are now sold at 20 to 25 cents each; a wire "carriage" and check spring cost 10 to 15 cents; a few cents' worth of No. 18 tinned wire, two or three triangles for changing the direction, costing 6 to 10 cents each, and a bell-pull costing anywhere from 25 cents for a slide, to a dollar or so for a japanned, bronzed, or plated crank with porcelain cap (*p*, **Fig. 1**,) are the requisites. Any mechanic, with common ingenuity, can hang one, though in a large house it is cheaper to employ a professional bell-hanger. The wires may run along the corners of a room, or behind the casing, or back of the lathing; but is far better when building a new house to use the zinc tubing prepared specially for the purpose, and costing only about 1¼ cents per foot. This is fastened outside the lathing and covered over with the plaster, with the end bent outward where the bell is to be hung. Bell boxes (**Fig. 2**,) are attached to the end where the pull is to be placed, and this is covered with the mortar, except at the opening for the pull. In this tubing the wire runs smoothly and with little wear, and is easily put through after a house is completed, if mortar has been kept out by the insertion of a wooden plug in the open ends. The bell boxes are kept by hardware dealers and sold for 6 to 8 cents each.

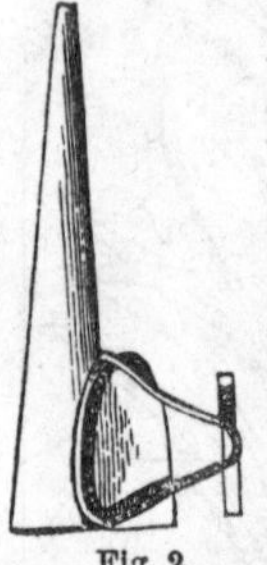
Fig. 2.

Speaking-Tubes.—Two persons standing at each end of a simple tin pipe, one inch in diameter, 50 to 100 feet or more long, with several elbows in it, and carried through half a dozen rooms, can still converse quite readily in a low voice. Such tubes may be carried between any two rooms in a house, however distant, and save a world of steps, and calling through the halls. (See pages 28 in our March number, and 88 in April number). A mouth-piece for inserting in each end, is shown in fig. 1. This may be of common tin plate, like the mouthpiece of a tin horn, but with the opening the same size as the tubing, or more elaborate. The three shown in fig. 1, have a porcelain rim, *m*, and a zinc shoulder, *z*. These retail at 30 to 40 cents each. Ready-made tin speaking-tube in 5 feet lengths, is retailed in the hardware stores for 4 cents a foot, and the elbows at 3 to 4 cents each. Fifty feet of tubing with two mouthpieces, and staple hooks for putting up the tubing, will cost only $2¼ to $3, and it can be put up by any expert carpenter. It is usually fastened firmly along the studding before lathing. It can be carried through any timbers in an auger-hole. An invalid unable to move upon a bed may have a flexible rubber tube extend from the tin pipe in the wall to the bed, with the mouth-piece in this, and thus be able to communicate with persons working in a distant room. Were we constructing a new house for ourselves, we should be tempted to run a speaking-tube from each to every other room. As they never wear out or need repairs, the annual interest on the cost of a tube 50 feet long would scarcely exceed 25 cents.

Fig. 3.

A Dumb-Waiter, is a great labor-saver, especially where the kitchen and dining-room are not on the same floor. It is simply a cupboard, or shelves, suspended on a cord passing over a pulley with a weight attached. **Fig. 3** shows a good form. Here the slide is suspended in the middle over a large wooden pully. It runs much better than the old-fashioned one with weights upon two sides. The weight is a piece of broad, flat cast-iron, running down upon one side. As the waiter does not need to rise more than 5 feet above the floor, there is room over it for the large pulley, or wooden wheel. A dumb-waiter may rise through the floor, its top being carpeted and forming part of the floor. In this case, two weights to slide down on two sides, are required, with the cords over small pulleys in the casing, and attached to the slide near the bottom.

Fig. 4.

Fig. 5.

Ventilators should be provided in every room of every house. We usually put in two, one (**Fig. 5**,) just over the baseboard; and another, (**Fig. 4**,) near the ceiling. The lower one is opened and closed by turning the ratchet-wheel, and the upper one by means of cords hanging from the two opposite sides. Where a cold room is being heated the upper ventilator is closed; the warm air rises to the ceiling and forces the cold air out through the lower ventilator. When the air becomes rarefied and impure, the upper ventilator will carry off that portion nearest the ceiling. Round, 5-inch ventilators, black or porcelain enameled, cost about $1.10 each, and the rectangular ones, 4×6 inches, cost about $1.60 each. An opening should of course be carried up through the wall from each ventilator or pair of them in one room, to the attic or to some point giving free exit to the air. Where a beam is in the way, it can be pierced with several small auger-holes, in sufficient number to allow free passage of the air. This can be done so as not to weaken the beam.

Fig. 6.

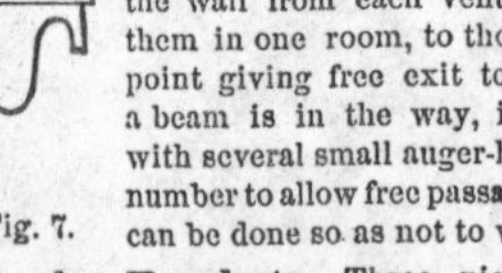
Fig. 7.

Cornice Brackets.—Three pieces are shown, copied from the houses described in March and April, which will serve as patterns. **Fig 6** is a portion of the cornice, and dentils under the eaves. **Fig. 7** is a neat form of a bracket for the head of a bay-window, or a small portico. **Fig. 8** is a tasteful pattern for brackets to be placed under and supporting a bay-window.

Fig. 8.

Egg-eating Dogs.—"W. C. G." of Boston suggests a specific for Egg-Eating Dogs viz: "Blow an ordinary hen's egg, expelling the entire contents, stop up one end of the shell with wax. Then fill it from the other end with strong spirits of ammonia, or 'Hartshorn.' Seal that end and then put it where the dog can get it. If he crushes it, he will never be desirous of repeating the luxury of egg-eating. After the dog has had one Ammonical feast, a little of the fluid poured into the nest, will remind him of the fact, that, he once was burnt, and also will serve to cleanse the nest from vermin."

Yellow-birds.—"M. B.," Williamsbridge, wishes to know how to keep yellow-birds out of his garden, as he finds them very mischievous. A small windmill, so arranged as to make a clatter, might answer. Who has had experience?

Raising Ducks.—Mrs. "S. A. P.," who had very poor luck with her ducks last year, asks how to raise them. We have had good success with the following method: The eggs are hatched under hens; when they come out, we confine them with the hen for

several days by placing boards to make a little yard around the hen-coop, as in the engraving. The coop is set on a board so that the hen cannot scratch, and water and food are placed where she can reach them. The ducks must always be called when fed; and at first allowed to get a little hungry before feeding, food should not be left before them. They will soon learn the call of the hen. When they do so and run quickly to her, and when they run to the coop for safety and when hungry, the boards are pulled up and the ducks let run. They have to be watched at first and driven back, when they wander far. Soon they may be trusted anywhere but in a brook or pond. If they get wet while in the down, and before two or three weeks old, they are in danger of being chilled. A heavy thunder shower often proves fatal from this cause, if they are caught by it far from home. After they are several weeks old they are in danger of being washed away and lost by the sudden rising of a brook or stream in which they may be allowed to forage. They need water only to drink, and to wash in, and a shallow trough or tub will hold enough. If rapid growth and large size are desired, ducks must be fed very frequently—as often as they will come and call for feed, is our rule—say once in three hours, and must have a range

Rock-Work.—"S.," Holly, N. J. Rock-work is best made of large stones laid up so as to represent a natural out-cropping of the rock. In arranging the stones, irregular spaces or "pockets" must be left to hold earth. Another way is to build up a structure of medium sized, irregular stones, using cement (water lime mortar) to hold them together; and afterwards paint the whole over with a wash of water lime to give it a uniform color.

Who Buys an American Watch?—Half a million watches have been made by the American Watch Co., at Waltham—No. 500,000 was finished a few weeks since.

Cranberry Culture.—A work upon the culture of the Cranberry, which should give the present approved methods of culture in a practical manner, has for a long time been needed. The publishers of the *American Agriculturist* are glad to announce a work of this kind, which is now in press and will shortly be ready. The author is J. J. White, Juliustown, N. J., a successful grower of cranberries on a large scale. There have been so many loose statements with regard to cranberry growing, and so much rash speculation in the business, that a well considered work like this will be timely.

The American Entomologist has changed its title to that of "Entomologist and Botanist." The botanical editor is Dr. George Vasey of Illinois, long known as one of the leading botanists of the West. Mr. Riley will continue to look after the "bugs." A popular botanical journal has long been wanted, and we shall be glad if the present one fill the gap. R. P. Shedley & Co., St. Louis Mo., Publishers, $2 per annum.

Salt and Lime Mixture.—Thos. Ford, Miss. Experiments indicate that there is a gain in mixing these substances, that is, in slaking quicklime with brine. The chemical reaction which takes place is not so definite as might be supposed. Lime is oxygen and calcium; salt is chlorine and sodium; an interchange of bases would convert lime and salt into chloride of calcium and soda. This action probably takes place, but to a very limited extent, depending upon the heat of the mass. The value of the article is probably chiefly due to the intimate mixture effected.

Laying Out a Vineyard.—"J. W.," St. Louis Co., Mo., finds laying out a vineyard by the use of a line to be too slow work, and asks if we can suggest a quicker one. By the use of flag stakes, a good plowman will run a very straight furrow. By running furrows in both directions, at the proper distances, the intersections would give the stations for the vines. We cannot say how much quicker this would be than the use of the line, but it would require fewer hands.

AMERICAN AGRICULTURIST

FOR THE

Farm, Garden, and Household.

"AGRICULTURE IS THE MOST HEALTHFUL, MOST USEFUL, AND MOST NOBLE EMPLOYMENT OF MAN."—WASHINGTON.

VOLUME XXIX.—No. 7. NEW YORK, JULY, 1870. NEW SERIES—No. 282.

"AMONG THE SAVAGES."—DRAWN BY WM. M. CARY.—*Engraved for the American Agriculturist.*

Wild horses are very clannish in their ways. If a strange horse enters a herd, he meets with a reception that is more demonstrative than friendly. The artist has represented a horse which has strayed from a train, and fallen in with his wild brethren. He fares as badly as would his owner, should he fall, defenceless, among a party of savages. The wild horse, or mustang, is the incarnation of viciousness. It roams the prairies with a certain rude grace, but submits sullenly to captivity, and though tough and serviceable, is seldom to be trusted. Few, who have not seen them, can have an idea of the immense numbers of wild horses upon the plains of Texas. The writer has seen them by thousands, in every direction, as far as the eye could reach. In some places they were annoying in their approaches, sweeping by the camp at full speed, and making imminent a stampede of the mules and horses. It was sometimes necessary to fire upon the animals to keep them away from the camp. Upon one occasion the train, while moving, was in great danger from the wild herds. Mules are very timid animals, and when once frightened, they become uncontrollable. The wild horses were so numerous, and dashed by the train with such a noise, that the drivers could not control their mules, and it was necessary to stop and secure them. As it was, one team of six mules escaped, and was only recovered after a long chase across the prairies. The direction of the wild herd was changed by a few shots, and after some delay the journey was resumed. The wild horses, now so numerous, are descended from domesticated animals, and are frequently made captives by the Mexicans and Indians, who catch them with the lasso. The murderous Mexican bit soon brings him under the control of a master.

MARK THE FACTS

Concerning

18,000 Cook's Evaporators.

Victor Cane Mills. 11,000

So many are in use all warranted and none returned.

THE FIRST PREMIUM

Has been given to

COOK'S EVAPORATOR

at

65 STATE FAIRS.

To the Victor Cane Mill, (since 1863,) at

42 STATE FAIRS.

TO BOTH

At the Louisiana State Fair of 1868, for working Southern Cane. All rivals fail to equal these machines on trial.

Farmers can't afford to risk crops of Cane on Mills that *break* or *choke*, or Evaporators that do *second-class work* and *only half enough* at that.

"While scores of new-fangled inventions have come up, had their day, and subsided, the "Cook" goes right along, constantly increasing in reputation."—*Prairie Farmer.*

Sorgho Hand-Book and Descriptive Circulars sent free.

For Farm and Neighborhood Use.

BUCKEYE

Thrasher and Cleaner.

This Machine is every way superior to anything of the kind ever introduced. It is made of the best material, is compact and handsomely finished, and does its work as well or better than the best of the large Thrashers. With our new PLANET LEVER POWER (4 or 6 horse) and four horses, it will thrash from 100 to 200 bushels of wheat, and from 200 to 400 bushels of oats per day; separating thoroughly, and delivering the grain fit for market and the straw in good condition.

The BUCKEYE is easily handled by the ordinary force of a farm, and is afforded at a price within reach of any thriving farmer. Send for Descriptive Circular.

BLYMYER, NORTON & CO., Cincinnati, O.; BLYMYER, DAY & CO., Mansfield, O.; BLYMYER, FEARING & Co., Chicago, Ill.

"Honor to whom Honor is Due."

TWO GOLD MEDALS AWARDED ONE MACHINE.

Harder's Premium Railway Horse Power and Combined Thresher and Cleaner, at the Great National Trial, at Auburn, July, 1866, for "*Slow and easy movement of horses, 1½ rods less than 1¾ miles per hour. Mechanical Construction of the very best kind, thorough and conscientious workmanship and material in every place, nothing slighted, excellent work, &c.,*" as shown by official Report of Judges. Threshers, Separators, Fanning Mills, Wood Saws, Seed Sowers and Planters, all of the best in Market. Catalogue with price, full information, and Judges Report of Auburn Trial sent free. Address

MINARD HARDER,
Cobleskill, Schoharie Co., N. Y.

The Victor Grain Drill,

a GREAT success; has positive FORCE FEED, light draft, no weight on horses' necks, can't choke. Best and handsomest. Sows all kinds of grain; and lime or plaster.

BLYMYER, DAY & CO., Mansfield, O.; BLYMYER, NORTON & Co., Cincinnati, O.; BLYMYER, FEARING & Co., Chicago.

HARRISBURG

PATENT SCREW POST

MANUFACTURING COMPANY.

Chartered by a Special Act of the Legislature.

PATENTS ISSUED FEBRUARY 16th, and AUGUST 24th, 1869.

W. O. HICKOK, Prest. EUGENE SNYDER, Treas.

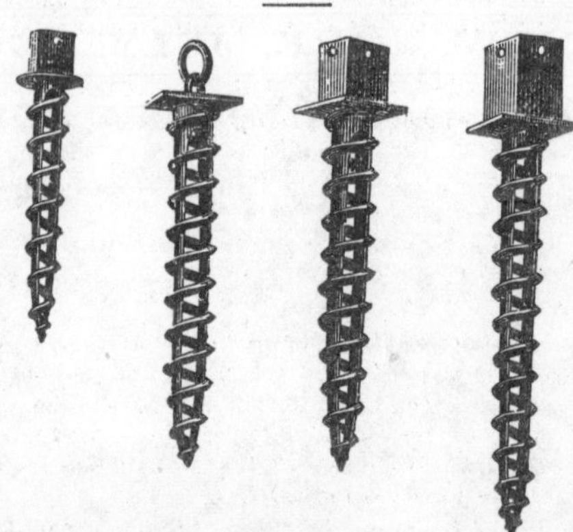

This new invention, the patent metallic Screw Post But, saves all digging in putting down posts for every purpose, and at same time it makes the only cheap and durable post but in use. It may be made of cast or malleable iron, being a hollow, gimlet-pointed screw with socket for the body of the post; is screwed directly into the ground by using a wrench; is suitable for all kinds of fences, grape arbors, vineyards, awning and hitching posts, trellises, seats for yards and gardens, &c., and wherever introduced takes the place of the old-style, wooden post. The body of the post, fitted into the socket may be either wood or wrought-iron or the entire post may be cast-iron. It has been thoroughly tested and is now in use in this city for various purposes, giving entire satisfaction for strength, durability and convenience.

Reliable parties wanted in all parts of the country to manufacture on royalty. Send for circular to No. 411 Market-street, Harrisburg, Pa.

JAMES R. PIPER, Sec'y.

COOPER'S STEAM ENGINE WORKS.

To keep pace with the growing demand for our Machinery, we are adding **$40,000** worth of new and improved Tools and Buildings to our present large Factories, and will continue to supply the following articles, after the best designs, on terms which will be found to be specially advantageous:

STATIONARY STEAM ENGINES,

For MILLS, FACTORIES, SHOPS, FURNACES, MINING, &c., of every required size, divided into three classes.

1. WITH SINGLE SLIDE VALVE, *Cutting off at two-thirds of Stroke by lap.*

2. WITH CUT-OFF VALVE, arranged so as to close at any part of Stroke and adjustable by hand-lever while engine is in motion.

3. WITH BABCOCK & WILCOX PATENT *Variable Cut-off, automatically adjusted by Governor.*

PORTABLE STEAM ENGINES,

Of 8, 10, 15, 20 and 25 Horse Power, combining all the improvements of the Slide Valve Engine. This is the only portable Engine and Boiler that has a *Combined Heater and Lime Catcher.*

Babcock & Wilcox Patent

Non-Explosive Tubulous Steam Boilers.

Grist Mill Machinery and Mills

Of any required size, with correct working drawings.

$1,500 WILL PURCHASE

A FIRST-CLASS COMPLETED TWO RUN GRIST-MILL, with erecting plans, &c., and guaranteed to give satisfaction and to be unsurpassed.

CIRCULAR SAW-MILLS

Improved in construction and combining all modern improvements.

☞ MACHINERY DELIVERED at New York, Philadelphia, Baltimore, Chicago, Saint Louis, or New Orleans.

☞ Full particulars and circulars on application. Address in full,

JOHN COOPER & CO.,
Mount Vernon, Ohio.

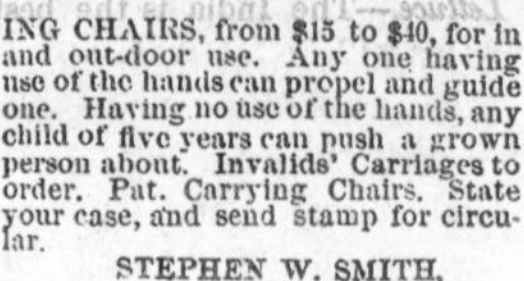

INVALIDS' TRAVELING CHAIRS, from $15 to $40, for in and out-door use. Any one having use of the hands can propel and guide one. Having no use of the hands, any child of five years can push a grown person about. Invalids' Carriages to order. Pat. Carrying Chairs. State your case, and send stamp for circular.

STEPHEN W. SMITH,
No. 90 William Street,
New York.

Over 2,000 PAIRS of CHAMPIONS

Sent to FARMERS last month, and we have received numerous testimonials of their adaptability to the FARMERS' WANTS.

NO HARD SEAMS TO HURT THE FEET. NO BOTHER WITH ROTTEN AND BROKEN SHOE-STRINGS. NO HOT AND CUMBERSOME BOOT LEG.

THOUSANDS of FARMERS wear HOT and HEAVY BOOTS all through the summer and fall, because they can not get a shoe to keep out the dirt. They have it in the "CHAMPIONS."

(Patented May 29th, 1866.)

NEW YORK, May 10, 1870.

We have examined carefully "BALLARD'S CHAMPION SHOE," and, without practical use, should judge it to be a capital article. The peculiar cut gives it the set and bearings of a boot, with the ease and lightness of a shoe; and the strap brings it as closely as desired around the ankle without the trouble of strings. Those which we have seen are of good stock and well made. ORANGE JUDD & CO.

DON'T FORGET THIS!

We warrant and guarantee the "CHAMPION SHOE" to wear longer and give better satisfaction than any other shoe you can buy. We make no cheap, shoddy, or inferior "CHAMPIONS," but every pair is warranted to be GOOD, HONEST, and SERVICEABLE in every particular.

FOR A PLOW-SHOE THEY ARE EXCELLENT. FOR BOYS' WEAR THEY ARE SUPERIOR. FOR THE HARVEST FIELD THEY ARE INVALUABLE, AS THEY KEEP OUT ALL DIRT, STUBBLE, AND THISTLE, AND YOUR FEET ARE EASY AND WELL AT NIGHT, NOT TIRED AND SORE.

It would be better for 10 or 12 to club together and send us *Registered Letter* or *Money Order.* It will be a saving in Express charges; and it is easier for us to send 12 pairs to one address than in single pairs.

INDUCE YOUR SHOE DEALERS TO KEEP A SUPPLY FOR YOUR NEIGHBORHOOD.

Address **A. BALLARD & SON,**
32 and 34 Vesey Street,
P. O. Box 5,000, New York.

THE JOHNSTON

Self-raking Reaper

is the MOST POPULAR MACHINE of the kind now before the public; it has more practical points that universally adapt it to the work of reaping grain than all others put together. It cuts

ANY GRAIN THAT GROWS,

whether high or low, thin or thick, lodged or standing, in the MOST SATISFACTORY MANNER; it is easily handled by an ordinary pair of horses, to cut from 15 to 20 acres between sunrise and sunset. It swaths, or lays the grain in gavels out of the way of returning team; is worked by almost any boy or ordinary field hand, or automatically, as you prefer. For further particulars send for Descriptive Pamphlets to the manufacturers,

C. C. BRADLEY & SON, Syracuse, N. Y.

H. KILLAM & CO.,

Chestnut St., New Haven, Conn.

We manufacture the finest class of carriages for city use, consisting of Landaus, Landaulettes, Clarences, Coaches, Coupes, Coupelettes, Barouches, Bretts and Phætons. Which we warrant equal in point of style, finish and durability to any built in this country.

Messrs. DEMAREST & WOODRUFF, 628 Broadway, are our Agents in New York City.

MONEY EASILY MADE

With our Stencil and Key Check Outfit.
Circulars Free.

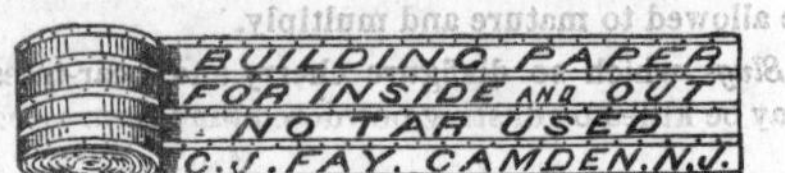

Improved Foot Lathes,

With Slide, Rest, and Fittings. Elegant, durable, cheap and portable. Just the thing for the Artisan or Amateur Turner. Send for descriptive circular.

N. H. BALDWIN, Laconia, N. H.

AMERICAN AGRICULTURIST.

NEW YORK, JULY, 1870.

July is the month which impresses farmers most with the desirableness of knowing how to forecast the weather by premonitory indications. No doubt could one "discern' the face of the sky" with anything like accuracy, injury to our grain or grass crop would be prevented, which might amount to thousands, yes, millions of dollars. The duty of the Government in this respect is clear—so far at least as to make use of telegraphic communication between the points of observation which exist all over the country, to convey to the community intelligence of approaching storms. It is not probable that thunder-showers which come up in an afternoon, and clear off before sundown, will be so foretold, that we shall not be in danger of getting our hay wet; but the great storms which do the real damage during haying and harvest, will be thus made known, and we hope the time may not be far distant.

Hints About Work.

Field Work for man and beast from early morning until late evening, leaves the farmer little time for anything else, so long as the weather remains fair. All the hands that can be got may now be employed, and it will pay to get only good men.

[Wor]king-Men need to be kept to regular hours of labor, eating, and rest. A little relaxation now and then is a grand thing too. Suppose all hands knock off work an hour or two earlier than usual of a Saturday afternoon and go to the brook or saw-mill pond for a frolic and a swim—depend upon it they will feel better and work better all the coming week. Good, satisfactory food is essential to having faithful work done.

Working-Animals should be very regularly fed and watered, and never over driven if it can be avoided.—Horses must be kept out of drafts, and not watered nor fed grass or grain until cool. Oxen when weary should have food and rest—be unyoked and allowed to lie down if they will. Many oxen will not willingly lie down in the yoke.

Orchard and Nursery.

Thinning has already been advised, and we refer to it again, as few have courage to thin sufficiently at the first going over. Fruit is likely to be very abundant, and it is all the more desirable to obtain large and fine specimens. When fruit is plenty, a poor lot is hard to dispose of at any price.

Grafts.—If they have grown vigorously, there is danger of their being broken by winds. Pinch the more rampant shoots.

Pinching and *Pruning* should continue; the first to regulate growth; and pruning is done in anticipation by removing shoots, that have started where they are not needed, while yet young.

Budding usually commences this month, but the condition of the stock is much influenced by the season. Whenever well-formed buds are to be had and the bark of the stock will lift, it may be done.

Insects.—We have in previous months noted the principal ones. The late broods of caterpillars will need watching. Borers have generally laid their eggs. Should the grubs have penetrated the tree, dig them out as soon as discovered.

Black Knot.—Cut it out at the first appearance and burn. No doubt insects do find a lodgment in it, as they will in any soft part of the tree, but they are not the cause. It is a fungus which should not be allowed to mature and multiply.

Slugs, which so disfigure cherry and pear-trees, may be killed by dusting powdered lime from a bag.

Seeds.—Collect cherry-stones and mix with sand to prevent them from becoming too dry.

Fruit Garden.

Grape-Vines.—Keep young growth well tied up, and rub out any superfluous shoot that may start. If mildew makes its appearance, use sulphur freely. Bellows for applying it may be had at the implement stores. Young vines should not be allowed to overbear; a bunch to the shoot is all that a vine should bear at its first fruiting. Beetles and caterpillars will need hand-picking. Layers may be made of the present season's growth, by bending a shoot down and burying a portion of its length.—The extreme end of the layered shoot should be tied to a stake in an erect position.

Currants and Gooseberries.—If the useless shoots are removed as they start, but little pruning will be required in autumn.

Strawberries.—To those who would make new beds, we cannot too strongly commend the method of striking runners in pots described on page 71 The plants, as soon as rooted, may be set out, and will bear a good crop next spring.

Raspberries.—Remove the old canes as soon as the fruit is off, and hoe off all suckers not needed to grow canes for next year.

Blackberries.—The new growth, which is to bear the fruit next year, is to be kept in proper shape by pinching. The canes should not be allowed to grow more than 5 feet high, and the side shoots, which they will throw out, are to be pinched when a foot or 18 inches long.

Dwarf Trees are to be kept in the shape of bushes, pyramids, or whatever style of training may be adopted, by pinching. If a shoot is disposed to grow too long, pinch the end, and if one does not grow as desired, there is probably a more vigorous one robbing it, which should be checked by the same means. What was said about thinning in the orchard is emphatically applicable to dwarf trees. They are particularly apt to overbear, and the best fruit is only to be had by care in thinning.

Kitchen Garden.

The soil should be constantly occupied by some crop. It is very poor gardening to allow a piece, after an early crop is off, to grow up with weeds. As soon as one thing is off, spade or plow up and put in a succession crop.

Beans.—The Limas, when they have reached the tops of the poles, may be pinched. Bush sorts may still be planted.

Beets.—If sown as late as the present month will, in ordinary seasons, make a crop. Thin early.

Cabbages and Cauliflowers.—Transplant the late sorts from the seed-bed. It is well to assort the plants and use only the most vigorous, rejecting any which show malformation of the root. Slugs are often very troublesome, and may be caught by laying cabbage leaves on the ground; they will hide under the leaves, and may be destroyed. Ducks will destroy them. These plants are much benefited by frequent hoeing.

Carrots are to be kept clear of weeds until the tops prevent working between the rows.

Corn.—The early sorts may be sown this month, and give a late picking.

Cucumbers.—Sow for pickles in well-manured soil. For seed, select the earliest and finest shaped.

Egg-Plants need all possible urging. Hoe frequently and give liquid manure. Place straw or hay under the fruit, to keep it from the ground.

Endive may be sown for late salad.

Herbs.—These are best grown as a second crop, transplanting from the seed-bed to occupy ground from which early things have been taken. They grow better late in the season than if put out early.

Leeks may be transplanted to rich soil, placing the rows a foot apart, the plants 6 inches in the row.

Lettuce.—The India is the best for hot weather, and this will do better if shaded a part of the day.

Melons.—Cultivate the ground as long as it can be done without injury to the vines. Remove all fruit that is not likely to ripen.

Onions need to be kept free from weeds. Those who live near cities usually bunch their onions and market the crop before it is ripe.

Potatoes.—Only the early sorts find a place in the garden, and these are usually dug while the tops are yet green. We open a trench and bury the tops, which decompose rapidly, and make an excellent manure for whatever crop may follow.

Rhubarb should have a rest as soon as fruit becomes plenty. Keep the flower-stalks cut off, as they exhaust the roots.

Sweet Potatoes, whether on ridges or in hills, should be kept clean until the vines cover the ground. The vines should not be allowed to root.

Squashes.—Keep the ground clean until the vines take possession of it, and then allow them to root at the joints. Look out for insects.

Tomatoes.—If training is followed, keep the plants tied up to the stake or trellis. They will bear cutting, and fruit all the better for having weak shoots cut out. If nothing better can be done, lay down some brush to keep the fruit from the ground. The "worm," as the large green caterpillar is popularly called, must be removed as soon as discovered, as it spares neither vines nor young fruit.—The notion that it is poisonous is an error.

Weeds are easily kept down if taken early enough. In many soils a sharp rake is the best implement to destroy them with. For more stubborn ones, the hoe-fork is preferable to the common hoe.

Flower Garden and Lawn.

Keeping is one of the chief things to attend to this month. Neatness in the borders, on the lawn, and in the paths, will make a small garden more pleasing than a large one badly kept.

The Lawn will require a weekly mowing. There are several excellent machines at moderate prices. The most of these scatter the grass, which wilts in a few hours, and is not noticeable. It serves as a mulch to the roots, and by its gradual decay adds vegetable matter to the soil. Annual weeds give but little trouble when the grass is mown frequently. The perennial ones should be pulled up or cut well below the surface by means of a knife, or a "spud," which is a sort of chisel with a long handle.

Margins of the turf where it borders upon a bed or a walk should be kept well defined, and no roots allowed to spread.

Climbers.—The new growth will often need directing. Do not allow the new shoots to get mixed inextricably with the old, as is apt to be the case with climbing roses and some others.

Dahlias will need tying to strong stakes, and the sluggish ones may be encouraged by liquid manure.

Bedding Plants, where planted in ribbons, need care to keep them effective. If those in front are inclined to outgrow the others, cut them back; and if the different lines intermix, cut out the straggling branches and keep the colors distinct. Coleus does all the better if cut back and made to grow dense. Some of the Cinerarias and other plants used for their gray foliage, will run up to bloom, and their buds must be pinched off.

Bulbs.—Take up when the foliage begins to turn yellow, and place them under shelter to ripen off; after which pack them away in a cool dry place until time to plant.

Seeds.—Secure as fast as they ripen. Some open their seed-vessels suddenly and scatter the contents; such should be gathered just before this occurs.—Seeds of perennials should be sown as soon as ripe.

Green-house and Window Plants.

There is not much to do to plants out of doors except to prevent them from suffering from dryness....Shade is necessary for Camellias and other plants of similar foliage....Look over the plants occasionally, and see that insects do not make a lodgment upon them....Pinch in unruly growth. ....Hanging baskets must have plenty of water in these hot days....Fuchsias in pots will do better in a partial shade....When plants are kept in the house the glass will need shading; whiting and skim-milk is the material usually applied to the glass....Make repairs while there is leisure.

A House Costing $1,300 to $1,900.

In accordance with our previous purpose, as well as in response to a great number of requests, we present well studied plans for a low-priced house. These were drawn by Mr. S. B. Reed, Architect and builder at West Flushing, L. I., with several modifications suggested by Mr. Judd, who will, by way of practically testing them, erect a couple of these houses in Flushing during this summer. We give the wide range of $600, for the cost of the same house. Where materials and labor are comparatively cheap, and little attention is given to ornament, inside or out, a house of this plan and size may be erected for $1,200 to $1,400. With labor and materials at the prices named below, the cost, exclusive of land, would run all the way from $1,300 to $1,900, or even more—according as one, or more, or all of the following things be provided. (1) Superior brick cellar wall, 7 to 8 feet high.—(2) The hall and one *cellar* finished off with floor, lathing and plastering the walls and ceilings.—(3) Use of the best "Novelty" siding, of good pine 10-inch boards, 1 inch thick, with groove in the center of each board, the same as shown in fig. 6, page 29 of our March number.—(4) More brackets on the cornice, and dentals added; and heavier, more ornamental window caps, than are shown in the engravings.—(5) Fine moldings in the lower rooms, and some fair moldings, instead of plain casings, in the second story.—(6) Thick plaster with fine hard finish and heavy cornice moldings, and center pieces in the ceilings, in the first story.—(7) All the walls filled in with brick and mortar, or back plastered, that is, put lath and plaster midway between the siding and

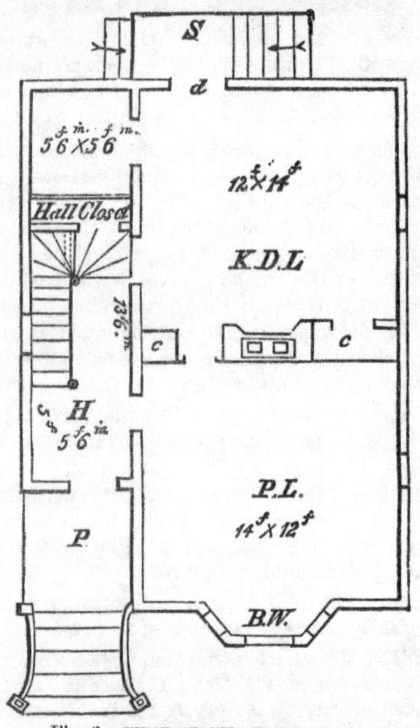

Fig. 3.—FIRST STORY—FLOOR PLAN.

inside lath and plastering.—(8) Water Tank over the hall stairs in the upper story.—(9) Range with water-back and upper boiler in *K D L*, fig. 3.—(10) Stationary Wash Tubs.—(11) Best quality of locks, with white porcelain knobs and plated shanks on all doors.—The addition of all these, or part of them, will add to the cost. After careful estimates of every part, in accordance with the below named prices, Mr. Judd is arranging to put up two houses in fair, comfortable style and convenience, stopping at just about $1,500 for the total cost, exclusive of land, but including cistern, fencing, privy, and, perhaps, a grape arbor. The 10-inch cellar wall will be of brick, with an air-space in the middle. The cellar to be built so as to be conveniently finished off if ever desired; Novelty siding; tin roof; moldings and cornice in first story, with center pieces; hard finished walls, etc., etc.—in short, a neat house, but without the extras. As will be seen, further on, the plan of the house, small as it is, admits of modifications, easily made, that will fit it for the wants of a large family. This is a noteworthy feature of this house. Put up cheaply at first, it can be extended at trifling expense and with very little "tearing up." The above-ground position of the basement renders the rooms in this quite light and cheerful.

Description.—Fig. 1 shows the **Front Elevation.** This gives a general view. The basement should be more above ground. As formerly explained, we believe in setting the first floor well up, so that the occupants shall always be as far above the ground as may be. For a basement 7 feet high, 2 to 2½ foot in the ground is deep enough, and 4½ to 5 feet above the ground—except in very cold regions where it is necessary to bank up the entire walls to keep frost out. A 10-inch wall, built hollow, or with air-space, will withstand pretty severe cold weather. The whole roof should be a few inches higher

Fig. 1.—FRONT ELEVATION.

than shown, so as not to appear so close to the upper windows. The cornice should be a little heavier than shown. Make the brackets a little shorter and broader. It adds greatly to the good appearance of a house of this kind, to have the roof project at least 20 inches all around. The cost is but little more. We have seen many, otherwise fine, dwellings deprived of all beauty by "sticking on" a very narrow and light cornice.

Fig. 2—Cellar or Basement.—(7 feet high in the clear.) The plan shows a division that may be made; but when great economy is studied these interior divisions may be omitted—until wanted. One or more chimney flues should start in the cellar for stove pipes there, if ever needed. An outside door and steps are shown at *C*. If the future wants and circumstances of the family require, *B* may be finished off for a Kitchen, with a dumb-waiter through the floor. *A*, may be fitted up as a Dining-room. In this case a small Cellar might be excavated under the piazza. We believe Mr. Judd intends to make the Cellar of the full size, piazza included. It will be seen that no more wall, and but little more digging, will be required. The girder will need supporting with locust posts, or brick piers, if the cellar division be not made at first.

Fig. 3—First Story.—(9 feet high in the clear.) The general arrangement is shown by the engraving. The piazza, *P*, may be lessened, if desired, and more room thrown into the hall, *H*, and into the room *I*, in fig. 4. The parlor, (*PL*), used somewhat as a living or sitting room, is of convenient size. A mantle-piece may or may not be added. An economical arrangement is to lead the stove pipe from *KDL*, through the wall into a drum in *PL*, and then turn it back into the chimney. This will keep the front room warm. The drum should have a small fire chamber in it for using wood during some of the coldest days. The closets or pantries, *cc*, should be made pretty deep to give plenty of room in them.—*KDL* is used for a kitchen, a dining-room, and partly as a living room. The back stoop, *S*, may be a simple

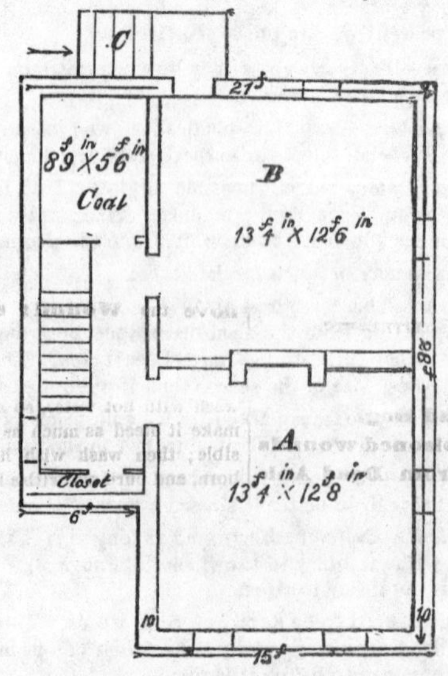

Fig. 2.—THE CELLAR OR BASEMENT—FLOOR PLAN.

platform with railing, or be covered in. *H*, may be lighted by a sash over the front door, and partly from the window over the open stairway in the second story. The window against the stairs at the left of *H* may be simply a blind one for the good appearance of the outside. A Pantry, 5½ feet square, is provided at the rear of the hall. The stairs to the cellar are in the hall *H*, under the other stairs. They should be wide enough for convenient descent to *A*, if that should ever be used as a Dining-room.

Fig. 4.—Second Story.—(8 feet high in the clear.) It will be seen that there are four sleeping rooms, *F*, *R*, *S*, and *I*, all large enough for a full-sized bed, except *I*, which will take in a ¾ bed. *F* and *R* may be heated by stoves, the pipes entering the chimney flues. *I*, may have a door also into *F*, if it is desired for a young or sick child. By having the door into the hall, *I*, may be kept as a "spare room," if the family require the other three sleeping rooms. We have slept comfortably in a "spare room," no larger than this, in many pretty well-to-do farmers' dwellings. Or *R*, or *S*, may either of them be used as the spare room. We don't believe in the frequent practice of living or sleeping *constantly* in a small room, and reserving the largest and best room for *occasional* visitors. That is sacrificing comfort and health to show. If friends, stopping for a night, or a few nights, cannot put up with such rooms and conveniences as may be readily furnished, their friendship is not worth a great deal. [And in this connection we will throw in our standing protest against the custom of giving up one's whole time to worry and excitement over getting up extra fine food for visitors—just as if *they* had no such things at home, and had come away to find "something good to eat." If *we* go to see our friends, it is to visit with *them*, and not with their larder. We do not want a hasty occasional word with them, while they are flying about to

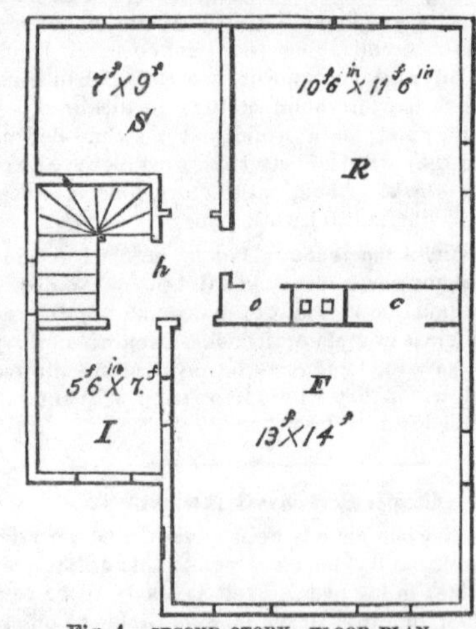

Fig. 4.—SECOND STORY—FLOOR PLAN.

feast our appetites. We prefer they should cook less than usual—give us plain, simple food, enough to satisfy the natural demands—and give us some of their *time* and attention for visiting. The venerable Bishop Hedding once called upon a good lady of our acquaintance awhile before noon. She was all excited, because no good preparations were made for dinner. "Sister," said he, "I have but three hours to stay, and I want to visit with you and your husband. You have good bread and milk: let us eat that, and *all* have the rest of the time to talk together"—and it was done. The writer happened to be one of the quartette, and we had a "feast of reason and flow of soul," such as a fashionable hotel table seldom supplies. The same principle applies to the general household arrangements for entertaining guests, especially where the house is small and means limited.]

Mr. Judd has another plan nearly perfected, for a house costing from $2,000 to $3,000, exclusive of land, which he is preparing to build the present season. It is designed to contain many conveniences, plumbing, etc. The plans of this will be described next month, probably.

Prices.—The following prices are used as the basis of calculation the present summer, here. The variations under the same items, refer to quality. Thus, while the best carpenters may receive $3 or more, per day, others less experienced, and less expert, will be dearer at $2 a day. There must be always some apprentices learning, at low wages, while they can do some kinds of work as well as full men. The same remark applies to other workmen.—*Wages* per day of 10 hours: Carpenters, $2.00@3.50; Masons, $3.50@$4.00; Laborers, $1.50@$1.75; Painters, $2.50@$3.50; Grainers, $4.00@$5.00; Roofers, $3.00@$4.00; Slaters, $3.50@$4.50.—Timber and Lumber per 1000 feet board measure: Timber, (beams, posts, etc.) $22.50; Siding, planed, $34@$40; Flooring, $32@$36; Joists, $18@$20; Clear Pine, $50@$55.

Poisons and their Antidotes.

BY OUR FAMILY PHYSICIAN.

[Even with the greatest carefulness, accidental poisoning will often occur. Children are much inclined to try bottles and parcels, before they are able to read the labels. Nearly every poison has its antidote, which will prove effective *if used at once;* loss of life often results from a delay of but a minute or two. A recent case of accidental poisoning, under our own observation, where a knowledge of chemistry, applied on the instant, averted much suffering, if not death itself, suggested the preparation of an article on this subject for the *American Agriculturist*, which should place against each poison some remedy available in almost every household, and Dr. M. R. Vedder, of Flushing, N. Y., has kindly prepared the following, and we have revised and arranged it with much care. Let it be carefully preserved where it can always be instantly referred to. Note well the closing remarks.]

Poison	Antidote
Arsenic. **Fly Powder,** ("Cobalt." **King's Yellow.** **Scheele's Green.** **Ratsbane.**	Stir 2 tablespoonfuls of ground **Mustard** in a quart of **lukewarm water,** and drink until copious vomiting is produced, tickling the throat with the finger or a feather. After vomiting, give large quantities of Calcined Magnesia.
LEAD POISONS. **Sugar of Lead.** **White Lead.** **Litharge.**	First, **Mustard** to vomit, as above, and doses of **Epsom Salts,** say a teaspoonful to a tablespoonful, according to the age of the patient, every half-hour for two hours.
MERCURIAL POISONS. **Corrosive Sublimate,** or (Bed-Bug Poison.) **White Precipitate** **Red Precipitate.** **Vermilion.**	**White of Eggs**—or Milk, or Wheat Flour—beaten up.—Administer all that can be got down in ten minutes, and then give mustard emetic as above.
COPPER POISONS. **Blue Vitriol.** **Verdigris.** **Food or Pickles Cooked** in **Copper or Brass Vessels.**	**White of Eggs,** or Milk taken very freely for ten minutes, to be followed with an emetic of Mustard as above.
IRON POISONS. **Copperas,** or **Green Vitriol.**	**Cooking Soda,** a teaspoonful to a tablespoonful, or more—according to age of patient, etc. — followed by plenty of Gum-Arabic water, or Flaxseed tea or Slippery-Elm tea.
ANTIMONY. **Tartar Emetic.**	**Powdered Nutgalls,** a teaspoonful or more in water; or tea of Oak bark or Peruvian bark. Give promptly.
SILVER. **Lunar Caustic.**	**Table Salt,** 2 teaspoonfuls or more, in a pint of water.
ZINC. **White Vitriol.**	**Warm water** to *relieve* vomiting; and 1 to 2 teaspoonfuls of **Baking Soda,** followed by Milk or White of Eggs.
PHOSPHORUS. **Matches.** **Rat Exterminator.**	**Mustard** and **Warm Water** to cause vomiting; then large draughts of water containing **Calcined Magnesia,** 2 tablespoonfuls to a pint, followed with **Flaxseed tea,** or Slippery-Elm tea.
ACIDS. **Acetic Acid.** **Citric Acid.** **Muriatic Acid.** **Tartaric Acid.**	**Baking Soda, or Saleratus, Lime,** or **Magnesia,** (a teaspoonful to a tablespoonful) dissolved in water and used freely. Powdered lime mortar from ceiling will do.
Oxalic Acid. **Nitric Acid.**	**Magnesia or White Chalk or Lime** stirred in water, drank freely and *quickly*.
Sulphuric Acid, (Oil of Vitriol.)	Drink much water quickly, and follow immediately with large doses of **Magnesia,** or Powdered White Chalk or Lime; or if these are not at hand, use Soda, or dissolved Soap. Follow with plenty of Flaxseed or Slippery-Elm tea.*
Prussic Acid. **Oil of Bitter Almonds.** **Laurel Water.**	Teaspoonful of **Hartshorn,** (aqua ammonia), in a pint of water. Drink the whole at once.
ALKALIES. **Pearlash.** **Salts of Tartar.** **Soap Lye.**	Drink freely of **Vinegar** in water, followed with Gum-Arabic dissolved in water, or Slippery-Elm tea, or Flaxseed tea.
Ammonia. **Hartshorn.** **Potash.** **Much Soda.**	Drink freely of **Vinegar** in water, or Lemon Juice, or Citric, or Tartaric Acid mixed with water. Sweet Oil, Castor Oil, Linseed Oil, or Cream, is also good, and should follow the other remedies above named—a tablespoonful first, and then a teaspoonful an hour for three hours.
ALCOHOL. **Any Spirituous Liquor.**	Two tablespoonfuls of **Mustard** in a quart of warm water. Drink till patient vomits freely, using a finger, or feather.
VOLATILE OILS. **Creasote.** **Carbolic Acid.** **Oil of Tar.** **Oil of Turpentine.** **Oil of Tobacco.** **Fusel-Oil.**	**White of Eggs,** or Milk, in quantity, followed quickly by a Mustard emetic.
MISCELLANEOUS. **Charcoal Fumes.** **Street Gas.**	**Fresh Air,** and artificial respiration.
Iodine.	**Starch,** or Wheat Flour beat up in water. Vomit with mustard and warm water.
Saltpetre. **Nitrate of Soda.**	**Mustard Emetic,** followed with Oil as for Ammonia.
VEGETABLE POISONS. **Strychnine.** **Nux-vomica.** **Opium.** **Laudanum.** **Paregoric.** **Morphine.** **Stramonium,** (or Thorn Apple, or Stink Weed). **Belladonna,** (or Deadly Night Shade.) **Croton Oil.** **Foxglove.** **Squirting Cucumber.** **Aconite,** (Monkshood). **Hemlock.** **Hyoscyamus or Henbane.**	**Emetic of Mustard** and warm water, as above; drink till patient vomits freely.—Tickle the throat with finger or a feather; or give a teaspoonful of powdered alum; or five grains of tartar emetic; or 20 grains (half a thimbleful,) of white vitriol, dissolved in half a tumbler of warm water, every ten minutes, till vomiting is produced. If the patient is drowsy, give the strongest cold coffee, or slap smartly on the back, and walk, or use electricity to keep him awake.
Arnica.	**Vinegar** and water.
Mushrooms.	**Emetic of Mustard** and warm water, until vomiting is produced; then frequent small doses of Epsom Salts.
Poisonous Fish.	**Emetic of Mustard** and warm water, tickling the throat.
BITES, ETC. **Serpents.** **Insects.** **Mad Dog.** **Poisoned wounds from Dead Animals.**	Tie a **String tightly above** the **Wound;** some one having no sores, broken skin, or exposed nerves in the mouth, suck out the blood, and wash with hot water, so as to make it bleed as much as possible; then wash with hartshorn, and burn out with a large red-hot wire or pointed Lunar Caustic; after this remove the string, & poultice with flaxseed.

We have endeavored to give those antidotes which are most likely to be at hand in every family, and can be readily administered by any one, but in all cases of poisoning, send for the family physician at once. It may be necessary to use the stomach-pump. The after treatment, to prevent or subdue inflammation, should be followed out with great care.

In poisoning by Arsenic, the *hydrated peroxide of iron* is an invaluable antidote, but as a physician would be required to administer this antidote, we have omitted it.

* [We have taken the liberty to change the usual directions for "Oil of Vitriol," which say, "use as little water or other liquids as possible." We should drink water very freely at once, to dilute the acid as much as possible. The heat produced by uniting sulphuric acid with water, will occur in any case, by the natural fluids in the stomach. No more heat will be produced by the large quantity of water, but the heat will be so diffused in the larger amount as to destroy its effects.—O. J.]

A ROLLER BARROW.—An implement in which the garden roller and wheel-barrow are combined, is in use in the public grounds and gardens in England. The engraving renders any description unnecessary. When used as a roller it may be brought to the desired weight by placing stones in the box, which is made of iron plate. It is emptied by lifting the handle and dumping the contents. It possesses advantages over a common wheel-barrow, as it is not, like that, liable to cut up the walks and lawns—a great annoyance, especially in wet weather. The

implement is found useful in carrying away cut grass and fallen leaves, and in bringing fertilizers on to lawns, or in general garden traffic.

Peach Crates.

Planters will now be making ready for the transportation of their crop to market, and we give the following timely directions for making a crate from Mr. Fulton's very thorough work on Peach Culture: "The standard dimensions of a crate are eight inches wide, fourteen deep, and twenty-three and a half long, outside measure. They are made of pine or other light wood. The ends and partition are sawed three-quarters of an inch thick, seven and a half wide, and fourteen long. The bottom and top twenty-three and a half long, six and a half wide, and three-eighths of an inch thick. The sides are composed of four slats, twenty-three and a half inches long, two and a half inches wide, and also three-eighths of an inch thick. Sometimes lighter stuff is used. The ends and partition are thicker, because to these all the other pieces are nailed. The whole crate consists of thirteen pieces. It is very simple in construction, and any intelligent hand, with a proper frame, can put it up without difficulty. The stuff is sometimes planed on the outside, which gives it a much neater appearance. Crates cost from thirteen to twenty dollars per 100.—They go with the peaches, and are never returned;

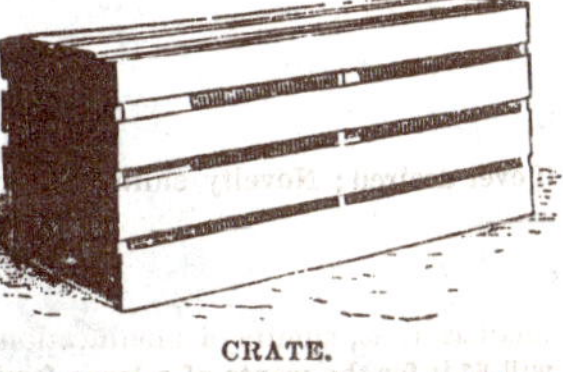

CRATE.

hence, those who ship in crates must provide as many crates as they have fruit to fill them. The reasons they are not returned, are two. First: they cannot be packed in each other like baskets, but occupy precisely as much room as when full. They are much more troublesome to handle, and the transporters will not return them free of charge as they do baskets. Secondly: they are usually reshipped or sent at once to a distant market, and sold with the peaches."

Our Common Tortoises or Turtles.

The structure of the turtles presents many interesting points. Externally they seem so unlike other vertebrate animals that it is difficult to see how they are classed with them. An inspection of the interior shows distinctly that the spinal column is present; the portions of it belonging to the neck and tail being movable, while the intermediate vertebræ, together with the ribs, are expanded and soldered together in such a way as to form the upper shell of the animal. The under shell, which is in part the breastbone or sternum, is joined to the upper at each side. Our streams and woods afford several species, the most common of which is the Spotted or Speckled Turtle (*Nanemys guttata*), which is readily distinguished by the yellow and orange dots upon its black shell. In the young the dots are upon the marginal plates only; but in the older ones the spots are on all parts of the back and present great variety in their size and distribution. The convexity of the shell varies considerably. It is probably the best known species, and may often be seen basking in the sun upon rocks or logs close to the water, from which they slip with remarkable rapidity when one approaches. They feed upon insects, frogs, and worms. When full grown the shell measures 4 or 5 inches in length. The young ones, an inch or two long, make amusing pets for the aquarium. Another very common species, and a most interesting one, is the Box Turtle (*Cistudo Virginica*), which is found upon dry land throughout the Atlantic States, and in some of the Western ones. The color is usually a dull brown, and marked with yellowish stripes and splashes; but some have been found entirely black. The plastron, or lower shell, is divided transversely into two unequal parts, which move upon the same axis and enable the animal to entirely conceal its head, tail, and legs. This peculiarity is recognized in the common name. This species is frequently kept in yards in a state of domestication, and it seems to be little disposed to wander. It conceals itself, by burrowing late in September, and reappears in spring. It feeds on insects and vegetables.

Muhlenberg's Tortoise (*Calemys Muhlenbergii*), found in New Jersey and Pennsylvania, and sparingly in New York, is a terrestrial species, 3½ or 4 inches long, with a marked keel or ridge upon its upper shell, and the plates of the shell neatly marked by ridges and furrows. It is readily distinguished by a large, orange blotch upon each side of its neck. In popular language these animals are all called turtles. Some writers restrict this name to those species which are found in the sea, and call all the others tortoises. The name tortoise is believed to be a corruption of the Spanish *tortuga*, and that of turtle, which originally belonged to the dove, is supposed to have been derisively applied by sailors to the sea-tortoise.

MUHLENBERG'S TORTOISE. SPOTTED TORTOISE. BOX TORTOISE.

THE VULTURINE GUINEA-FOWL.

The Vulturine Guinea-Fowl.

A bird which promises to be an acceptable addition to our list of domestic fowls has just now an interest for a large number of persons. We give an engraving of a species of Guinea-fowl, which has recently attracted the attention of the naturalists and poultry fanciers of England—the Vulturine Guinea-fowl. The specimen arrived at the London Zoological Gardens from Africa via the Suez Canal. It is said to be common upon the east coast of Africa near the Equator, and though used there as an article of food, it has heretofore only been known in England by a single museum specimen. It is described as "peculiar in having the head and greater part of the neck devoid of feathers, and without any caruncles, and in the possession of long ornamental hackles surrounding the base of the neck and breast." The description gives no idea of the size of the bird, nor does it state the color. We are informed that arrangements have been made for a supply of the birds, and before long we may be able to learn something more of its history and habits. Those who have eaten the wild bird in Africa, speak of its good qualities as a table fowl. If capable of domestication, it may prove to be valuable.

Curing Hay in the Cock.—We are not only in the habit of cutting our grass too late, but we give it too much sun, in our haste to finish the job in one day. This course may save labor, but it loses nourishment. It is not very much more laborious to cure hay in the cock. Cut the grass after the dew is off, and let it lie until the middle of the afternoon, and then bunch it in the usual way. Let it remain in this condition until the third day. Then shake up the cocks a little with a fork, bringing the bottom to the top, and let it remain until the next day. If it storms the hay will not harm much to lie over. If it is good weather, give it an hour or two of sun, and put in the barn. This course saves much of the labor of tedding and turning. There is but one raking, and the quality of the hay is very much better than the sun-dried and hastily cured article. All the sweetness and aroma of the grass are retained; and the cattle eat it with a higher relish, and thrive better upon it. If one is furnished with hay caps there is very little risk in curing hay in this way. The capped cocks might stay in the field a week without harm. A farmer may use his hay-tedder quite too much for the good of his fodder.

Half-hardy Passion-flowers.

The most beautiful species of Passion-flower (*Passiflora*) can only be grown in the green-house; but there are two which can, with a little care, be kept in the ground through the winter, and though not so showy as the more tender ones, are sufficiently beautiful and interesting to repay the little trouble they require. The species referred to are the Blue Passion-flower (*P. cœrulea*), from Brazil, and the Flesh-colored (*P. incarnata*), which is native from Virginia southward. These are vines, climbing twenty feet or more by means of tendrils, and producing lobed leaves and pretty flowers that are remarkable in structure. The one figured is the Blue Passion-flower, and shows the general characters of the flower in all the species. There are five petals and as many divisions of the calyx, which last are colored on the inside and give the flower the appearance of being composed of ten petals. Within the petals is a conspicuous crown of rays, or thread-like bodies, which adds much to the showiness of the flowers, as it is usually more strikingly colored than the petals. Then we have five stamens and one pistil, which bears three styles with large button-like stigmas at their extremities. The showy crown, which is composed of one or more rows of slender threads, has been a puzzle to botanists. Some regard the rays as modified stamens, and it is stated that anthers have been produced upon them. The fruit is a large berry with a tough rind, containing numerous seeds imbedded in a pulp. The fruit of many species is edible; that of our native Flesh-colored one is known as "Maypops," and that of a South-American species is the much esteemed Granadilla. The name Passion-flower was given by the early South-American missionaries who saw in different parts of the plant symbols of the crucifixion. The three styles were to them the three nails, two for the hands and one for the feet; the five stamens were the five wounds; the rays the crown of thorns; the ten parts of the calyx and corolla the Apostles; Peter who denied and Judas who betrayed, being absent; the tendrils represented the scourges, and the lobed leaves the hands of the persecutors. It does not need any such exercise of the imagination to make the Passion-flower beautiful; and it deserves to be classed with our most desirable climbers. Plants of the Blue species are to be had of the florists; it may be kept as a green-house plant, or, planted out to ornament a veranda or trellis. The stem is killed by the frost, but if the root has a good covering of leaves over it, it may be preserved through the winter. In warmer countries it is hardy, and forms a trunk as large as one's wrist. The native species is rarely seen in cultivation, probably for the reason that it, like many other native plants, is more difficult to procure than the exotic ones. It is, however, well worthy of a place in the garden. Another native species, found as far north as Pennsylvania, is the Yellow Passion-flower (*Passiflora lutea*), which has greenish-yellow flowers about an inch broad.

BLUE PASSION-FLOWER—(*Passiflora cœrulea.*)

Pot-layered Strawberries.

The propagation of Strawberries by striking their runners in pots has already been recommended in these columns, but another season's experience has so impressed us with the superiority of this method, that we desire to call attention to it with more emphasis. It is not a process that commends itself to those who grow strawberries by the acre, but for the amateur it is of the greatest convenience; as it not only renders him independent of the season, but saves him a year's time. Let us suppose a runner roots at the present time in the soil of the bed. If taken up and planted out in the fall, it may bear a few berries next year, but not a full crop until the spring of 1872. If put out in the spring of '71, it will not bear until the spring following. If the same layer be rooted in a pot, and as soon as it is fairly established, it is turned out of the pot, say in July, it will make a strong plant before winter, and bear well the next season. One great advantage in using pot-layers is, that the roots are undisturbed, and the planting may be done in the hottest weather.—The operation of pot-layering is a very simple one. Prepare a compost of good garden soil, and well-decomposed manure, letting it be light, rich, and fine. With this fill small pots; some use them as large as 4 inches across, but the size known in the trade as verbena pots are sufficiently large. These may be had at the potteries at about a cent a piece, if many are taken. The pots of compost are sunk in the soil of the bed under a runner, before it has taken root and become fixed in the bed. The pots are plunged quite down to their rims. It will be found necessary to lay a small stone or clod on the runner to keep the wind from moving it out of place, or to fix the runner by pressing it into the soil of the pot.—Roots will be formed rapidly in the rich compost in the pots, and when the plant has become well rooted, it may be separated from the mother plant, and set in the bed where it is to fruit. The illustration shows the manner in which the pots are placed. After a runner takes root, it will throw off another, and this another, and so on. When but few plants are wanted, it is well to pinch off the secondary runners, and allow only those from the mother plant to take root.

STRIKING STRAWBERRY RUNNERS IN POTS.

☞ *For other Household Items, see "Basket"*

A Jelly Bag.

The brilliancy of jellies is secured by careful filtering through a flannel bag. The old chemists under the name of Hippocrates' sleeve made use of the same thing that we call a jelly bag. It is a conical bag of stout flannel, shaped as in the figure, stretched upon a hoop. One 14 inches deep, and 7 inches across the mouth, is a convenient size for ordinary use. The seam should be double stitched to make sure that no liquid will pass through it. As the bag requires to be washed, it ought not to be permanently attached to the hoop. A broad hem may be made at the top into which a wire or whalebone hoop may be inserted, or the mouth of the bag be finished by sewing a cord to it to make a strong edge. It can then be easily sewed to a hoop each time it is used. In straining jellies which solidify when cold, the bag should be hung near the fire. The first portion which runs through will be turbid, and should be poured back into the bag.

About Canning Fruit.

So much has been said about preserving fruit in air-tight cans or jars, that we are surprised that the inquiries still come for descriptions of the process. As our circle of readers widens, it each year includes new ones who do not seem to have practised this admirable household economy. Only a short time ago a letter came from Africa, saying that the writer had seen peaches from America put up in a thin liquid, and asking if we could find out how it was done. In our own families the canning of fruit is a matter of course, and we advise all who have never tried it to do so once; they will be sure to keep up the practice afterwards.

THE PRINCIPLE should be understood, in order to work intelligently. The fruit is preserved by placing it in a vessel from which the external air is entirely excluded. This is effected by surrounding the fruit by liquid, and by the use of heat to rarify and expel the air that may be entangled in the fruit or lodged in its pores. The preservation does not depend upon sugar, though enough of this is used in the liquid which covers the fruit to make it palatable. The heat answers another purpose; it destroys the ferment which fruits naturally contain, and as long as they are kept from contact with the external air they do not decompose.

THE VESSELS in which fruits are preserved are tin, glass, and earthenware. Tin is used at the factories where large quantities are put up for commerce, but is seldom used in families, as more skill in soldering is required than most persons possess. Besides, the tins are not generally safe to use more than once. Glass is the preferable material, as it is readily cleaned and allows the interior to be frequently inspected. Any kind of bottle or jar that has a mouth wide enough to admit the fruit and that can be securely stopped, positively air-tight—which is much closer than water-tight—will answer. Jars of various patterns and patents are made for the purpose, and are sold at the crockery and grocery stores. These have wide mouths, and a glass or metallic cap which is made to fit very tightly by an India-rubber ring between the metal and the glass. The devices for these caps are numerous, and much ingenuity is displayed in inventing them. We have used several patterns without much difference in success, but have found there was some difference in the facility with which the jars could be opened and closed. The best are those in which atmospheric pressure helps the sealing, and where the sole dependence is not upon screws or clamps. To test a jar, light a slip of paper and hold it within it. The heat of the flame will expand the air and drive out a portion of it. Now put on the cap; when the jar becomes cool the air within will contract, and the pressure of the external air should hold the cover on so firmly that it cannot be pulled off without first letting in air by pressing aside the rubber or by such other means as is provided in the construction of the jar. When regular fruit jars are not used, good corks and cement must be provided.

CEMENT is made by melting 1¼ oz. of tallow with 1 lb. rosin. The stiffness of the cement may be governed by the use of more or less tallow. After the jar is corked, tie a piece of stout drilling over the mouth. Dip the cloth on the mouth of the jar into the melted cement, rub the cement on the cloth with a stick to break up the bubbles, and leave a close covering.

THE PROCESS.—Everything should be in readiness, the jars clean, the covers well fitted, the fruit picked over or otherwise prepared, and cement and corks, if these are used, at hand. The bottles or jars are to receive a very hot liquid, and they must be gradually warmed beforehand, by placing warm water in them, to which boiling water is gradually added. Commence by making a syrup in the proportion of a pound of white sugar to a pint of water, using less sugar if this quantity will make the fruit too sweet. When the syrup boils, add as much fruit as it will cover, let the fruit heat in the syrup gradually, and when it comes to a boil ladle it into the jars or bottles which have been warmed as above directed. Put in as much fruit as possible, and then add the syrup to fill up all the interstices among the fruit; then put on the cover or insert the stopper as soon as possible. Have a cloth at hand dampened in hot water to wipe the necks of the jars. When one lot has been bottled, proceed with more, adding more sugar and water if more syrup is required. Juicy fruits will diminish the syrup much less than others. When the bottles are cold, put them away in a cool, dry, and dark place. Do not tamper with the covers in any way. The bottles should be inspected every day for a week or so, in order to discover if any are imperfect. If fermentation has commenced, bubbles will be seen in the syrup, and the covers will be loosened. If taken at once, the contents may be saved by thoroughly reheating.—Another way is to prepare a syrup and allow it to cool. Place the fruit in the bottles, cover with the syrup and then set the bottles nearly up to their rims in a boiler of cold water. Some wooden slats should be placed at the bottom of the boiler to keep the bottles from contact with it. The water in the boiler is then heated and kept boiling until the fruit in the bottles is thoroughly heated through, when the covers are put on, and the bottles allowed to cool. It is claimed that the flavor of the fruit is better preserved in this way than by the other.

WHAT MAY BE PRESERVED.—All the fruits that are used in their fresh state or for pies, etc., and Rhubarb, or Pie-plant, and Tomatoes. Green Peas, and Corn, cannot be readily preserved in families, as they require special apparatus. *Strawberries.* Hard-fleshed sour varieties, such as the Wilson, are better than the more delicate kinds. Directions for these, as well as for *Raspberries* will be found in a Basket item.

Currants need more sugar than the foregoing. *Blackberries* and *Huckleberries* are both very satisfactorily preserved, and make capital pies. *Cherries* and *Plums* need only picking over. *Peaches* need peeling and quartering. The skin may be removed from ripe peaches by scalding them in water or weak lye for a few seconds, and then transferring them to cold water. Some obtain a strong peach flavor by boiling a few peach meats in the syrup. We have had peaches keep three years, and were then better than those sold at the stores. *Pears* are pared and halved, or quartered, and the core removed. The best, high-flavored and melting varieties only should be used. Coarse baking pears are unsatisfactory. *Apples.* Very few put up these. Try some high-flavored ones, and you will be pleased with them. *Quinces.* There is a great contrast between quinces preserved in this way and those done up in the old way of pound for pound. They do not become hard, and they remain of a fine light color. Tomatoes require cooking longer than the fruits proper. See directions for these among Basket items. Any intelligent person who understands the principle upon which fruit is preserved in this way, will soon find the mechanical part easy of execution and the results satisfactory

Fruit Juices.—The juices of raspberries, pineapples and other fruits, are useful for flavoring ice-cream and similar purposes. The juice can be readily preserved by bottling. Express the juice and put it in bottles; set the bottles in a cold boiler with a board or grating under them to prevent contact with the bottom of the boiler. Heat up the water and continue at the boiling point until the contents of the bottles are heated through. Cork the bottles while hot, seal and keep in a cool place. The bottles should not be so large as to contain more juice than enough to use at once, as it will not keep long after being opened.

Aromatic Mustard.—Our request for recipes for preparing mustard after the manner of the Germans and other Europeans, has been replied to by several. The following is from Mrs. "W. S. K.," Rochester, N. Y.: "To 1 quart pure cider vinegar, add 2 tablespoonfuls each of ground allspice and cinnamon, 1 do. of cloves, 3 do. brown sugar, 4 do. salt, and 3 large onions cut fine. Boil until the strength is well extracted, say ½ or ¾ of an hour; then strain the vinegar, boiling, on to 1 lb best English mustard, and stir until it is perfectly smooth. If not thin enough, put more vinegar to the same spices, boil, and mix." This keeps well in a fruit jar, and improves by age. Boiling the vinegar for the length of time directed above would weaken it very much. The flavor of the spices would be extracted equally well by keeping the mixture nearly boiling hot in a covered vessel.... Another formula is given by Mrs. "E. D. C.," Meadville, Pa.: "4 tablespoonfuls of ground mustard, 1 do. flour, 1 do. sugar, 1 teaspoonful salt, 1 do. black pepper, 1 do. cinnamon, 1 do. cloves. Mix smoothly with boiling vinegar and let stand several hours before using. It may be thinned with cold vinegar. Will keep any length of time."

containing a great variety of Items, including many good Hints and Suggestions which we throw into smaller type and condensed form, for want of space elsewhere.

Transplanting and Watering.—"W. S. B.," Cass Co., Ind. In setting out cabbages, if the soil is very dry, make holes with a trowel, pour in a pint or more of water, and set the plant. We do not like to water in a dry time, unless it is necessary to save the life of the plants, as, if commenced, it must be continued until a rain comes.

Sparrows.—W. B. Christopher. The European Sparrow is thoroughly established in New York and vicinity. Imported birds are sold at about $4 a pair. They need to be furnished with small box houses, and in winter should have food scattered where they can find it. We do not know that they will destroy the Canker-worm, but have no doubt of it.

Trouble with Vines.—"D. D.," Winfield, Ind., in training Concords on the horizontal arm system, finds that some of the buds did not start. The Concord is usually very tractable. It may be that the fall pruning was too close, or that the Vine Flea-beetle has been at work at the buds.

Thomas' Smoothing Harrow.—This implement was mentioned in the Ogden Farm Papers for June, and several have inquired where it can be had. We cannot inform them. It is probably designed for a select few, or the makers would advertise.

Trial of Farm Machinery in Wisconsin.—A circular from the Ripon Farmers' Club informs us that arrangements have been made for a great trial of Farm Machinery. The time will be announced hereafter. Particulars may be had by addressing D. T. Glaze, Secretary of the Club.

AMERICAN AGRICULTURIST

FOR THE

Farm, Garden, and Household.

"AGRICULTURE IS THE MOST HEALTHFUL, MOST USEFUL, AND MOST NOBLE EMPLOYMENT OF MAN."—WASHINGTON.

VOLUME XXIX.—No. 8. NEW YORK, AUGUST, 1870. NEW SERIES—No. 283.

THE MAPLE-SHADE COTSWOLDS IN THEIR FLEECES.—Drawn from Life by E. FORBES.—*Engraved for the American Agriculturist.*

Our readers will call to mind a picture we presented of this flock in October of last year, and it would be well for them to refer to that number of the *Agriculturist.* There, they were shown as they appeared soon after shearing —the wool being perhaps an inch long. Shortly after this engraving appeared, the entire flock was purchased by one of the proprietors of the *American Agriculturist*, as was previously noticed. It has had the best of care, and being in fine condition this spring, several of the animals stood for their likeness to Mr. Forbes, who has succeeded in presenting us beautiful and accurate pictures. The portraits now given show the animals just before shearing, covered with their long, silky fleeces, the staple of which varies from 10 to 14 inches in length. The change in the appearance of the sheep is very striking; and the two pictures exhibit the excellencies of long-wool mutton sheep very satisfactorily. We see in the one the real "Shorthorn carcass" —long, deep, broad, compact, with nothing superfluous,—nothing, which is not either essential to the sheep's well-being, or profitable to the butcher. In the other, the one now shown, we see the immense fleeces of valuable wool which they carry, and the noble lambs which they bear, and which are a source of such great profit in the spring. The above pictures were taken late in April, just before the shearing, which took place early in May. Such wool, unwashed, is quoted in the open market at 40 to 45 cts.; and grade Cotswold lambs sell in April and May for 20 to 25 cts. per pound on their feet, netting their breeders often over $10 a head. Now-a-days certainly no sheep are more profitable, or better worthy, either the attention of the breeder of thorough-bred stock, or of the farmer who raises mutton and lambs for the market.

CHEESE-MAKERS AND DAIRYMEN ATTENTION!

If you would have a quality of ANNATTO always reliable and uniform in color and strength, buy only that known as **large "F,"** and be sure you get the Genuine.

Always on hand by WARD, SOUTHERLAND & CO.,

Wholesale Druggists, 130 William Street, New York.

Also, an assortment of the other brands that are best known.

Holbrook's Regulator Seed Drill.

A perfect hand-machine for sowing with *regularity and in proper quantity*. Beet, Carrot, Onion, Parsnip, Spinach, Sorghum, Turnips, Peas, Beans, etc. It is very simple, compact, durable, easy to operate, and *shows the seed as it drops*, to prevent any mistake or failure. Price $12.00. Address

F. F. HOLBROOK & SMALL, Boston, Mass.

THE PARLOR

Air Target Pistol,

Hawley's Patent, June 1, 1869.

No dirt, dust, nor danger of fire, uses compressed air. *No cost for ammunition*; from one charging, it shoots from five to ten shots, from thirty to fifty feet, accurate as any pistol. A great favorite with Ladies and Gentlemen for practice and amusement. Price $3.50. Sent C. O. D., and return charges.

AGENTS WANTED. P. C. GODFREY.

119 Nassau Street, New York.

THE PLANET DRILL.

For Garden Seed or Guano, the best: the most simple, compact, largest, easiest regulated, lightest, cheapest; no gearing, no slides; sows in open sight, and *evenly, whether full or not*, all seeds ordinarily sown; also salad, salsify, broom-corn, nursery seed, &c.; spreads fertilizers in the row at any rate, *without loss from winds*. No. 1, 5 lbs. seed, $12; No. 2, 12 qts. of seed or 25 lbs. guano, $20.00. Manufactured by S. L. ALLEN & CO., Forest Building, Philadelphia, Pa.

Hexamer's Prong-Hoe.

Is the most useful implement for Farm and Garden. No one can afford to do without it. Price $1.50. Send for Circular.

Address REISIG & HEXAMER, New Castle, Westchester Co., N. Y.

Or, B. K. BLISS & SON, 41 Park Row, New York.

SEND $1.25 for **"O'Hara's Pocket Giant Corn Sheller,"** (awarded first premium at Miss. State Fair, beating the Crowfoot Sheller two to one), and how to make **$20 per day.** Nothing like it, does not scatter, is simple, durable, cheap.

CHAS. MELSOM O'HARA, Bolivar, Tenn.

NO LOTTERY.

EVERY INVESTMENT DRAWS A PRIZE. INVALUABLE TO FARMERS, STOCK-RAISERS AND OWNERS, AND TO ALL CLASSES OF PEOPLE. SEND FOR A DESCRIPTIVE PAMPHLET OF **BUCHAN'S**

CARBOLIC SOAPS,

FOR CURE OF FOOT ROT, SCAB, MANGE, SCRATCHES, AND SORES OF ALL KINDS, DESTROYING THE SCREW WORM, TICKS ON SHEEP, FLEAS ON DOGS, LICE ON CATTLE OR POULTRY, FREEING HOUSES FROM INSECTS, CLEANSING, DISINFECTING, AND PREVENTING THE SPREAD OF INFECTIOUS DISEASES, THESE SOAPS AND COMPOUNDS OF *CARBOLIC ACID* ARE WITHOUT A RIVAL.

BOWMAN & BLEWETT, Sole Agents,

52 BARCLAY ST., N. Y.

$100 to $250 *per Month guaranteed. Sure pay.* Salaries paid weekly to Agents everywhere selling our *Patent Everlasting White Wire Clothes Lines.* Call at, or write for particulars to the Girard Wire Mills, 261 North Third-st., Philadelphia, Pa.

INVENTORS Who wish to secure **PATENTS** should write to **MUNN & CO., 37 Park Row, New York, for Advice and Pamphlet,** 108 pages, **FREE.**

A GREAT CHANCE! AGENTS WANTED!

$1000 a year sure made by agents, male or female, selling our *world-renowned Patent Everlasting White Wire Clothes Lines.* Cheapest and best clothes lines in the world; only 3 cts. per foot, and will last a hundred years. Address the *Hudson River Wire Co.*, 75 William St., N.Y., or 16 Dearborn St., Chicago, Ill.

LOCAL and TRAVELING AGENTS WANTED for the New, Revolving, Brass Wheel, Dating Stamp. Rapid Sales and large Commission. For Circular and particulars, address JOHN GARRETT & CO.,

201 Broadway, New York.

MAPLE LEAVES

Is the CHEAPEST, the BEST, and the MOST POPULAR Magazine published. It contains Stories, Sketches, Useful and Scientific Articles, Stories for Boys and Girls, Merry Moments, Puzzles, Illustrations, etc., etc., etc.

Only Fifty Cents

for one year. Specimen copies six cents. Subscribe now and get all your friends to do the same.

O. A. ROORBACH, Publisher,

102 Nassau Street, New York.

Magic Photographs, wonderful and amusing, 25 cts. a package. Library of Love, Courtship and Marriage, 4 Books for 50 cts. MAGIC LANTERNS, with a doz. views, 2, 3, 5, 8 dollars each. W.C.WEMYSS, 3 Astor Place, N. York.

PREMIUM CHESTER WHITE SWINE.—T. B. SMITH & Co. has the best stock in America. See adv't on page 116.

FLORA. BOTANY.

The latest, best, most popular Botanical Text Books.

By ASA GRAY, M. D.

Of Harvard University, and Botanic Gardens, Cambridge, Mass.

SALE GREATER THAN THAT OF ALL OTHERS COMBINED?

THEY HAVE NO EQUALS IN ANY RESPECT?

Gray's "How Plants Grow,"	$1.20
Gray's Lessons in Botany. 302 Drawings	1.40
Gray's School and Field Book of Botany,	2.50
Gray's Manual of Botany. 20 plates,	2.50
Gray's Lessons and Manual. One vol.,	3.00
Gray's Manual, with Mosses, etc. Illustrated.	2.50
Gray's Structural and Systematic Botany,	3.00
Flora of the Southern States,	3.50
Gray's Botanist's Microscope, 2 lenses	2.00
" " " 3 "	2.50

IVISON, BLAKEMAN, TAYLOR & CO.,

47 and 49 Greene Street, New York.

PUBLISHERS OF THE

American Educational Series of School Books,

AND MANUFACTURERS OF THE CELEBRATED

SPENCERIAN DOUBLE ELASTIC STEEL PENS.

For Berry Baskets, Grape Boxes, Crates, &c., address AMERICAN BASKET Co., New Britain, Ct., East. NEWFANE BASKET Co., Newfane, N.Y., West.

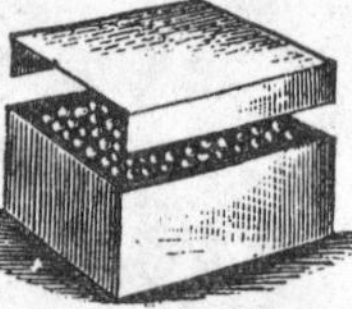

HOW TO MAKE THE FARM PAY

The value of this work is no longer questioned. More than 20,000 farmers, in every State in the Union, have used it the past year, and from all these comes one universal voice of approval. Thousands of dollars have already been made, and hundreds of thousands can yet be made by following its directions; the materials are on every farm without expending a dollar. Every Farmer, Stock Raiser, Gardener and Fruit Culturist can easily double all his profits. Published in English and German. 5,000 copies ordered the first 20 days of 1870. The sale is immense wherever introduced. Farmers and energetic young men wanted in every township to act as agents, on very liberal terms. Send name and address for Circular to ZEIGLER, McCURDY & CO.,

16 South Sixth Street, Philadelphia, Pa.

CHAS. A. DANA, EDITOR.

The cheapest, smartest, and best New York newspaper. Everybody likes it. Three editions: DAILY, $6; SEMI-WEEKLY, $2; and WEEKLY, $1 a year. ALL THE NEWS at half-price. Full reports of markets, agriculture, Farmers' and Fruit Growers' Clubs, and a complete story in every Weekly and Semi-Weekly number. A present of valuable plants and vines to every subscriber; inducements to canvassers unsurpassed. $1,000 Life Insurances, Grand Pianos, Mowing Machines, Parlor Organs, Sewing Machines, &c., among the premiums. Specimens and lists free. Send a Dollar and try it. I. W. ENGLAND, Publisher, New York.

NEW SUNDAY SCHOOL MUSIC.

The Song Garland, or **Singing for Jesus.** An entire *new* collection of beautiful Sunday-school music, pronounced by all who have used it the *best* book for Sunday-schools ever published. Price, 35 cents, or $30 per hundred. Specimen copies mailed to any address on receipt of price. S. BRAINARD & SONS, Publishers,

Cleveland, Ohio.

WE WANT AGENTS EVERYWHERE.—To canvass for Henry J. Raymond and New York Journalism, an Octavo volume of 500 pages, beautifully illustrated and richly bound; being a life and history full of deep interest to all. The author, Mr. Maverick, Managing Editor of the New York Evening Post, has in this book revealed many scenes of stirring interest never before shown to the public. *Canvassers for this work will derive great benefit from gratuitous editorial advertising.* Send for descriptive circular and see our extra inducements.

A. S. HALE & CO., Publishers, Hartford, Conn.

LUM'S EVERBEARING RASPBERRY.—Very fine, plants by mail, $2.50 per dozen. Address H. B. LUM, Sandusky, Ohio.

WHOEVER wants Wilson's Early Blackberries, Philadelphia Raspberry, *or other* Small Fruit Plants, should send for *my prices*, as I will sell extra good plants *very low*. Address DAVID BAIRD, Manalapan, N. J.

WATER-PROOF PAPER FOR OUT, AND inside of buildings. C. J. FAY, Camden, N. J.

Differing widely from the usual monthly magazines.

OLD AND NEW

A MONTHLY MAGAZINE.

EDITED BY REV. EDWARD E. HALE.

January and February numbers now ready.

THE welcome given to the first numbers assures the Proprietors and Editor that their plan for a VIGOROUS, COMPREHENSIVE AND POPULAR Magazine is well founded.

We mean to make the **Stories** in OLD AND NEW attractive alike to its youngest readers and to its oldest readers in their moments of recreation. We mean, also, to discuss the gravest themes, of **SCIENCE**, of **BUSINESS**, of **POLITICS**, and of **RELIGION**. We do not mean that these discussions shall be childish, any more than we mean that our stories shall be dull. Our desire is, not controversy, but to state the highest and best results arrived at in fearless science, in Christian morals, in free government, and in Liberal theology.

The **EXAMINER**, in each number of OLD AND NEW, will contain such reviews of books recently published in all parts of the world as shall guide buyers of books, and put them on their guard against what is worthless.

The January number of the **RECORD OF PROGRESS** was devoted specially to improvements in **Education**; succeeding numbers will touch on the **Homes of the People**, on **Sanitary Improvement**, on **Co-operative Experiments**, and other enterprises for the improvement of the methods of daily life.

The Editor has distinguished assistants. The journal must, of course, be printed in one place; but it does not pretend to be a local organ. There is no State in the Union, and scarcely is there an important city in the Old World, in which we have not active correspondents.

OLD AND NEW is a royal 8vo of 128 pages, handsomely printed.

TERMS.—Yearly subscription, $4; single numbers 35 cts. *Specimen number mailed on receipt of 35 cents.*

H. O. HOUGHTON & CO., Publishers,

135 Washington Street, Boston,

HURD & HOUGHTON,

459 Broome Street, New York.

Unparalleled Success!!

THE SOUTHERN FARMER, published at Memphis, Tenn., enters upon its 4th Vol., with a circulation of 10,000! Increase in the month of Dec. over 3,000. It is the Standard Agricultural and Horticultural paper in the six South-western States. Thousands at the North would do well to take this paper to see the rare openings for profitable investments monthly offered. Advertisers will find no better medium in the South. Terms, $2.00 per annum.

Address M. W. PHELPS & CO.,

361 Main St., Memphis, Tenn.

THE WIDE WORLD Sent Free! Best Story & Family Paper in United States. Agents Wanted. WIDE WORLD, Boston.

THE PIANO BOOK,

Which cannot be Excelled; is Richardson's New Method. Regular Sale,—30,000 a year.

Sold by all Music Dealers, Price, $3.75. Sent post-paid. **Oliver Ditson & Co.,** Publishers, 277 Washington St., Boston. **Chas. H. Ditson & Co.,** 711 Broadway, New York.

SMALL FRUIT RECORDER AND COTTAGE GARDENER.

A monthly paper at only 50 cents per year. The Jan. No. (alone worth that amount) will be sent to all applicants *free*. It copies the "*cream*" from all other papers on these interesting subjects, and is admitted by all who have seen it to be the most valuable paper of the kind printed in the United States. Address

A. M. PURDY, Palmyra, N. Y.

TO TEACHERS, COLLEGE PROFESSORS, AND CLERGYMEN.

A Rare Chance to learn Phonography.

JAMES E. MUNSON, the author of the COMPLETE PHONOGRAPHER, 117 Nassau Street, N. Y., will give, by mail, to any Teacher, College Professor, or Clergyman, a complete Course of Lessons in Phonography, giving a thorough, practical knowledge of the Art, for Five Dollars. Send for Circular.

10,000 AGENTS WANTED FOR

PRIEST and NUN.

Apply at once to CRITTENDEN & McKINNEY, 1,308 Chestnut Street, Philadelphia, Pa.

GAMES, PUZZLES, PARLOR FIRE-WORKS, Books, &c., &c. Catalogues sent to any address, FREE, by O. A. ROORBACH, 102 Nassau St., New York.

A Beautiful Full Gilt Photograph Album, holds 20 photographs. Sent free for 25 cts. Pat. March 16th, 1869. Address J. J. HAMMOND, Holliday's Cove, W. Va.

THE HOLIDAY GUEST: a collection of Stories, Games, and Amusements for Winter Evenings. Illustrated. Price 10 cents, Sent on receipt of price, by O. A. ROORBACH, 102 Nassau St., New York.

AMERICAN AGRICULTURIST.

NEW YORK, AUGUST, 1870.

Relief from the hardest of the summer labor and care comes to almost every farmer in August. After spring grains are harvested, and the buckwheat and turnips are sown, and the latter hoed and thinned out, and the early apples are marketed, and preparations are made for winter wheat, though still there is work enough for his men and teams, he has time to draw a long breath and look about a little. Even the farmer may take his family for a visit to a neighboring county, to the springs, or to the sea-side; and some are wise enough to do it. The good it does cannot be measured. If it took every cent of the profits of the entire year it would often pay. Who knows when this incessant drive and worry, which seems to be the very life of many farmers, and into which they inconsiderately force their poor, work-worn partners, will end in a fit of sickness or in the quiet of the church-yard? Better rest, and enjoy life a little, now and then.—You will enjoy the final rest as well, and add essentially to the grand sum total of human happiness.

Hints about Work.

Compost Heaps.—While the warm weather lasts, and other more important work is not pressing, make compost heaps; laying up manure from the hog-pens with weeds, sods, etc., from the fence rows, freshly mown reeds, brakes, and swamp-grass. Compost makes very fast with a little good manure as a basis, and the quantity may be increased six-fold in a few months. Where no manure is at hand, use lime or ashes, sprinkled upon layers of dry muck and any vegetable matter. Dead animals are an admirable addition to such composts, and if used, the heap should be covered with soil.

Compost for Pastures.—Top-dressing old pastures with a compost made of such vegetable or peaty matter as can be collected in the vicinity, with lime, to which is added as much bone-dust as one can afford, thoroughly worked over and mixed, spread and harrowed in, will make a wonderful difference in the yield of grass.

Working Oxen are ill adapted to severe labor during the hottest part of the day, but if put to the hardest work in early morning, and after the heat of the day is passed, they will be capable of doing a great deal without falling off in flesh. Hired men will gladly accede to a proposition to break their day's labor by a three-hours' nooning and rest.

Orchard and Nursery.

Marketing.—The early fruits will now be coming, and will need attention if sent to market. Apples and pears should be picked when fully matured, but before they begin to mellow. Recollect—what has been so often repeated—it does not pay to send poor fruit to market. It costs as much for freight and packages for poor stuff as it does for good. Superior fruit will always sell, no matter if it is a season of plenty. In such seasons it is difficult to dispose of an indifferent or a poor article in our city markets at any price. Assort the fruit.

Peaches are to be picked in just that condition of ripeness in which they are firm enough to endure transportation, and will be in eating condition by the time they reach the consumer. If too hard, they will fail to ripen up properly, and if too soft, they will bruise in the carriage. Experienced pickers soon learn to recognize the proper condition by the appearance of the fruit, without subjecting it to the test of touch. Crates are becoming each year more popular for packing peaches. Directions for making them were given last month.

Thinning should have been done before; but few thin the fruit sufficiently at the first going over, and it is better to look over the trees again and remove inferior specimens.

Insects are to be watched. As soon as the nests of the fall web-worm appear, destroy them. This month the red spider is often troublesome, not only on fruit-trees, but on ornamental ones. It shows itself by a browning of the leaves; and a close examination will detect the minute, red pest. Frequent syringings with soap suds, or even clear water, will dislodge it, if persisted in.

Budding.—The time to bud is when the bark of the stock will "run," or part readily from the wood, and well-formed buds of the same season's growth can be procured. The maturing of the buds may be hastened by pinching the end of the shoot. When sticks of buds are cut, the blade of the leaf should be cut off, leaving the leaf-stalk attached to the twig. Keep the twigs moist until they are used.

Fruit Garden.

Thinning the Fruit is easily performed upon dwarf trees, and such should never produce other than fine specimens. The trees are so small that they can be petted and cared for in a manner not practicable with standards.

Strawberries.—Beds may be set by the use of potted runners; directions for making were given last month on page 71. Keep the runners cut off from vines grown in hill culture. A dressing of guano or ashes just before a rain will help the plants.

Blackberries and Raspberries.—The new growth should not be allowed to reach higher than 5 feet; some stop as low as 4 feet. The tender point is pinched out, or a knife is used. This stopping of the growth causes the side shoots to push, and these are in turn pinched when they are 18 inches long. Keep all suckers hoed down, and when the fruit is off, remove the canes that have borne.

Black Caps differ in their manner of growth from ordinary raspberries; they do not throw up suckers from the root, but multiply by rooting at the ends of the canes, which naturally bend down and touch the earth. If it is not desired to propagate plants, the new growth is pinched at the hight of about 30 inches, which causes lateral shoots to push.

Grapes.—After the first appearance of mildew there is no time to be lost in applying sulphur. Grayish patches on the fruit stems or upon the leaves call for an immediate and thorough dusting with sulphur, hand-pick beetles and caterpillars; keep the new growth tied up and pinch the laterals back to one leaf as often as they push.

Kitchen Garden.

Beans, if desired for salting or pickling, may be sown and will give a supply of green pods.

Cabbages and *Cauliflowers.*—Give thorough culture, using the hoe frequently. If any are disposed to be slow, give a dressing of guano or hen manure. Lime or salt is used for destroying slugs.

Carrots.—Thin the late sowings and keep cultivating until the tops are so large as to prevent.

Celery.—Keep the soil well stirred and free from weeds. Plants for a late crop may yet be set.

Corn.—Those who save their own seed should select the earliest and best formed ears for the purpose. It is very easy to keep the variety improving, and it is equally easy to deteriorate it by taking the leavings of the crop for seed.

Cucumbers.—Save the finest and earliest for seed. Pick for small pickles every day.

Egg-Plants.—Continue to forward them by cultivation and manuring. The "Tomato-worm" will often attack the plants and do much mischief.

Endive.—Set out the plants a foot apart each way in rich ground; cultivate well to induce a rapid growth. They may be blanched by tying, covering with a flower-pot, or by laying a board upon them. The operations should be performed only when the plants are quite dry.

Melons.—Remove such fruit as will not ripen before frost. In saving seeds select the earliest, and those grown at a distance from any other variety.

Onions are ready to harvest when the tops lose their stiffness and fall over to one side. When a good share have done this, the crop may be pulled. If the onions are to be stored, they should first be thoroughly dried and then spread in a cool loft where they will have ventilation.

Radishes.—Sow the winter sorts at the proper time for sowing round turnips. The Chinese Rose-colored Winter is the best, though some fancy the white and black.

Spinach, if sown in rich ground, will give a supply this fall. It is too early to put in the crop intended to keep over winter.

Sweet Potatoes, whether in hills or upon ridges, must be kept free from weeds until the vines completely cover the ground. Do not allow the vines to take root, as they are apt to do when they come in contact with the soil; move them occasionally.

Tomatoes. — Those who have a choice variety should take pains to preserve and even improve its good quality. A dozen vines of the same variety planted side by side will show marked differences in earliness and productiveness, and in the shape and solidity of the fruit. By attending to these points and sacredly reserving the most desirable for seed, each grower can establish a style, if not a variety of tomato, that will suit his purposes. If training is practised, whatever form has been adopted should be faithfully followed up. Give the tomato-worm no quarter. If left to itself for a day it will make havoc. Its droppings seen upon the leaves or upon the ground are warnings to be heeded.

Turnips.—Dust the young plants with lime or a mixture of plaster and ashes, if insects trouble them.

Weeds.—Keep at them. Those cut up this month die quickly and are quite sure to stay dead.

Flower Garden and Lawn.

Lawns.—This is a trying month for new lawns; if, as is often the case, the green was produced in good part by annual grasses, these will die out, leaving brown and bare spots. Encourage the formation of a close turf by rolling after a rain, and by frequent mowing.

Edgings and *Margins* will require the use of the edging knife to keep them neat.

Bedding Plants.—See that they do not grow out of shape and get mixed. Most of them will require the use of the knife to keep them in shape.

Dahlias usually suffer at this season for want of water. They should be kept in a growing condition.

Roses.—To keep the ever-blooming sorts up to their name, as soon as the flowers drop, cut the stem back to a strong bud, and it will soon push new shoots and flower again.

Gladioluses.—Tie up those that need it; the flowers are so heavy that the stalk is frequently unable to support them.

Lilies will also need staking. Cut away the flower as soon as it fades, unless it is desired to save seed. Keep the foliage free from caterpillars, which are apt to attack them at this season. The future condition of the bulbs depends upon healthy foliage.

Chrysanthemums.—Bring into desired shape by pinching. Do not let them become crowded, or the lower leaves will decay. Tobacco water will kill lice; and the caterpillars, often so troublesome upon them, must be pinched off by hand.

Last Month we gave several hints upon general care and keeping, which are still timely.

Green-house and Window Plants.

The plants out of doors continue to need the care mentioned in last month's notes. The earlier repairs are made to the house, the better. When this and the heating apparatus are in good order early in the season, much hurry and confusion will be saved at the time of taking in the plants. Propagation by cuttings from plants out of doors can go on. It is well to partially sever or "tongue" the cuttings of geraniums and other very soft-stemmed plants before removing them entirely. The cut portion becomes callous while still upon the plant, and when removed is ready to strike root.

Cranberry Culture.

"Cranberry Culture, by Joseph J. White, a Practical Grower," is the modest title of a work recently published by Orange Judd & Company. Cranberry culture has developed wonderfully since the time Mr. Eastwood wrote the only considerable treatise we had upon the subject.

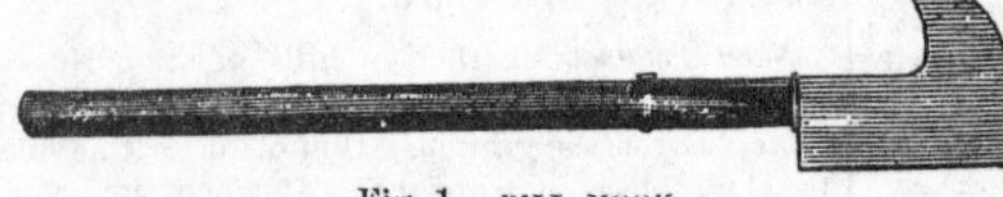

Fig. 1.—BILL HOOK.

The present work not only gives instructions drawn from the author's own experience, but presents the practice of the most advanced practical growers. Some who have written upon the subject have left it to be inferred that cranberries would grow almost anywhere. The author states distinctly, and various other growers sustain him—that successful culture can only be carried on where certain conditions are present. Where these are wanting, vines may grow, but their culture will not be profitable. The

Fig. 2.—TURFING A SWAMP.

requisites are muck, sand, and ability to drain the meadow for a foot below the surface, and to flow it at will. Natural bogs are sometimes found which may be improved at a moderate outlay, but the great majority of plantations are made by clearing the swamps and setting out the plants. In many swampy places there are 5 or 6 inches of muck, underlaid by sand, which may be prepared for cranberries by simply plowing under the muck and turning up a layer of sand. In swamps where the muck is deep, sand

Fig. 3.—SANDING A MEADOW.

must be at hand to spread upon the surface. Mr. White gives full directions for preparing the different localities that are naturally available. As summer is the most favorable time for preparing a swamp, we give an outline of Mr. White's directions. The first step is to open a ditch to the depth of 2 feet below the surface of the muck. The bushes are cut down by means of a bill-hook like that shown in figure 1, collected in heaps, and when the leaves have dried, burned. Turfing, or "scalping," is next in order; the whole grassy and weedy surface is to be removed. The turf is cut into strips a foot wide, by the use of a cleaver made for the purpose, and then removed in pieces by means of a turfing-hoe, as shown in figure 2. The pieces of turf are used to make a fence or wall around the plantation. The cost of turfing is $40 per acre, and the hauling off and building the fence costs about as much more. Any stumps which remain are removed. A sufficient number of ditches, to drain all parts of the meadow a foot or 18 inches, are cut, and the meadow is ready for sanding. Sand without any admixture of clay or loam, is spread regularly over the surface of the meadow to the depth of 2 to 6 inches, according to the depth and character of the muck. The sand may be carried on in wheelbarrows running upon plank, or a portable railroad may be used, like that shown in figure 3. Planting is done in the spring, and the different methods of setting the plants are fully discussed and illustrated, as are the after treatment of the meadows, gathering the crop, overcoming insects—in short, all that one needs to know about cranberry culture seems to be contained in this work. As we are frequently asked about flooding, we may state that it is done after vegetation has ceased in autumn, for the protection of the plants during winter, the water being kept on until spring has fairly opened. The meadows are again flooded just before the blossoms open, to destroy the vine-worm. At this time the water is kept on only 24 hours.

Camellias in Window Culture.

Of all the green-house plants grown in window culture, there is none with which the amateur is so generally unsuccessful as the Camellia. All goes well until the buds are beginning to swell, and then they begin to blast and drop, carrying with them the anticipations of the grower. The trouble is a too hot and dry atmosphere. A correspondent, whose abundant bloom last winter we had the pleasure of seeing, writes: "I have had capital success with Camellias this winter. These are generally failures in the dwelling-house, as the buds drop just about the time they should open. The window of an upper hall was devoted to camellias and a few other plants; twice during the winter the earth in the pots froze, and twice they were removed to a warm room to prevent freezing. In February the buds began to show signs of opening, and such plants as were ready to flower were taken to the dining-room, where they rapidly came into full bloom. One later than the rest has over a dozen flowers on it now (in April). Those who have no green-houses can have Camellias if they will only take a little trouble; and so magnificent a flower is worth a great deal of trouble. Nice, moderate-sized plants can be bought at from $1 to $3, and upwards. Keep the plants out of doors in a shaded place all summer, taking care that they do not get too dry and that no insects eat the leaves. I don't know what one it was, but some "bug" disfigured the leaves of one of my plants. When frosty nights come, remove the plants to the house, to a room without any fire, and keep the window open whenever it is not too cold. Water as needed, and during the winter, wash the leaves with a sponge or soft cloth every week or two. A moderate freezing will not hurt the plants, unless the buds are too far advanced; but it is easy to prevent it altogether by removal to a slightly warmer place during very cold nights. As soon as the buds swell enough to show the color of the petals, the plants may be removed to a sitting-room where their flowers can be enjoyed. Soon after the flowers have fallen, the camellia makes its wood growth. At this time it may be repotted, giving it fresh soil, in the same pot, if a cramped condition of the roots does not show that a larger one is needed. Fresh soil, such as decomposed sods, is as good as any of the mixtures of peat, sand, and other things recommended by the books. Prune into shape, if necessary, and the plant will push its new shoots and grow rapidly. During the time it is growing, give it plenty of water and all the light possible. Set the plants in the open air as soon as frosty nights are over. The treatment is simple enough, and the success most gratifying."

The Twelve-Spotted Squash-Beetle.

Squashes, melons, cucumbers, and in fact, the whole Gourd Family, have a hard time of it in the way of insect attacks. The little "flea" attacks them while in the seed-leaf, then comes the striped-bug, and after these, the disgusting squash-bug; all of these feed upon the leaves, unless the borer comes along, and by going to the root of the matter, destroys the vine outright. To this list we have to add the Twelve-spotted Beetle—*Diabrotica 12-punctata.* We first became acquainted with this pest some ten years ago, in a garden in Michigan. We had, at that time, never heard of it at the East, nor is it mentioned in Harris' Insects. Now it is disastrously abundant in the vicinity of New York, and we have heard more of it this year than in previous ones. The beetle presents a general appearance to the "Lady-birds," or Coccinellas, but it is of a dark yellow color, with twelve roundish black spots upon its wing cases. The engraving shows the insect of about twice its real size. It is closely related to the striped-bug, and a great many times more mischievous; while the plants will soon get strong enough to not mind the striped-bug, this one will attack them when in full vigor, and wherever it goes it makes as clean work as if fire had been among the leaves. This insect is said to be very destructive to the flowers of Dahlias and Asters. We find but little account of its habits in the Entomological works, but know that, excepting the borer, it is the worst enemy that the grower of melons, etc., has to contend with. We know of no remedy save hand-picking.

Fig. 1.—FRONT ELEVATION.

No. 9—A House costing $2,200 to $3,000.

This house we designate as No. 9 of the series mainly designed by MR. JUDD, and already tested or being tested by him. It will come within the range of a great number of families. The variation in cost will depend partly upon location and price of labor and materials, and more upon the style of work, cornices, moldings, plumbing, whether basement be finished or not, etc., etc. The variation may indeed go further, say from $1,800 to $3,500, without materially changing the general plan or size. A careful estimate, at the prices given in our July number, page 68, makes this house, in fair style and finish with good moldings, and the plumbing described below, cost about $2,800, without the land. In none of the plans thus far given, has special provision been made for a sleeping room on the first floor, as is customary in most farm dwellings. The stratum of air near the ground is always unhealthful, especially at night. Fogs, dampness, and malaria abound near the soil. We see heavy fogs along the ground when the air is clear at the hight of 10 to 15 feet. The same reasons that lead invalids and health seekers to go to the mountains or elevated localities, should always take lodgers to a second or third story at night. An invalid, too feeble to be daily moved, and requiring constant attention, may be necessarily placed upon the first floor—though by no means the best place for a feeble person—and a first floor bed-room may be indispensable in some cases, though it would usually be preferable to give up the parlor or sitting-room to such a person, temporarily.—All the outer walls are filled in with brick between the studding. The siding is the 1-inch thick "Novelty," described in fig. 6 of our March number.

Fig. 1—Elevation.—The engraving gives the general style and appearance, as nearly as the artist can do so from the builder's minute drawings, which are plain lines (not in perspective). The basement is mainly above ground, for reasons here and previously given. Three or four extra steps elevate the occupants at all times above the damp soil. For flat prairies, or other localities exposed to very violent winds, the house is proportionately too high. In such cases the basement may be lowered, and the attic be omitted, though the latter is always desirable, and adds but little to the cost, while it furnishes much store room and two or more sleeping rooms if needed. The basement, floors, ceilings, roofing, doors, windows, etc., cost no more for a house 20 to 25 feet high, than for one only 15 to 20 feet high. The cornice, it will be seen, extends well out—a great help to the appearance of most houses. The appearance can be materially improved by more brackets, and by dentals on the cornice; by heavier moldings, window caps, etc. The roof is covered with tin, which is nearly as cheap as shingles now, (8½ to 9½c. per foot,) and cheaper in some places. For some styles of brackets, etc., etc., see our June number, page 64.

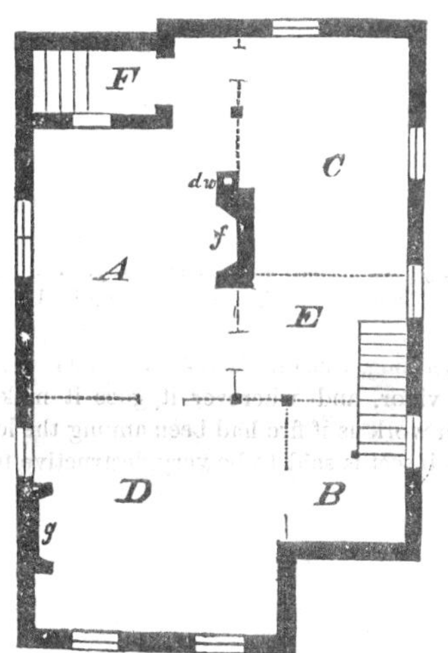

Fig. 2.—BASEMENT.—FLOOR PLAN.

Fig. 2—Basement.—Hight in clear 7½ feet. Walls of brick, 8 inches, with an air-space all through the middle, except where tie-bricks are thrown across. The whole Basement may be left in one room, as a cellar, with three or four piers of brick or locust posts firmly set on sunken, flat stones, to support the girders and partition walls above. The dotted lines show a good arrangement of basement rooms that may be made at first, or which can be put in readily afterwards if the means permit, or if the enlargement of the family require more room. The stairs and windows and fireplaces (*f* and *g*) should be adapted to this possible or probable use of the basement. In this case *D* would be the Dining-room, *A* the Kitchen, *C* the Cellar, *E* a large Pantry, *B* a Hall, *F* the rear entrance. Or, if desired, simply the room *A* may be finished off as a laundry or wash-room, or as a kitchen with a dumb-waiter, to be carried up, say at *dw*, or to whichever one of the rooms above is to be used as a dining-room.

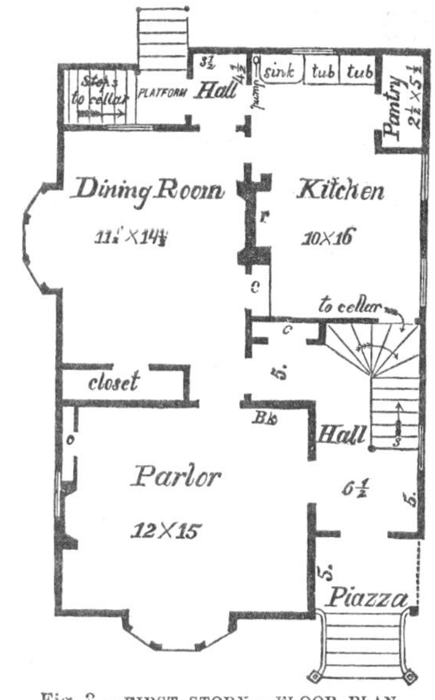

Fig. 3.—FIRST STORY.—FLOOR PLAN.

Fig. 3—First Story.—Hight in clear, 9 feet. A careful examination of the engraving will show the general arrangement, the economy of space, the "saving of steps" in the cooking, and other items of work, the provision for closets and pantries, *c, c, c, c*, etc. The Bay-windows add greatly to the apparent as well as real size of the rooms, and are ornamental to the exterior. The main hall and kitchen would be improved by adding 6 to 12 inches to the width of the house, where great economy is not studied, but they are of fair and convenient size, as indicated.——It will be noted that the doors, sink, pump, stationary wash-trays, and large pantry are placed near together to save steps. A cooking range, *r*, is intended, but a place for the copper boiler was overlooked. It was intended to put it by the side of the chimney where the door into the dining-room is now placed, and have that door the other side of the chimney where the closet is located; but that made the distance too great from the dining-room table to the kitchen sink. The chimney can well be placed a foot or so nearer the front of the house, and leave room between that and the door for the copper boiler. The exact location of every thing should be considered and decided upon before starting the foundations.——The rear, outside platform may be covered, or not, as desired. The stairs to the basement may be entered from the kitchen, or from the front hall when the basement is finished for a dining-room. There is sufficient head-room *under* the stairs to the second story.——*Bk*, in the Parlors, indicates a bell-pull to the kitchen. By passing the wire down and along the basement ceiling, a bell may be also attached in the basement kitchen (*A*, fig. 2). The stationery wash-tubs and sink are supplied with cold-water cocks from the pump, and hot-water cocks from the boiler. The force-pump draws water from the cistern or well, and when needed forces it into the tank placed in the Attic or elsewhere.

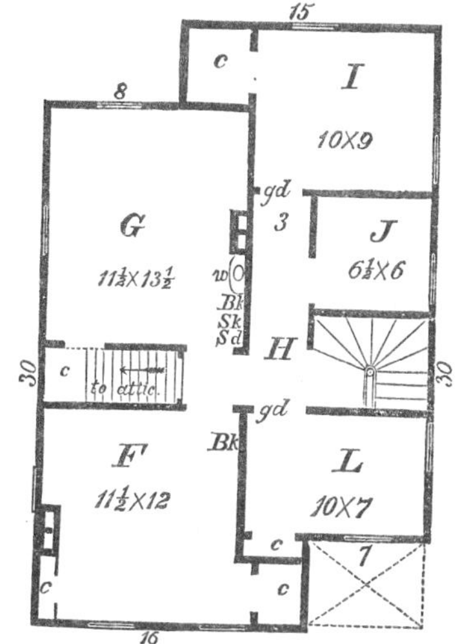

Fig. 4.—SECOND STORY.—FLOOR PLAN.

Fig. 4—Second Story.—Hight in clear, 8¼ feet. The Hall, *H*, is lighted either by sash in the upper half of the doors, *gd, gd*, or by head-lights over these doors, which, if swung on hinges or pins, also serve as ventilators. (Ventilators, by the way, are to be placed in all rooms in both stories, with flues in the walls to carry the air to the attic.) Where needed, the flues may be carried through beams or girders, by piercing these with sundry auger holes for the passage of air, boring so as not to weaken the timbers too much. With the "Novelty Siding," firmly nailed on, a house is very strong, even though the timbers be very light, or weakened by cutting or boring. ——It will be seen that of the four larger chambers each has one or more closets. The room *J* will admit a 3-4 bed, or, it may be used for a bath-room or store-room. *F* and *G* may have mantels and fireplaces, or simply holes into chimney-flues, for stoves.——In *G*, the wash-sink, *w*, may be supplied with hot and cold water.——The stairs to the Attic are necessarily quite steep to afford head-room under the roof. The water-tank to catch water from the roof or receive it from the pump, for supplying the boiler, etc., is to be placed in the Attic, or if that be not raised high enough, it may be placed *over* the stairs, occupying the upper part of the second story. It may be of any desired length and width, and any where from 20 inches to four feet in depth. The larger it is, the less pumping up of water will there be. Water comes in from the roof, and when the tank is full, the surplus runs over through a pipe to the cistern. Wherever a Tank is placed, stronger or double studding should always be carried up from the basement, to support it firmly and safely.——Bell-pulls, *Bk*, to Kitchen, are located in *F* and *G*, and in *G* a speaking tube, *Sk*, to the kitchen, and, *Sd*, to the front door, as described previously, for other houses.——Blind windows are necessary back of the chimney, at the left of the Parlor, and of the Chamber *F*... In order to fully appreciate this plan, one will need to consider the many items taken into account, such as the least possible building of outer walls for the same space, the saving of steps in the arrangement of the working-rooms and apparatus, the securing of an abundant supply of closets or pantries, the provision for finishing off the basement in whole, or in part, whenever it may be desired, without much disturbance to the rest of the house and at small cost.

A Native Crinum.—(*Crinum Americanum.*)

The exotic species of Crinum are well-known, showy ornaments of hot-houses, and we have one, a native of Florida, which, if not so brilliantly colored as the others, is a very pretty and interesting plant. This spring we received from Mr. C. L. Allen, of Brooklyn, a few bulbs of this Crinum, which allowed us to see it in flower. The flowers are from 6 to 8 inches long, and borne from two to four together at the end of a stalk, which is a foot or more high; they are white, delicate in texture, and fragrant. The engraving gives a representation of the plant, reduced, of course, in size. This Crinum is a native of the river swamps of Florida, and it will probably be necessary to take up the bulbs at the approach of winter. We shall be very glad if Mr. Allen succeeds in adding this to our list of popular bulbs; aside from the fact that it is a native plant—which we are always pleased to see brought into cultivation—it has merits of its own which will commend it to those who love delicate and modest flowers.

THE AMERICAN CRINUM.

Notes on Strawberries.

In conversation with an experienced strawberry amateur a short time ago, he remarked that we had made no progress in strawberries in ten years—indeed, that our present varieties were not so good as those that were popular ten years ago. Our friend was nearly right; we have allowed many excellent varieties to fall into neglect, and instead, accepted fruit that was either of large size or very productive. The ease with which the Wilson can be raised has caused fruit which required careful cultivation to be discarded; and the enormous size of the Seth Boyden has made us forget the flavor of such varieties as Hovey's Seedling and Hooker. Then again the introduction of such musky varieties as the Triomphe de Gand, which has but little strawberry taste about it, has created a false standard of flavor. We would not be understood as condemning large berries, but size cannot compensate for lack of flavor. The grower for market needs a productive fruit, one that is sufficiently firm to carry well, and of good color and size. The amateur who grows for his own table will be content with fewer berries of a high flavor, and the ability to bear transportation is of no consequence to him. We append a few notes on some of the newer varieties we have grown or have tested this season.

Nicanor.—We fruited this variety on young plants put out last fall, and have seen it on two-year-old plants under the best treatment, and wonder why so little has been said about it. It is vigorous, hardy, and in productiveness it is remarkable, if not unequalled. The berry, though not of the largest, is of good size, a bright scarlet, and of excellent quality. Our illustration, (fig. 1,) is from only a medium-sized berry. Should this variety, in other parts of the country do as well as in the instances referred to, it must become a popular fruit.

Fig. 3.—LATE PROLIFIC.

Seth Boyden.—This was first exhibited as Boyden's No. 30. Three years ago we described and figured it as "Seth Boyden," and since the death of the originator, this name has

Fig. 1.—NICANOR. Fig. 2.—RUBIS.

been generally adopted. It stands at the head of the large berries, being a good bearer, of a fine shape, and little disposed to cockscomb, and its conspicuous polished neck makes it showy and attractive. It unfortunately lacks flavor and is rather soft, but it is the best large berry yet introduced into general cultivation.

Colfax.—This variety was heralded a year or two ago, as something remarkable. Dr. Hexamer recently exhibited it as the "poorest strawberry ever grown," a statement that both its appearance and taste confirmed.

Late Prolific.—E. W. Durand, Irvington, N. J. This variety took the prize at B. K. Bliss & Sons' Exhibition, as the best variety not before exhibited. It is a cross between Haquin and a seedling of Mr. Durand's, which was itself a cross between Green Prolific and Triomphe de Gand. Hermaphrodite. The foliage is very vigorous, of a dark green, and endures the sun. The fruit of the largest size; obtusely conical, and not inclined to cockscomb. Seeds in medium depressions; surface brilliant, bright crimson; flesh crimson with a few white streaks, very juicy, with a rich, sprightly flavor. On Mr. Durand's grounds this is certainly a very promising berry, but Mr. D. wisely declines to disseminate it until he is thoroughly satisfied that it retains its good qualities. (Fig. 3.)

Rubis.—Mr. Louis Ritz, of San Souci Fruit Farm, Plainville, Ohio, has a remarkable collection of novelties, especially of foreign varieties. He sent us a number of varieties for trial, and though small plants, some of them fruited. One of them, Rubis, pleased us so much that we have figured it, (fig. 2,) though being the product of a plant set out this spring, it cannot do justice in size. This variety originated with Dr. Nicaise, in 1868. The berry is bright scarlet, and very glossy; seeds but little imbedded; flesh rosy-white, solid, juicy sweet, and of excellent flavor. It is not safe to commend a berry with so slight an experience with it, but it is a variety of which we have strong hopes.

Her Majesty is another new sort from Mr. Ritz. The few berries we had were large, of a fine carmine color, and possessing excellent characters, being fine, sweet, and of delicious flavor.

Mexican Everbearing. It will be expected that we shall say something of this bone of contention. We have it growing, and if it is in any important character different from, or better than, the Monthly Alpine, we are unable to see it. Mr. Meehan finds a difference in the young leaves, but as we do not grow the plant for leaves, that has no practical bearing. There are some who fancy Alpine strawberries, and these will be pleased with the Mexican, Autumn Galande, or a half-dozen others, the chief difference in which consists in the name. A pot each of the Mexican and Alpine was exhibited at Bliss & Sons' show, without labels. There were many persons present who are practical strawberry growers, and none of them could see in what particular the plants differed.

Black Defiance.—(Fig. 4.) Another seedling by Mr. Durand, which received the premium at the N. J. State Fair in 1868. It was exhibited at B. K. Bliss & Sons' show, where it attracted much attention by its large size and very dark rich color. This year it has not proved as

Fig. 4.—BLACK DEFIANCE.

abundant a bearer as heretofore. Its richness of color and flavor will commend it to the amateur.

The Mezquit Tree.

Those who have read accounts of travel in Western Texas and the adjacent regions, have noticed the frequent mention made of the Mezquit. In many parts of the country it is the prevailing tree, and, without which, these regions could not be traversed with any comfort; as it is often the only resource for both fuel and forage. The tree grows some 30 feet high, and sometimes, where it has a chance to develop, forms a regular spreading head, and resembles at a distance a large apple tree in its outline. More generally, probably from the attacks of insects and the injuries received from the Mistletoe, the tree is irregular in its form. Upon a closer inspection the foliage reminds one of the Honey-Locust, and the presence of strong spines, a pair at the base of each leaf, increases the resemblance. The shape of the leaflets varies considerably, as does their shade of green. The engraving gives the leaves, flowers, and fruit. The flowers are very minute and clustered in small, greenish-yellow spikes. The fruit is a bean, six inches or more in length, straight or somewhat curved. This pod at a certain stage of ripeness, is pulpy. The pulp is very sugary and often of pleasant flavor. At full maturity the pulp becomes dry and spongy, and still retains its saccharine character. The pods form a most acceptable food for horses and cattle, and are found of great service by travelers. The Indians and the poorer Mexicans make use of them to form a kind of *pinole*, a meal obtained by pounding and sifting the pods; this, though sweet, is accompanied by a taste which is nauseous to the civilized palate. The tree exudes a gum, which, in its qualities, is very much like Gum Arabic. It is used as a substitute for that gum by the Mexicans. The sapwood of the Mezquit is yellowish, and the heartwood of a reddish-brown color, as dark as black walnut. It is very hard, and in durability seems to be equal to the wood of the locust. For fuel, nothing can exceed it in excellence; and a bed of live Mezquit coals makes in intensity of heat a near approach to anthracite. Upon the plains, where frequent fires sweep over them, the Mezquit appears only as a shrub a few feet in hight, and an inexperienced person would be in a straight for fuel. The old traveler, however, regards the appearance of a Mezquit bush, however small, as a sign that a good fire is in prospect; and by the use of a grub hoe he soon unearths roots as large as his leg, which make the most perfect fuel. The Mezquit has been proposed, in Utah, we think, as a hedge plant. It certainly possesses many qualities which adapt it to that use, as any one who has ever had the misfortune to lose himself in a thicket of it can testify. Experiments are needed to ascertain how far north it will prove hardy. It will probably be found, as is the case with many other trees, to endure the winters much farther north than where it grows naturally. The Canadian River is the northernmost locality that we have seen mentioned for it. The subject is one worthy the attention of those who live in the South-west, and we shall be glad to hear of the result of any experiments that have been made in this direction. The tree is related to the Honey-Locust, though it belongs in a different suborder. Its systematic name is *Prosopis glandulosa.*

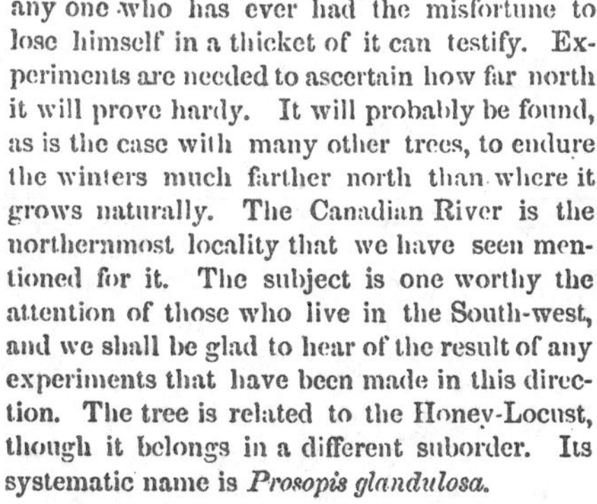

MEZQUIT TREE.—(*Prosopis glandulosa.*)

CURRANTS.—A well-known fruit grower once said to us, as we were discussing varieties of currants: "I can take the old Red Dutch, and treating it in my way, get $200 an acre more from it than from any other variety." This seemed a pretty large statement, but upon seeing the result of severe pruning and liberal manuring upon some of our own bushes of Red Dutch, we think it was not much out of the way. This variety is superior to all others in flavor, and with proper treatment will rival them in size. Our friends' "way" is judicious pruning, liberal manuring, and thorough cultivation. With these, this old sort, as well as the White, will attain a development and excellence in marked contrast with the fruit as ordinarily seen.

Bowman's Root, or Indian Physic.

In a large collection of herbaceous perennials, there was nothing which attracted more attention in June than a native plant which has received the rather inelegant name of Bowman's Root or Indian Physic. Its botanical name is *Gillenia trifoliata;* and we think that those who dislike botanical names will prefer to call this plant Gillenia, rather than by either of the English names, which have a strong flavor of the shop. The plant is closely related to the Spiræas, and in its foliage resembles some of them. The stems are two to three feet high, several from a root, of a pleasing reddish color, and clothed with neat, dark-green foliage. The flowers form a large, loose cluster, one of which, much reduced in size, is given in the engraving. The buds are rose-colored, but the flowers are white, with sometimes a tinge of rose color. The appearance of a plant when in flower is extremely pleasing: the colors of foliage and stem produce a fine effect, and the quaint-looking flowers are supported by such slender stalks that at a little distance they appear as if floating in the air. The root is a powerful emetic, and having been more or less used as such, it has received the names we have already given as well as several others having reference to its medicinal properties. The Gillenia is found from New York southward, in woods and on the borders of rivulets. It flourishes better in the soil of the border than we have ever seen it when growing wild, as not having to struggle with other plants, it has a chance to develop its form. It is readily multiplied by dividing the roots, or by seeds, which should be sown as soon as ripe, as they will otherwise be long in germinating.

BOWMAN'S ROOT, OR INDIAN PHYSIC.—(*Gillenia trifoliata.*)

Cross-breeds and Grades.

The distinction between cross-bred animals and grades is so seldom insisted upon, that we are inclined to define the terms when used in connection. "Grades" among neat-stock, sheep and swine, are animals which have thorough-bred sires, with more or less, or altogether common blood on the side of the dam; while a cross-bred animal has both sire and dam thorough-bred, but of different breeds; or it traces its blood on both sides to thorough-bred stock of different breeds. Thus, if a common cow has a heifer calf by a Shorthorn bull, the calf is a half-grade Shorthorn, and *her* calf, by a bull of the same pure breed, would be a three-quarter-grade Shorthorn. The next grade would be seven-eighths, the next fifteen-sixteenths, etc.—Any pure-blood cow, crossed with a pure bull of another breed, produces a cross-breed, which, crossed with a full-blood or another cross-breed, would represent no gradation of blood, but be a cross-breed still, combining and exhibiting with considerable distinctness, the characters of its different parent breeds, or the predominating ones.

CROSS-BRED HEIFER, "LIZZIE HUYCK."

In the case of grade animals, the common or native characteristics are often almost lost in the half or three-quarter grades, except perhaps some long-cultivated points, like the milking qualities of good, old, native cows, which are often intensified in their grade offspring. This is supposed to show the strength of the vital power of the breed, which has, as it were, accumulated through many generations. By the use, then, of thorough-bred sires, we are enabled to reproduce their valuable characteristics in their grade stock with great certainty. Inferior specimens always occur among herds of pure stock. These must be sold at low prices, or sent to the shambles. They may, however, if females, be used to great advantage often in breeding cross-breeds. As a rule, none but excellent animals should be used as *sires* of either thorough-breds, grades, or cross-breeds.

THE SMALL HARVEST MOUSE.—(*Reithrodon humilis.*)

To illustrate the successful combination of the traits and points of widely differing breeds, we introduce the three-year-old heifer *Lizzy Huyck*, bred by Robert Trimper, of Valatie, N. Y. She is the result of a preconcerted plan of crossing, whereby Mr. Trimper designed to produce good size, feeding capacity, and constitution, and a great yield of milk of high quality. His success in this and several other experiments has been very uniform. Lizzy Huyck is out of a pure Shorthorn cow, (Lady Gifford), a deep milker; and by a cross-bred bull, which was out of a Shorthorn and Ayrshire cow, and by a pure Jersey bull. This makes the heifer, Lizzy Huyck, $^1/_2$ Shorthorn by the dam, $^1/_8$ Shorthorn, $^1/_8$ Ayrshire, and $^1/_4$ Jersey by the sire—or, to state it differently, $^5/_8$ Shorthorn, $^1/_4$ Jersey, and $^1/_8$ Ayrshire. She calved January 3d, 1870. The first trial was begun January 11th; the week following she gave an average of 14 quarts of milk a day, and made 11 pounds of butter. The next trial was begun February 8th; she gave 16 quarts of milk a day, and made 12$^3/_4$ pounds of butter. The cost of food, which consisted of "hay, corn-stalks, and ground feed" for the first week, was about 33 cts. a day, which, with butter at 45 to 55 cts. a pound, paid very well. For the second trial, the feed cost 49 cts. a day. Other trials were not made, as the milk was required for use. The engraving is from a photograph, and represents the points of the Shorthorn with those of the Ayrshire, while the quarter of Jersey blood seems to disappear, to show in the pail.

Our Native Mice

The mouse which is best known, on account of its frequency in our houses, is a native of the old world, and belongs to a different genus from any of our native species. We are accustomed to speak of wild mice, or field mice, as if there were but one, whereas there are over fifty different species described by naturalists. Some of these are found from Labrador to the Southern States, while others are of very local occurrence. The explorations of the far West and of the Pacific Coast have, within a few years, greatly increased the number known to science. The genera are mainly distinguished by the structure of the teeth. There are some twenty species of Field Mice belonging to the genus *Arvicola*. About fifteen of White-footed Mice, *Hesperomys*—and four of Harvest Mice, *Reithrodon*. The Harvest Mice have short, hairy ears and tail, and the upper incisor teeth have a longitudinal channel along the front face. The one figured is the Little Harvest Mouse of South Carolina and some other of the Southern States, *Reithrodon humilis*. It is about 2$\frac{1}{4}$ inches long from the nose to the tail, which is shorter than the head and body. The color is reddish-gray above, yellowish-white below, with a buff-colored line separating the colors of the back and the under surface. It builds a nest upon the surface of the ground, among the long grass, using as a material, soft and withered grass. This species is not considered particularly injurious to the farmer, as the stores in its nests have been found to be mainly the seeds of wild grasses. Some of the wild mice are very destructive to young fruit and other trees, by girdling them during the winter.

Irrigation.—Whoever has brooks or springs which may be led in small streams over grass land, ought not to neglect doing it. One properly irrigated acre is worth two or three dry acres; and the art of thorough irrigation and the best way to manage the water is easily learned.

THE HOUSEHOLD.

For other Household Items, see "Basket"

Graham Mush.—Faith Rochester says: Does anybody want to know how to make Graham mush? There are people who raise the best of wheat, year after year, and live on fine flour always, and have no idea how sweet, as well as healthful, the unbolted, or Graham flour is. Make Graham mush as you do corn-meal hasty-pudding, sifting the meal with your hand slowly into boiling water, stirring briskly meanwhile. A few minutes' boiling seems to cook it sufficiently, though many cook it longer. Sweetened cream is an excellent dressing for it, and then if you add fresh berries!

About Keeping Cool.

BY FAITH ROCHESTER.

Amid the fierce heats of July and August, it seems impossible to "keep cool" always, but with proper care much may be done to make life more tolerable than we usually find it. The whole house should be opened and well aired in the cool, fresh morning, letting in the sunshine for a little while. Everything thrives best with plenty of fresh air and sunshine; but we have to choose between evils in some of the hottest days, and I think we shall prefer shaded rooms that are comfortably cool, to a brighter light and oppressive heat, during the hottest part of the day. Rooms may be kept comparatively cool if they are shaded, beginning early in the day, on the side where the sun's direct rays beat in. If there are no wooden blinds at the windows, thick paper curtains answer very well, and may be drawn down close, if light can be admitted from another side, or they may be raised only while the room is actually occupied. During the busy part of the day our sitting-rooms are often vacant a large part of the time, and they should be guarded from the heat of the kitchen stove. Our comfort depends a good deal upon the "looks" of things. Clean rooms seem warmer in winter and cooler in summer, than untidy ones. The sight of clean muslin or lace curtains refreshes one in summer; and so does the appearance of a fresh muslin or a calico frock, or a "spick-and-span" linen coat and trowsers. A room full of buzzing flies never seems comfortable, but it is not well to keep darkened rooms all summer for the purpose of keeping out flies. Better have netting in the doors and windows, and set fresh fly-traps every day. With clean breezy rooms, netting is hardly *necessary* where mosquitoes are scarce, especially if there are no bad breeding places for flies in drains, compost heaps and out-houses. I don't know much about fly-poisons. We have tried cobalt and fly-paper, but nothing serves us so well as traps, made thus: half fill a tumbler with strong soap suds, and cover it with a thin slice of bread, spread on the under side with molasses; cut a round hole in the middle, smearing its edge with molasses. Set a trap near each window, and knock upon its cover every little while, and down go dozens of victims to a soapy death. Don't let the glass stand long enough to become like the "ointment of the apothecary." Straw matting is more suitable than woolen carpeting for summer. Many prefer it all the year round, with large rugs to lay over it in winter. An instructed taste will prefer clean, bare floors to greasy carpets with untold horrors of dust beneath them. I think Art has not yet done its best for us in the way of floors—at least for dwelling-houses. When it does take hold of the matter, I suspect that *nailed-down* carpets will go out of fashion, to the great advantage of human lungs. It is common in stores and offices, and country schoolhouses to sprinkle the bare floor frequently with cold water, which has a cooling effect on the temperature. A sheet rung out of cold water and hung in a room has the same effect. I suppose the pitcher of ice-water on the table comforts us not only by its cool suggestions to the imagination—as the reading of Kane's Arctic Explorations makes life endurable to some imaginative people when the mercury touches 100° above zero,—and by its effect upon us as we drink, but it constantly produces a cooling effect upon the atmosphere of the room; as nature seeks to produce an equilibrium of heat.

It is on account of this constant tendency to equalize the temperature of different things in the same neighborhood that we choose fabrics for summer wear which will best conduct away the heat of our bodies. The human body, when healthy, varies little from a temperature of 98°, the world over. The atmosphere is seldom so warm as this, so the body is almost constantly parting with its heat; but the internal fires keep up the supply in proportion as they are fed by our diet, pure air, and exercise. When the atmosphere is almost or quite as warm as our bodies, we are uncomfortable, because we cannot part fast enough with the extra warmth produced, and we seek the aid of rapid currents of air, using fans to create artificial breezes. Linen garments, being good conductors of heat, are the coolest when the atmosphere is not higher than blood-heat, or 98°. But for keeping *out* external heat when working in very hot rooms, or in the hottest days of summer, flannel serves best. For this reason we wrap ice in flannel to keep it from melting. How people do sometimes burden themselves with unnecessary clothing! Are the demands of fashion so very imperative? or are the poor human creatures who go loaded down with broadcloth, or heavy silks and rich shawls, so stupid that they don't know how to be comfortable? Thin lawns and muslins, white or of light tints, are the coolest ladies' fabrics—comfortable alike to wearer and beholder. A fresh calico frock is cooler than one of silk or worsted, and is always in good taste. Close bands at the neck and wrists add to the warmth of a garment. A lace collar, or bit of lace sewed in the neck of a dress, is more comfortable than a linen collar. Dressed in clean, airy garments, and sitting at leisure in rooms from which the glaring sunshine is excluded, with cool lace or muslin curtains at the windows, and with sprays of delicate flowers and trailing vines in vases on table and mantels, with a dewy pitcher of ice-water, or iced-lemonade ready for any thirsty one, and with fans conveniently disposed, a reasonable person ought not to complain of the heat.

But, in the kitchen, and in the hay-field? Our good old friend, the cook-stove, has the aspect of a demon in such weather. If possible, the cooking ought to be turned out of doors with the coming of summer. If the regular occupant of the kitchen, with its reservoir and roasting oven, is too heavy or too precious to be removed to the wood-shed or back "stoop," a cheap second-hand stove could be made to do most of the ordinary work, and a small charcoal furnace would boil the tea-kettle and heat the flat-irons. One summer, our cook-room was a rough shed, with a wall on only one side (the wall of the house) and we never enjoyed the necessary work over a cook-stove so well before nor since. There are few women that would n't like to hear the birds sing while getting breakfast. It is bad enough to have a hot stove in a room where the exercise of house-work is mainly taken; but it is quite too bad if those who have been busy in preparing the dinner, and those who have just come in from the sweltering harvest-field, must sit down to their meal in proximity to a raging fire. Better dine in the parlor if that is the only alternative.

If the poor people who can't get a comfortable breath of air while they eat, attempt to satisfy their hunger with the most heating kinds of food—as fat, starch, and sugar, or their compounds, how can they reasonably pray for comfort? The experience of all observing people corroborates the testimony of Chemistry and Physiology with respect to the effects of different kinds of food in heating or nourishing the body. In animal food the fat contributes little except heat; the lean part being that which builds up the body and gives strength. In the wheat kernel the inner portion which makes the fine, white flour, is largely carbon, in the form of starch; while the muscle-making, strength-producing portion of darker color lies between this and the woody bran, and is nearly all sifted out in the bolting of the flour, and fed to the cattle. Unbolted wheat flour contains all the elements needed by the body. Fine, white flour, so highly prized by ignorant house-keepers, contributes scarcely anything but fuel for the burning-up of worn-out particles of the body. Just think then of the absurdity, on a hot July or August day, of such a dinner as is common among farmers, consisting of fried pork (fat, nearly all carbon or *fuel*); potatoes (nearly all starch, or fuel for combustion); bread made of fine, wheat flour (starch, or fuel); butter (which is the oily, or carbonaceous portion of milk, that, if taken whole, contains all the elements of the body); pie (whose crust is indigestible fine flour and fat); and suppose we add to this hot coffee!

Think of this combination for a "tea!" fine, flour bread, perhaps "hot, shortened biscuit;" butter; fine crackers; rich cakes and cookies; sweetened preserves, or highly sweetened sauces of any kind; hot tea. How unreasonable! Bread made of unbolted, wheat flour, lean meat, fresh vegetables, and fresh fruits, are the best articles for summer diet. Butter is not a necessity, but do not think to substitute molasses, for that is quite as heating. A great many sensible folks give sincere thanks for a supper of unbolted wheat "gems," with good apple-sauce, or eat with keen relish and an intelligent expectation of benefit, "wheaten grits" boiled soft, with strawberries and other fruits. The melons and berries and tree fruits that come with the season, are wholesome and comforting articles of summer diet, *if taken as food at meal times.* Hot food and hot drinks always raise the bodily temperature. Hearty dinners might often be taken cold as well as hot in summer. Cold meats sliced, cold beans or peas, cold puddings (not too starchy nor too sweet), and if tea or coffee must be had, they may be taken cold. Use *very* cold drinks with moderation; they are not best taken with food; as they lower the temperature of the stomach below the point necessary for good digestion.

Some folks shut all their doors and windows at nightfall, because they think the night air is injurious. Such a supposition implies a doubt of the Creator's wisdom; for what air is it natural for us to breathe in the night if not the air of night? The experience of soldiers shows how groundless is this whim. It is a great comfort to let in the cool, evening air, after a very warm day.

Bitter Butter.—J. G. Caulkins, Dutchess Co., N. Y., says: "When the milk is brought in and strained, set the pans, one at a time, over a kettle half full of boiling water, and let them remain until the milk is *thoroughly scalded;* this is to be repeated the next day, and the milk then set aside in the pantry adjoining the sitting-room or kitchen, and kept comfortably warm until fit to skim; the cream is to be kept in a loosely covered jar, in the same temperature, and well stirred every time fresh cream is added, and churned at least once a week; the butter will be as sweet, and almost as rich as in June or October. If an Orange Carrot be grated fine, a little warm water poured on it, and the juice pressed out, strained, and stirred in the cream before churning, the butter will be of a beautiful golden yellow."

Cooking Tomatoes.

The tomato is a vegetable that is difficult to spoil, and it is generally acceptable even when rudely cooked. It is capable of so much change in the cooking as to afford a pleasing variety. One way of stewing tomatoes is to choose very ripe ones, skin, and slice, rejecting any hard parts. Put in a pan with salt, butter, and pepper, and cook very slightly, not more than ten minutes. Another way is to stew the tomatoes until thoroughly soft, rub them through a sieve, and then cook them down to the desired thickness. Butter, salt, and pepper, are the usual seasoning. Those fond of the flavor of onions will find the addition of chopped onions while cooking, to make an excellent variety. Baked tomatoes are fine. Choose large fruit, and cut out a cavity at the stem end; fill this with a mixture of powdered cracker or bread crums, butter, salt, or other seasoning, set on a pan and bake until done. If managed carefully, the tomatoes retain their shape. Tomatoes may be broiled; cut them in halves crosswise and put them cut-side down, upon a gridiron over the fire. When the cut surface is seared, turn them and put butter, salt, etc., on each, and cook with the skin side down until done.

How to use Canaille.

The request of a correspondent for information upon the use of Canaille, has brought out several answers. We give the following from "J. G. C.," Knoxville, Tenn.—There are several ways in which Canaille, or wheat shorts (or midlings, as it is termed by some), can be prepared as food, each of which will be found both palatable and acceptable.

Bread and Biscuit.—Mixed with an equal part of wheat flour it makes a nice, wholesome bread, a little dark colored perhaps, but light, spongy, and sweet.—It also makes most excellent biscuit, which, if served up before they are entirely cold, and with butter and honey, will be found very fine.—Mixed with a little Indian meal it makes superior griddle cakes, if served hot.

Canaille Pudding.—Take sour milk, a little saleratus, and if the milk is not pretty sour, a little cream of tartar, a couple of eggs, and canaille stirred in until the batter is a little too stiff for plain cake, bake in any convenient dish, in a rather slow oven, serve hot with hard or wine sauce.

Hints on Cooking, Etc.

Water-Ices.—These are an agreeable change from ice-cream, and are by many considered preferable to it. They may be made from any juicy fresh fruit, or various flavors may be given without the use of fruit. Pine-apple, Orange, Raspberry, and Currant, are the fruits most commonly used. A syrup is made in the proportion of one pound of the best white sugar to a pint of water. If the sugar is clean, this will need no straining or clarifying. Equal parts of this syrup and fruit juice are used and frozen in the same manner as ice-cream. In freezing, water-ices require more thorough stirring than ice-cream, to prevent them from becoming lumpy. The addition of lemon juice will improve the flavor of pine-apple and other water-ices from fruits that do not contain much natural acidity.

Raw Tomatoes.—The almost universal popularity of the tomato has led to a great variety of ways of eating it. Probably more are consumed raw than in any other way. The manner of dressing them varies greatly. Some use only vinegar and salt. Others, vinegar, salt and oil, and others again, vinegar and sugar. We often see them served with a regular salad dressing, but the general way is to place them upon the table plain, and allow each one to prepare them to suit his own fancy. They are usually sliced with their skins on, but those who are very particular about such matters scald and peel them before slicing, and the slices are put with pieces of ice to cool them and restore their firmness. Tomatoes when eaten raw show the qualities of the different varieties better than when cooked.

Pickles.—J. M. Fleming sends the following: "Pick each morning, and put into weak brine, allowing them to remain three or four days, or long enough to become sufficiently salt for use, putting in mustard pods and horseradish leaves to keep them green. Then take out and drain, and cover with good vinegar for a week, at which time take out again, drain, and put into fresh vinegar, adding mustard seed, ginger root, cloves, pepper, and red pepper pods—about one or two ounces each to the barrel, or to suit taste. The pickles will be nice and brittle, and pass muster at any table. Put on the vinegar cold, and add the spices as desired; but the vinegar must be changed once, as the large amount of water in the cucumbers so reduces the vinegar that this change is absolutely necessary; and if they should lose their sharp taste, just add a little molasses or spirits, and they will be right again. Mrs. F. put hers up in this manner last season, and we never had so good pickles; they were always ready for use, and kept well.

containing a great variety of Items, including many good Hints and Suggestions which we throw into smaller type and condensed form for want of space elsewhere

Substitutes for "Muck."—Swamp muck is vegetable matter in a state of slow decomposition. In the swamp the decomposition is exceedingly slow—when dried and mingled with the soil, much more rapid. Any vegetable matter may be used in a manure heap as a substitute for swamp muck or peat. Wood mold is the closest approach to it. The parings of wet meadow land, or any grass sods taken from the fence rows or sides of the road, laid in a heap to decay, resemble it closely, and both these articles may be used when dry in the stables or yards, as absorbents. They will make mud, however, if they get very wet.

Apples for Minnesota.—The State Horticultural Society, at its last annual meeting, *Resolved*, That the following varieties of apples have been found worthy to be recommended for planting in Minnesota.—Duchess of Oldenburgh; Haas; St. Lawrence; Price's Sweet; Fameuse; Golden Russet; Red Astrachan; Talman's Sweet; Tetoffsky; Saxton, or Fall Stripe; Perry Russet; Ben Davis, or New York Pippin.

Mixing Soil with Manure.—Heavy loam or clay mixed with manure in a heap, has a tendency to retard fermentation, and may, consequently, be used to advantage with horse or sheep manure, which, when placed in a loose heap by themselves, ferment too rapidly. The clay will also hold the ammonia, and prevent its escape from the heap. On the other hand, sandy loam, or sand mixed with manure, regulates fermentation, and may be used to advantage with hog and cow manure, which is of a sluggish nature. The better way, however, is to mix all these manures together as made.

Sore Mouth in Cattle.—A correspondent in Arlington Co., Mo., writes that a disease is there prevalent called Black-tongue, Sore-tongue, and Sore-mouth. Several of his neighbors have lost cattle. The symptoms described are foaming at the mouth, and, inability to eat. Our friends of the Veterinary College advise a weak solution of carbolic acid—say 1 to 5 drops to the ounce of water—washing the mouth every few hours, allowing a little to be swallowed, and following this with mild tonics and food that will not irritate the mouth. Fluid preparations of gentian and iron in small doses, given with the feed, is a good tonic; for food give a warm gruel of oil meal or wheaten flour.

Joint Corn.—A variety of corn has been shown to us which is at least a curiosity, as it produces an ear at each joint. Each of the several stalks we saw had from 8 to 12 ears upon them. It is small in the ear and grain, like pop corn, but if, as is claimed, it will yield 200 bushels to the acre, it will prove valuable.

How to Feed Fowls.—Fowls are not fed for the mere sake of keeping them alive and healthy on the least possible amount of food. We wish to convert the food into flesh, or into eggs. In feeding for quick fattening it is understood that the poultry should be made to eat as much as possible. Our rule for feeding is to throw out the feed twice a day as long as the fowls will run after it and no longer. We are told, and it is our own experience also, that fowls thus fed will eat considerably more than if they can go to a feeding box and help themselves at all times. We want the fowls to eat; the more they eat, within reasonable bounds, the more eggs they will lay, the longer they will lay, and the better condition they will be in. Laying fowls should take exercise. If they can go to a trough and eat at any time they wish, they will take next to none. If they are fed but twice a day, they will hunt insects and wander much more. If fed soft feed such as wheat bran mixed with corn meal or ground oats, they will be hungry again in two hours after feeding, and be off after insects, etc. Give feed, then, only to adult fowls while they will run after it—soft feed morning, whole grain at evening. Keep them supplied with gravel, lime (plastering, or, better, oyster shells), ashes to dust in, and fresh pure water, some meat in winter, and they will be healthy and prolific.

Worms in Horses.—J. W. Bruce of Mass., asks for a cure for "the small intestinal worms in the horse." Give drachm doses of tartar emetic twice a day for two or three days, and follow with a mild purgative, say one of epsom salts, or four or five drachms of aloes.

Arbor Vitæs.—A lady wishes to know if pruning will prevent her Arbor Vites from turning brown in the fall. It is one of the faults of this tree that it becomes unsightly in winter. No pruning or other treatment will remedy it.

Four-leaved Clover.—J. R. Carter. The clover with four or more divisions of the leaf instead of the usual number of three is not a species, but a not very unusual development of common clover.

Wild Lilies.—A correspondent sends us a specimen of the Turk's-Cap Lily, and asks if it can be transplanted to the garden. This species (*Lilium superbum*) makes a fine plant in cultivation, and is highly valued in Europe. The most favorable time for taking up the bulbs is when the leaves have turned yellow.

Onion Seed.—A. N. Curtis. Our seed-growers thrash with a flail and winnow. Some wash the seed to clear it of the light particles, but our best growers do not approve of the practice.

Geranium Sporting.—"F. K. M.," Philadelphia. Some of the geraniums will sport in the way you mention, and by propagating the sporting branches the peculiarities are perpetuated.

Pear Blight.—C. H. Kent. The blight is probably a minute fungus, the presence of which is not suspected until the mischief is done. No preventive is known. The best way is to cut back the limbs to sound wood and burn the trimmings. White lead applied to the wounds of a tree has no special curative properties. It protects the part from the action of the weather, and is as easily applied as anything.

Characteristics of American Horticulture.—Under this title the editor of the Horticulturist writes to the English Gardener's Magazine. He says: "You do not find here a true American son of the soil touching anything high or low (much less horticultural) unless there is *some money* in it."—"There's an opinion as is an opinion."—Neighbor Williams should publish this article in his own journal, if he wishes to get up an excitement among American horticulturists.

Tan for Hot-beds.—"Weymouth." Tan is sometimes used for producing bottom heat. It takes much longer to heat up than stable manure, and the heat continues longer. It is sometimes mixed with stable manure. It is but little used in this country, and we cannot say how it [illegible]d do for winter work.

Plants Named.—E. C. Breed, Waupacca Co., Wis. The Twin-flower, *Linnæa borealis*. A beautiful little vine and doubly interesting on account of bearing the name of the great Linnæus.... W. S. Williamson, Coffee Co., Kansas. *Callirrhoe involucrata*. We know no

Harris on the Pig.—Mr. Joseph Harris, of Rochester, has long been a raiser of thorough-bred pigs, and has experimented largely in crossing these animals on the common swine of the country. Mr. H. now presents, in a neat volume of a little over 200 pages, an excellent manual upon the subject. The various foreign breeds are described and their qualities discussed, as are those breeds which are considered to be purely American. The principles of breeding, the care of young pigs, feeding, fattening, the construction of piggeries, and other pertinent matters are treated with satisfactory fullness. The work is illustrated by numerous engravings of the different breeds, plans of houses, troughs, etc. We have heretofore, in this department of agricultural literature, been mainly dependent upon reprints of foreign works, and we are glad to be able to present a work so well adapted to the wants of the American farmer and breeder. Published at this office. Price, $1.50, post-paid.

Bad Smelling Cisterns.—Mrs. "R. B. J.," Delaware.—The bad odor does not arise from the cement, but from the organic substances washed into the cistern from the roof. The remedy is to clean the cistern and wash it out, and then to have all the water that flows into it filtered. For the present a filter may be prepared thus: Take a keg holding half a barrel or more, put in 6 inches of clean coarse gravel, upon this 4 inches of sand, then several inches of freshly heated charcoal, pounded small, with the *dust* blown out, to be covered with 4 inches of sand and 6 of gravel. The water is poured in at the top and drawn off at the bottom as wanted. A coarse bag, filled (while wet) with hot charcoal broken up somewhat, and suspended over a cistern or well, will often destroy bad odors and flavors in the water.

AMERICAN AGRICULTURIST

FOR THE

Farm, Garden, and Household.

"AGRICULTURE IS THE MOST HEALTHFUL, MOST USEFUL, AND MOST NOBLE EMPLOYMENT OF MAN."—WASHINGTON.

VOLUME XXIX.—No. 9. NEW YORK, SEPTEMBER, 1870. NEW SERIES—No. 284.

THE SOCIETY OF FRIENDS.—FROM A PAINTING BY HERRING.—*Drawn and Engraved for the American Agriculturist.*

Mr. Herring's pictures of horses are widely known. They have not the dash and spirit of those of Rosa Bonheur and some other celebrated animal painters, but they are exceedingly pleasing on account of their domestic character. In the many farm scenes which this artist has painted, he has given the animals a thoroughly home-like expression. An artist, to be able to give in a picture the varied expression of which a horse's face is capable, must be not only a close observer, but a real lover of the animal. This, Mr. Herring was; for in early life he was the driver of a stage-coach, and in his later years was constantly surrounded by the pet animals, which he has introduced into his celebrated picture in so many pleasing relations. In the above picture he presents horses as affectionate friends. That horses do exhibit remarkable attachments to one another, as well as to man, is within the observation of almost every one. Mules, usually, unjustly considered inferior to the horse in every "moral" quality, form attachments even more readily than horses, and show them in the most positive manner. The writer knew of a mule that was so fond of a particular horse that, no matter how hungry it was, it would not eat if the horse was in sight, and would be perfectly contented if it could only be allowed to stand near him. We are all familiar with the accounts of the manner in which the Arabs treat their horses; they not only make friends of them, but even admit them into their families. The artist calls his picture the "Society of Friends"—we do not know if he had our friends the Quakers in mind, at any rate, the animals have that air of content and inward satisfaction, which comes from kindliness and well-doing, which is one of the characteristics of that estimable people.

THE PRACTICAL POULTRY KEEPER.

A Complete and Standard Guide to the

MANAGEMENT OF POULTRY,

FOR DOMESTIC USE, THE MARKETS, OR EXHIBITION.

BEAUTIFULLY ILLUSTRATED.

BY L. WRIGHT.

CONTENTS:

CONTENTS:

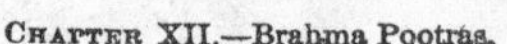

Price, Post-paid, $2.00.

ORANGE JUDD & COMPANY, 245 Broadway, New York.

TEGETMEIER'S POULTRY BOOK.

THE BREEDING AND MANAGEMENT OF POULTRY, THEIR QUALITIES AND CHARACTERISTICS.

To which is added "THE STANDARD OF EXCELLENCE IN EXHIBITION BIRDS."—Authorized by the Poultry Club.

BY W. B. TEGETMEIER, F. Z. S.

English Edition, 8vo. With colored Illustrations by HARRISON WEIR, and numerous engravings on wood. **Price, Post-paid, $9.00.**

SAUNDERS' DOMESTIC POULTRY.

A PRACTICAL TREATISE ON THE

PREFERABLE BREEDS OF FARM-YARD POULTRY, THEIR HISTORY AND LEADING CHARACTERISTICS.

With complete instructions for breeding and fattening, and preparing for exhibition at Poultry Shows, etc., derived from the author's experience and observation.

BY SIMON M. SAUNDERS.

FULLY ILLUSTRATED. Price, Post-paid, paper covers, 40 cents; cloth, 75 cents.

ORANGE JUDD & CO., 245 Broadway, New York.

AMERICAN AGRICULTURIST.

NEW YORK, SEPTEMBER, 1870.

The distressing heat and dryness of the summer have been in a measure relieved by the rains of August. Serious losses in important crops are felt by many farmers, and these should, by all means in their power, try to make good the deficiency, especially of fodder. Hay will be high. Corn-stalks are the best substitute we can get, and if these are well secured, we may be able to sell hay, which it will be quite worth while to do if we can get a sufficiently high price, and have enough left to winter the stock. Farmers should contract early, if they can make favorable terms for feed. Bran, linseed and cotton-seed meal may be bought better now than later. European farmers have suffered from the drouth more than we have; and prices there are advancing, both on that account, and on account of the Franco-Prussian war, which, at the time we write, is so deeply interesting the whole civilized world. We shall probably soon feel the effect of such heavy drafts upon the productive labor of both countries engaged. The crop of wheat is excellent, and far more than is needed in this country; while the amount of corn, notwithstanding the drouth, will be immense, so far as we can now form a judgment. It is therefore to be hoped that our farmers will realize good prices for those things which they may have in abundance.

September is the seventh month by name, though the ninth of the Calendar. It is the seventh month of plowing and sowing; the seventh of growth and of tillage. It is the month of golden corn and pumpkins, of golden apples and fair pears. 'Tis the month of Farmer's Festivals, of Cattle and Implement Shows, where there is so much to see and learn that the mind wearies at the thought, as will also back and limbs before the days of sight-seeing are over. Fairs are to be very numerous and very fine this year. We have never known more interest manifested in them, and the list we prepare is longer and more complete than usual. Every farmer should try to go with his family, and to give his hired men a chance besides.

Hints about Work.

Full Barns.—The constant destruction of grain and grass by a few hundred mice is enormous. As soon as the weather begins to cool they will come in from the fields and make great havoc. Several traps, well baited, and well watched, will thin their numbers, as also those of rats and other vermin; but the farmer should especially guard his premises from fire—which the ashes of a pipe, the end of a match, and many another bit of carelessness will easily kindle. Smokers must do their puffing, away from barns and woodsheds. There is no more terrible thing than a fire in the country, and the insurance of house, barn, and stock against it should never be neglected. Country insurance rates are moderate, and few farmers can afford not to be well insured. This subject naturally suggests

The Water Supply.—During such a drouth as we have had, the wells and springs have stood a severe test. Pumping water for cattle is hard work, but there are pumps which save a vast amount of labor, and to which a hose may be attached, and water thrown over garden and lawn, and over house and barn in case of fire. Barns may be supplied by

Cisterns.—If these can be placed underground, and at a higher point than the cattle floor, it is a great convenience; and it will pay to carry water several rods from the eaves to find an elevated spot.

Draining should be in the plan of fall work, for some thoroughly done every year will not be much felt as a heavy expense. Living springs are often tapped by drains, and thus a supply of water is secured at the outlet, which may be turned to good account. If not useful as a supply of water for the stock yard, it may serve a good purpose in

Irrigation.—When ground lies lower than an abundant source of water, there is no excuse for not irrigating. Irrigated ground must be well worked and drained so as not to remain water-soaked. When irrigation is well done, it is the most economical way possible of manuring grass.

Manure.—Everything that will rot and that is not full of weed seed should be thrown into the hog-pen, or made into compost with lime or yard manure. We prefer to subject all such material to be worked over by the hogs.

Grass.—Manure at this season if it has not already been done. Leave the aftermath; neither cut it nor feed it off unless it is heavy. Sow grass and clover on thin spots, and harrow in lightly with some good compost, rolling the surface afterward.

Orchard and Nursery.

Planting may usually be done this month, and in localities where the autumns are mild, it is preferable to spring for all except stone fruits. The work is done more thoroughly, and the trees usually succeed quite as well as those set in spring.

Nursery Trees should be ordered in good season that the planting may be done before heavy frosts.

Stripping the Leaves.—We are frequently asked if stripping the leaves from trees taken up in the fall is injurious to them. It is a common practice among nurserymen, and we see no objection to it if the leaves have finished their functions in ripening the wood and perfecting the buds. It merely anticipates what would take place at the first storm.

Picking Fruit.—It is easy to lose money by the careless handling of fruit. If there is neglect in the picking and packing, much of the care expended in raising goes for nothing. Our most successful fruit men are those whose fruit opens in the market in the best shape. Autumn varieties should be picked when fully mature, but before they show signs of softening, as they should reach the retailer before they are in eating condition. Use new barrels. Half-barrels are preferred for pears, though many use crates. In either case the package must be closed by pressure, so that the fruit will not shake. All good fruit should be hand-picked; some ladders for the purpose are figured on page 88

Fallen and *Refuse Fruit.*—It does not pay to send poor fruit to market; all the cullings as well as that which falls from the trees should be utilized at home. Apples, ground and pressed to make vinegar, pay a good price; pears may have the sound portions cut out and dried, and peaches may be dried; and those portions of all fruits not available for other uses, may go to the piggery.

Budding.—The stocks budded last month will probably need to have the ties cut. Pear upon quince, and peaches may be budded.

Seeds are to be secured. Peach and other stone fruits are to be mixed with earth and exposed to frost during winter. Peach stones are usually strewn thickly upon a bed and spaded in.

Fruit Garden.

A rich, deep soil should be secured by spading or subsoil plowing. Where the season is mild, fall planting is advisable.

Blackberries.—As soon as the crop is off, remove the old canes. Top the new growth at 5 feet if not already done, and keep the side shoots to about 18 inches. Three or four canes are enough to a stool, and these should be tied to a stake with strong twine. Hoe off all suckers.

Raspberries should have superfluous suckers kept down and the canes tied up, unless they are made self-supporting by timely shortening.

Black-caps.—If it be desired to propagate these, throw a little earth upon the pendant tips to keep the wind from moving them about. The tips will soon strike root.

Strawberries.—Plants rooted in pots may be set at any time. The spring is the favorable season for planting strawberries, but many are obliged to do it in the fall. The plants at this season should not be allowed to dry at the roots, as they with difficulty recover. If those set this fall make runners, pinch them off and keep the beds clear of weeds.

Grapes.—Use scissors in gathering and avoid unnecessary handling, in order to preserve as much as possible the bloom upon which the beauty of the fruit depends.

Currants and *Gooseberries.*—Make cuttings as soon as the wood is thoroughly ripened, and set them.

Kitchen Garden.

Clear up the ground after each crop, and manure and plow or spade for winter crops.

Beans.—Gather and salt string beans, and shell and dry Limas for winter use.

Cabbages and *Cauliflowers.*—Seeds for the spring crops are sown here this month, from the 10th to the 20th, and later further south. The object is to get a good, strong, young plant before the ground freezes. These plants are set out in cold-frames, where they are kept as nearly dormant as possible, until time to set them out in early spring. The seed is sown in well-prepared seed beds, and the plants thinned and weeded as usual.

Corn.—Dry for winter use, taking that which is just fit for the table. Old sweet corn makes a very poor dried article. It is boiled long enough to set the milk, before cutting from the cob.

Cucumbers.—Gather for pickles every other day.

Endive.—Blanch by tying, or by covering the plants with mats in dry weather.

Melons.—Choice specimens may be made to ripen evenly by putting a bit of board under them. Use for mangoes those which set too late to ripen.

Onions to be stored for winter, need to be thoroughly cured. They should not be in large enough masses to heat. Onion sets of all kinds need to be spread in thin layers in cool lofts.

Radishes.—The winter varieties may be sown. The Chinese Rose-colored we consider the best.

Sweet Potatoes.—By carefully opening the rows, some of the largest roots may be removed for use, and the others allowed to grow. It is necessary to dig the crop at the first touch of frost, otherwise the potatoes will not keep.

Tomatoes.—Can and make catsup while in their prime. The late ripening is apt to be watery.

Turnips.—The round varieties may still be sown. Give the Swedes good culture, and phosphate.

Manure.—Garden refuse should be turned to account in the piggery or compost heap. Burn weeds with ripened seeds, and spread the ashes.

Flower Garden and Lawn.

Now that the intense heats are over, the garden should be gay with late-blooming plants.

Cannas.—These fine ornaments to the garden are very sensitive to frost. If the foliage be injured, it is with difficulty that the roots can be preserved. When frost is threatened, lift the roots, and lay them under a shed to mature.

Chrysanthemums are so brittle that they are liable to be broken by storms, and should be staked. Pot those intended for house blooming, as soon as the buds are well formed, shading them for a few days.

Dahlias will need careful staking and tying, to keep them from injury. Side stakes, in addition to the central one, will be useful. Remove misshapen buds and flowers that have passed their prime.

Bulbs.—All of the spring-blooming ones that were taken up in summer, are to be put out at the end of the month.

Herbaceous Perennials.—The majority are better removed this month, than in spring. Sow seeds.

Pæonies, if they require moving, should be transplanted this month.

Pits and Cellars for the reception of half-hardy plants, will allow one to preserve many things usually kept in a greenhouse. A dry, well-lighted cellar, which is safe from frost, will keep plants in a dormant state admirably.

WOOD DUCKS—MALE, FEMALE AND YOUNG.—DRAWN BY HERRICK.—*Engraved for the American Agriculturist.*

Greenhouse and Window Plants.

The houses should be in readiness to receive the plants as soon as the weather requires that they should be taken in.

Potting of plants that have been turned out into the borders, will require early attention. It is better with most things to start with young plants from cuttings, as it is often difficult to bring the old ones into satisfactory shape.

Bulbs for winter blooming are to be potted as early as they can be procured; plunge the pots in a dry place, and cover them with coal ashes.

The Summer, or Wood Duck.

This familiar, native species is universally admitted to be the most graceful and beautiful of all ducks. We can hardly compare its delicate beauties, its gay and richly tinted feathers, its vivacious motions, whether upon land or water, or on the wing, with the deliberate movements, and snowy plumage of the majestic swan, but it is not too much to say that with this single exception it is the most beautiful of all water fowls. It is notable as being the only wild duck which remains naturally with us during the summer. A few Mallards and Black Ducks on the prairies, and belated pairs of several other species breed within the United States, but the Wood Duck is here at home. It seeks retired wooded streams or lakes, and in a quiet nook finds some hole within a few hundred feet of the water, where a dead limb has decayed, which may easily be cleaned out and made fit for a nest, where it may rear its young. The pair are devoted nurses, and a prettier sight can hardly be imagined than such a family, as they scoot about among the leaves of aquatic plants hunting for seeds, water insects, and minute shell-fish. They breed but once, and the young are able to fly soon after midsummer. Their geographical range is from Texas to the St. Lawrence and Columbia Rivers. This duck seldom winters further north than Maryland, and it is protected by law from reckless sportsmen during the summer months. As the country becomes settled, we must expect to see it gradually disappear, but its extinction need not be anticipated, especially as it is easily domesticated; where its natural habits are studied and accommodated, it breeds and does well in confinement, if pinioned, but has a permanent tendency to roam and return to the wild state unless its wings are cut. It loses its brilliant plumage also, to a considerable degree, unless it be bred in a half-wild state, and can get its natural, wild food. When these conditions are provided, though it may be tame enough to feed from the hand and run with other ducks in the yard during the winter, it will breed freely if unmolested.

Ornamental Shrubs.—The Bladder Sennas.

The Bladder Sennas, or *Coluteas*, are desirable in a collection of shrubbery, as they are of rapid growth, and though not particularly showy, are pleasing both in flower and in fruit.—

BLADDER SENNA.—(*Colutea Haleppica.*)

They belong to the Leguminosæ, or Pulse Family, and are nearly related to the Locust. The flowers are borne in small clusters in the axil of each leaf, and are of the shape and size shown in the engraving. The flowers are succeeded by thin, bladdery pods, which look as if they had been inflated; these, when suddenly pressed, burst with a small report. The pods being usually of a reddish color, the shrub is quite attractive in fruit. The leaves have been used as a substitute for senna, a circumstance which, together with the bladder-like character of the pods, has given the popular name. The origin of *Colutea*, the botanical name of the genus, is unknown. Several species are given in the books, most of which are probably varieties of *Colutea arborescens*, the common Bladder Senna of Southern Europe. Our engraving is taken from the one known as the Aleppo Bladder Senna, *Colutea Haleppica*.

The Elegant Humea.

In the ornamentation of the Flower Garden at Central Park, the Humea is introduced with fine effect. The engraving which we present, while it gives the form and habit of the plant, cannot convey a proper idea of its real beauty. One must imagine a plant of from four to eight feet in hight, with its gracefully pendent branches stirred by the breeze, and crowded with minute crimson or purplish flowers, to have an idea of its elegance. The plant belongs to the Composite Family; its heads of flowers are very small, and covered by a colored involucre, which is the showy part of the flower. The flowers have the dry, papery character of the Rhodanthe and others of the "Everlastings," and like them may be used for winter bouquets. This Humea is no novelty, as it was introduced into cultivation from Australia in 1800; still it is, in this country at least, very rarely seen. The reason it is so seldom grown, is probably due to the fact that it is a biennial. Our people are generally too impatient, to grow a plant a whole year in hope of its flowering the next; and in the desire for immediate returns, they overlook many beautiful things. The seeds are sown in September, and the young plants, when large enough, are potted singly, and are kept in the greenhouse all winter. In spring they are turned out into a rich border or bed, and usually come into flower in July. The plants may be put in groups of several, but well-grown specimens placed singly upon the lawn produce a fine effect. At flowering time the leaves die away, and the plant looks naked at the base, a defect which may be concealed by setting some other plants around it. The young Humeas, during the first year, have so strong a resemblance to tobacco, that several have mistaken our young plants for "the weed." Unlike that, the leaves of the Humea have a spicy fragrance, compared by some to that of cinnamon. The genus was named in honor of Lady Hume.

The Culture of Spinach.

BY PETER HENDERSON.

As September is the time in this region for sowing the winter and spring crop of spinach, a few suggestions may be useful to your readers. Any soil that will grow a good corn crop, will grow spinach, though, as is the case with all other vegetables in which the leaf or stem is the part used, the land can hardly be made too rich. Our practice is to grow it on our best soils, applying not less than 50 tons of well-rotted stable manure to the acre, or, in lieu of stable manure, one ton of bone-dust; or about 1,200 lbs. guano, sown after plowing, and deeply harrowed in.—The rows are made with the ordinary garden "marker," at the distance of 12 or 15 inches apart. The seed is sown rather thickly, we prefer to do it always by hand, using about 10 to 15 lbs. per acre; when thickly sown, the plants can be thinned out so that a much larger yield will be given. We sow here from the 5th to the 15th of

THE ELEGANT HUMEA.—(*Humea elegans.*)

September, and quite frequently sell, by thinning out, 50 or 75 barrels from an acre, which usually, in October and November, sells for $2 per barrel. This thinning out, which is done by cutting out the plants where thickest with a knife, if carefully performed, does not at all injure the main crop, which is to stand over winter until spring. I may here caution the inexperienced of the necessity of treading down the soil on the seed, if the land is dry; the crop is often ruined by the want of this precaution, in continued hot, dry spells that are frequent with us during September. If the soil is left loose, the hot air shrivels up the seed so that it will never germinate. If a heavy roller is not at hand, the best way is to tread in each row with the feet. The same precaution is necessary in the sowing for cabbage and lettuce plants; at this season these are often lost from the same cause. There has some question arisen of late whether the round or prickly seeded Spinach is the best; as far as I can judge, it makes but little difference which kind is used, though we use the round almost exclusively, as it is the easiest to sow. The price of Spinach in the New York market, last year in February, for a few days, reached $10 per barrel, although the average for the spring months of April and May, (its regular time of selling,) was not more than $3 per barrel, but fair crops give a yield of 200 barrels per acre; at a high estimate, the expenses will not exceed $300 per acre, so that it is safe to claim a net profit of $300, although extraordinary crops often do much more than this. A near neighbor of mine realized $900 from three-quarters of an acre last spring, getting the land cleared early enough in May to succeed the Spinach with a crop of Flat Dutch Cabbage.—Spinach is hardy enough to grow in almost any part of the country; but in districts where the thermometer falls below zero, it is necessary to cover it up about Christmas with hay, straw, or leaves, to the depth of 2 or 3 inches; it is best done just as a snow storm is setting in, as the snow settles down the covering, and keeps it from blowing off. Spinach is yet comparatively little grown for our Northern markets at the South, but no doubt soon will be; in such latitudes as Charleston and Savannah, it should be sown in October and November, and would be in fine order for use in February. At this cool season it could be shipped with safety, as it will remain in good condition for three or four days if packed, and would command a rapid sale at the time when our northern crops are still frozen solid. There is another vegetable but little known outside of New York, which is called

German Sprouts, properly *Siberian Kale;* its cultivation is identical with that of Spinach, except that from 2 to 3 lbs. of seed are sufficient for an acre. It sells at nearly the same rates, and is used in the same way; hundreds of acres of Sprouts are grown in the vicinity of New York City (on Long Island and New Jersey). It is mainly used by our German population, who show their good taste in preferring it to heading cabbage, of which it is only a variety.—Sprouts, like Spinach, might also be successfully grown at the South; though I doubt if they would bring as high a price early in the season.

Fruit Ladders.

In years of plenty it is only the best fruit that brings good prices. One step towards having good fruit is to secure careful picking, and to this end it is necessary to have a sufficient supply of convenient ladders. For low trees, step-ladders will be found serviceable. Fig. 2 gives the ordinary step-ladder used in the peach orchards of Delaware. It is made of two boards 10 feet long, 6 inches wide, and 1 inch thick for the sides; the steps are of the same material, let into grooves in the side pieces. At the top is a board about 10 inches wide, upon which the basket stands. The support is of two narrow strips, strengthened by cross-bars; this is attached to the steps by an iron rod passing through its ends, and through the side pieces.—

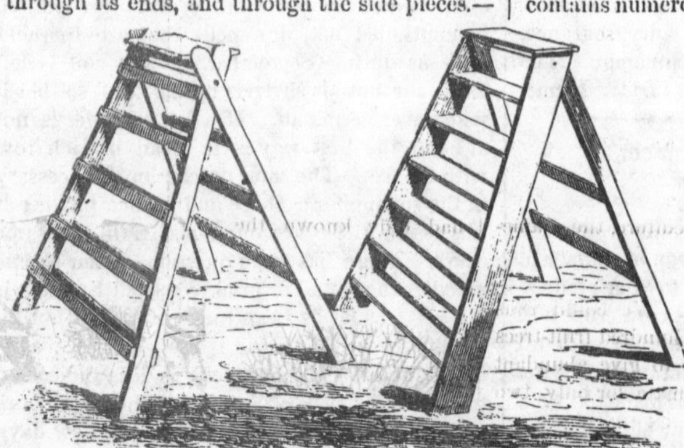

Fig. 1. STEP-LADDERS. Fig. 2.

A cheaper ladder is made of, four pieces of shingling lath. Two of these have strips of the same material nailed opposite each other for steps, as in figure 1. The top step has a board nailed to it to receive the basket. The other two pieces are made to serve as supports as shown in the figure. In both these ladders the bottom is wider than the top, in order that they may stand firmly. It is often the case that ladders much taller than these are required. A common ladder, which should have iron points

Fig. 3.—EXTENSION LADDER..

at the bottom of each side piece, may be so guyed with ropes as to be perfectly safe. It takes considerable time to change the position of such a ladder, and it will be found more convenient to support it by means of wooden stays of a proper length. In September 1868, and January 1869, we illustrated two very efficient methods of doing this. Another form of ladder is proposed by Mr. George H. Russel, which is shown in figure 3. It is really two ladders hinged together, and may be used by two persons at once. The dimensions are: side pieces 9 feet long and 3 by 1½ inches. Width of ladder 1 foot 9 inches, distance between the rounds, 1 foot 4 inches. Long hinges of malleable iron are used to fasten the two parts together, as shown at the right hand of the engraving. The basket can be placed upon the upper two rounds. By opening out this, and using it with the hinges down, it will answer as a single ladder.

In Our Garden.

[The following letter from Galen Oderkirk contains numerous good suggestions, but in presenting it we must dissent from the writer's views in regard to the introduction of the Onion Slug. We do not think it could have easily been introduced with the seed.—Eds.]

During the long drouth we have experienced here in Wayne Co., N. Y., nothing in our gardens has excelled the cucumbers. I attribute this to the fact that we transplant to rows rather than hills. The seed was started upon sods three inches square. It germinated in the hot-house about the twenty-fifth of April, and the plants have so flourished that we had cucumbers upon the fourth of July. The rows are five feet apart, and the vines are at intervals of three feet. We prefer sprinkling our fertilizers along the rows, to manuring in hills, or beneath each plant. Had we reared the vines upon hills in the usual manner, they would have dried out in such a drouth as they have experienced.

For early cabbage we plant the Winnigstadt, with a few Early York, and some new varieties for trial. The Winnigstadt being larger and tighter-leaved than any other variety, brings a better price, or at least finds a more ready sale than other early varieties. Between our rows of cabbage plants we set lettuce and early radishes, which mature without molesting the cabbages, and can be out of the way by the tenth of July. Since that period we have harrowed the cabbages and by aid of the rain they seem beyond injury from any insects. The latter have not troubled us much this season; perhaps owing to a superabundance of lime and salt in the ground.

In the current season we have had among our onions a pest which defied our lime and salt, or suds and tar-water,—an Onion Slug. We have heard much of this noxious mollusk, but we have raised onions very successfully for fifteen years, and never experienced this pest. Why? Without doubt for the reason that we had *pure seed* to begin with, and have raised our own ever since, until last year. The present season we bought seed from Mass. containing these slugs in *embryo*. Perhaps if we had fumigated the seed it would have been different; but as it is, as much labor has been spent for nothing as would have resulted, other things being as usual, in four hundred bushels of onions more than we will have. We have sown turnips in the long gaps where the slugs worked ruin to the onions. The turnip seed had no larvæ in it; the slugs are done with their work, and we live in hope of a crop from the rich, moist soil. But we caution purchasers to soak in copperas water or fumigate their seeds when there is the least apprehension of results like the one we experience. We almost always raise our own seeds. Onions as a specialty, we rely upon, as they usually have resulted in good profits. They have always brought fifty cents per bushel, and sometimes $2. They are more profitable at fifty cents than potatoes. Early peas and beans have been very prolific with us; so also have salad, beets, and early onion sets. The raspberries in this locality have been as plenty as last season. Prices have averaged less than seven and a half cents. The time, toil, and expense of producing and marketing a quart of berries are more than this sum, with the majority who raise raspberries. The Doolittle variety has been a drug on the market, selling as low as at five cents a quart. Our fruit-growers here have as yet, I blush to own it, not associated together, either in keeping prices to a profitable level or in assembling to interchange ideas upon the subjects connected with their business. Enthusiastic writers,—who often judge of the fitness of a locality from one unexceptional season,—have praised our district as of the highest excellence for small fruits. Nearly every one has therefore entered the business. If there were association among us sufficient to demand and require profitable results from investment and labor, perhaps no better section exists for fruit growing than this. But I think many have been induced to spend labor and means upon a business which must prove, as it has again and again proved, unremunerative. Indiscriminate praise of locality, conceit, and entire lack of association, among horticulturists, are productive of failure everywhere. I hope those who read this will do something towards association in every section lacking in this respect. Experience universally teaches that "union is strength." The only question is how to unite. What I have said in relation to raspberries is equally true of strawberries, and other small fruits in this section.

Gardens or Greenhouses upon the Roof.

BY SAMUEL B. PARSONS.

Some ten years ago we suggested an idea of roof gardens, which was conveyed by the *Agriculturist* to its numerous readers. Since that time the almost universal adoption of the Mansard roof has increased the availability of our plan, and has led to numerous inquiries which make necessary an improved description. Time has not diminished our sense of the value of an adaptation of this idea to city buildings where land is almost fabulous in price, and the air in which to build is free of cost.

Fig. 1.—MANSARD ROOF CONVERTED INTO A GREENHOUSE.

Marvellous accounts have come down to us of the beauty of the hanging gardens of Babylon, and the lavish expenditure upon them by the monarchs of that magnificent and wonderful city. We think of their beauty as something unattainable now, and scarcely realize that in any of our cities they can be rivaled without extraordinary expenditure. It is within the means of any man who builds a house to rent for $800 per year, to have a garden on the roof which, during the summer can be filled with the most luscious grapes, peaches, plums, etc., and in the winter with plants, the beauty of the flowers of which will afford a charm far beyond the trifling cost of their maintenance.

A glass roof costs very little more than a tin or slate one. Let the roof, therefore, be covered with glass, and let the garret floor be covered with concrete, sloping gently from the center to the sides, around which a slight depression in the floor can carry the moisture or drip into the leaders which pass from the roof of every house to the ground. With this slight expense, a perfect greenhouse may be had. A Mansard-roof glazed in this manner is shown in figure 1. Now for heating: Every one knows that the upper rooms of his house are so warm from the ascending heat of his furnace that registers are scarcely needed. Let the doors be kept open and the waste heat of the house will keep the garret at the highest desirable temperature. Thus the greenhouse is heated without any extra trouble or expenditure.

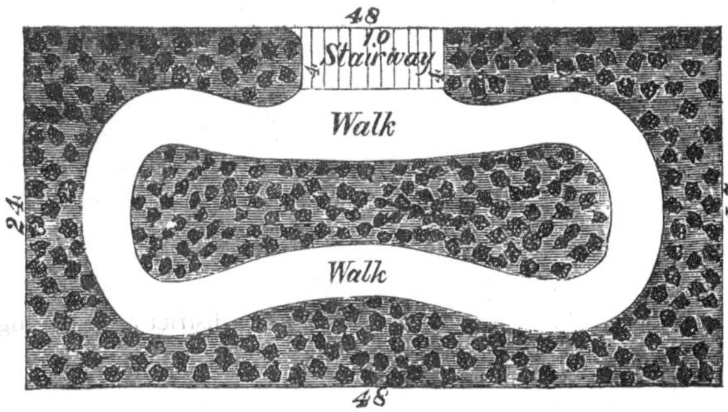

Fig. 2.—PLAN OF GREENHOUSE WITH PLANTS BEDDED OUT.

Its care would be a pleasant recreation for any of the family, provided they enjoy working among plants as much as a friend of ours, who would leave the counting-house and its engrossing cares at a reasonable hour in the afternoon, and after leisurely dining, would put on an apron and go among his camellias, potting and trimming, and enjoying their beauty while giving them those nice manipulations which only a true lover of flowers knows how to administer. A lady fresh from such occupations would lend new charms to the evening hours; and the memory of her children in the upper air would always have power to bring a sparkle to her eye or a glad expression to her lip. And then the pleasure of cutting one's own flowers or sending to a friend one's own roses or camellias or Black Hamburg grapes, is not to be despised. In case the demands of the counting-house or the drawing-room are too engrossing to allow any attention to flowers, there are numerous florists in every city who would be glad to keep such a place in perfect order for a very moderate compensation.

If a little extra strength is given to the beams which sustain the upper story, sufficient earth could be placed there to lay out the whole space of twenty-five by fifty feet as a garden, with winding walks, delightful carpets of moss and roses, camellias, etc., planted in the soil, as in figure 2. By this mode the illusion will be complete, and in the middle of winter one may have a tropical landscape. Those who have visited the greenhouses of Prince Demidoff, at Florence, will have some conception of the beauty of such an arrangement of the plants.

But for fruit as well as flower culture the use of pots will be preferable. Let us see what can be done with these. The superficial area of nearly every good city house is more than twelve hundred square feet. Deducting the space required for the walks and the stairway, there will be more than eight hundred square feet on which pots can be placed.—By the most approved mode of pruning for pot culture, the diameter of foliage in a fruit-tree should not exceed the diameter of the pot. We could thus place eight hundred fruit-trees in the garden, but in order to give abundant room and air, we will estimate for only two hundred. If one wishes no flowers, but fruit only, he can have forced peaches and nectarines at a season when he cannot buy them of the confectioner for less than a dollar each. The two hundred trees ought readily to yield a crop of a thousand peaches. If one's taste runs on strawberries these will yield a good return.

But if it be desired to have the house filled with flowers through the winter, we cannot cultivate forced fruit. We can, however, have flowers, stone fruit, and Black Hamburg grapes in succession. If the house has been filled with flowering plants in the winter, and there is plenty of yard room, they can be taken out and arranged in groups in the yard as soon as all danger of frost is over. The house can then be filled with peaches, plums, and nectarines in pots, which can be obtained of the nurserymen ready for fruiting, or prepared the previous year by the florist having charge of the house, and kept in the cellar during the winter. These can remain in the house until the fruit has attained sufficient size to be safe from the curculio, when they also can be grouped in the yard where they will grow, and ripen early and well. Their place in the house can then be supplied with grapes in pots which have been retarded by being kept in a cool, dark place in the cellar. These will then bear abundantly during the summer, and, before the flowering plants require to be taken in the ensuing autumn, will duly respond to the tiller in Black Hamburgs and Muscats. Two pounds to each vine, or four hundred pounds of grapes would be a moderate estimate for the space mentioned.

Both stone fruits and grapes are easily managed, and a man of ordinary intelligence could soon learn to grow them even if his life has been passed in the midst of dry goods or hardware; if, however, his own skill fails him, florists are always attainable. Here then are new luxuries—flowers, peaches, and grapes—within the reach of every man of moderate means.

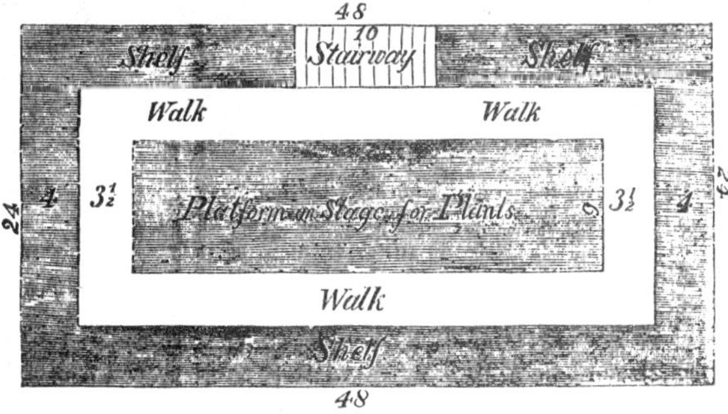

Fig 3.—PLAN OF GREENHOUSE WITH SHELVES AND STAND.

If the capabilities of this plan and its economy were thoroughly understood by architects and proprietors, the time would soon come when a roof garden would be considered just as essential an appendage to a house as a bathroom. The demand for care takers would bring forward a host of candidates for this new branch of industry, and it might furnish an excellent and remunerative vocation for women.

THE TWELVE-SPOTTED SQUASH-BEETLE.—Last month we gave on page 76 an engraving and some account of the Twelve-Spotted Squash-Beetle—*Diabrotica 12 punctata*. Heretofore we had only known the perfect insects, and finding them so destructive we had picked them off before they had a chance to propagate. This year the in-

Fig. 1. LARVA. Fig. 2.

BEAR-GRASS.—(*Yucca filamentosa.*)

WHITE-SPOTTED CALLA.—(*Richardia albo-maculata.*)

sects came in too great numbers to be disposed of by hand-picking, and they produced larvæ in abundance, which are quite equal in their ravages to their troublesome parents. The full-grown larva, or "grub," is about half an inch long, of a lemon-yellow color, and covered with spines, which are yellow near the body and black above. These are themselves armed with short bristles. As the larva has six of these spines upon each ring of its body, it presents a formidable appearance under the magnifier. The engraving (fig. 1,) shows the larva of twice the natural size. It eats the cellular tissue of the leaves, in circular patches, an inch or more in diameter, leaving the net-work of fibres, as shown in fig. 2. They seem to work upon either side of the leaf indiscriminately. We made an application of powdered white hellebore to the vines, without any marked effect, and have since resorted to hand-picking.

Tying Material—The Bear-grass, or Adam's Needle.—(*Yucca filamentosa.*)

The commonest of the Yuccas is the *Y. filamentosa*, commonly known as Bear-grass and Adam's Needle, and in some localities called Cliff-Lily and Gardener's Twine. It is a native plant, being found wild as far north as Virginia, and is quite common further south. It has long been known in our gardens, and of late years has become quite popular; fine old specimens being highly ornamental for their leaves, and surpassingly beautiful when they throw up their enormous panicles of white flowers. Our purpose is, to speak of it as a useful rather than as an ornamental plant. An article on tying materials has induced Mr. Jas. T. Worthington, of Chillicothe, O., to present the claims of the Yucca as a superior article for bands and strings. Mr. W. considers it of great value, not only to the gardener but to the farmer, and advocates its cultivation for its useful leaves. He states that when the plant is well established, or the third year after planting out a shoot or sucker, hundreds of leaves may be removed without injury. He sets the plants four feet apart each way. Mr. Worthington, in a communication to the Scioto Gazette, upon the same subject, says: "For all purposes where a string is needed, from a bouquet to a bacon ham, including bands for grape-vines and shocks of corn, and for garden vegetables, it has no equal and requires only to be generally known to be universally cultivated." We have known the Yucca to be used in some of the Missouri vineyards, and thank Mr. Worthington for calling attention to its availability for other purposes. Plants are sold by most nurserymen, the catalogue price being 50 cents each.

The White-spotted Calla.
(*Richardia albo-maculata.*)

Last year we received from Messrs. Bennett & Davidson, Florists, of Flatbush, L. I., a specimen of a Calla quite unlike the well-known Ethiopian Calla, and which was ascertained to be the White-spotted Calla (*Richardia albo-maculata*). The present one seems to be but little known, and having grown it for two seasons we think it merits more attention than it has received. The plant is smaller than the common Calla, growing about two feet high, and though it has a general resemblance to that, it will be seen from the engraving that the leaves are different in shape and are distinctly marked with numerous white spots. The spots, though they appear quite white by reflected light, when held between the eye and the light are found to be translucent. The flower, or more properly, spathe, the shape of which is given in the engraving, differs in form and in size from that of the common species. It is white without, but purple at the base within. The plant is a summer-flowering one; planted out in the open ground it has flourished in ordinary garden soil, without any care in watering, and has thrown up an abundance of foliage and a succession of flowers. It produces seed freely, though it multiplies so readily by offsets there is no need of growing it from the seed. The thick rootstock, or "bulb," may be kept through the winter, in a dormant state in a box of dry earth, in a greenhouse or other place where there is no danger of frost. The plant is from Natal, and was first described by Sir Wm. Hooker, in 1860. A recent French work mentions it as a spotted variety of the *Richardia Africana*, the common Calla, which is an error, as it is distinct from that species in several important particulars.

Shallots.—(*Allium Ascalonicum.*)

The early "green onions" which appear in our city markets are properly Shallots—a different species from the common onion. The mature bulbs are smaller, oblong in shape, and made up of several smaller bulbs, which are enclosed in a skin like that of an onion. They keep with difficulty, as they are very apt to start into growth. In flavor they are stronger than the onion. With us they are chiefly used in the green state; but abroad they are employed when mature for flavoring soups, etc. Being perfectly hardy, shallots are best planted in September. The bulbs are broken up, and the separate sets, or "cloves," are planted in good soil, in rows a foot apart, the sets being six inches apart. In planting they need to be simply pressed into the soil with the fingers. They require no protection during the winter.

Work in the Horticultural Departments.

In September we not only harvest many crops, but we sow and plant for another year. The true horticulturist will not only do this in his orchard and garden, but will make it with himself a season for harvesting ideas, and acquiring facts to be of use to him hereafter.

Labor Saving in Churning.

Contrivances to lessen labor in churning have been, and are, many and various, from the European peasant's plan of lashing a jug of cream on each side of a wheel of his cart, in which the cream was found to be butter on returning from market, to the most improved "Blanchard." Mr. A. Kemler, with a commendable desire to relieve his wife of some of the labor of churning, made the contrivance here exhibited, which works well. The affair is easily understood by the accompanying engraving. Two bent levers of iron, (*a*), to the long arms of which are attached wooden "heads," (*b*), which receive the ends of the two dasher handles, and in which they are fastened securely by thumb screws, are suspended from the ceiling by a piece of joist, (*c*). From the short arms of the levers, wires are attached, connecting them to treadles. Thus the churn is worked by the feet. The weight of the heads is sufficient to cause the dashers to go down, but were this not the case, as it might not be, were the cream very thick, cords extending from each treadle to the long arm of the lever, raised by the other, would obviate the difficulty. It would be necessary for such cords to be attached at points as far from the fulcrum as the length of the short arm.

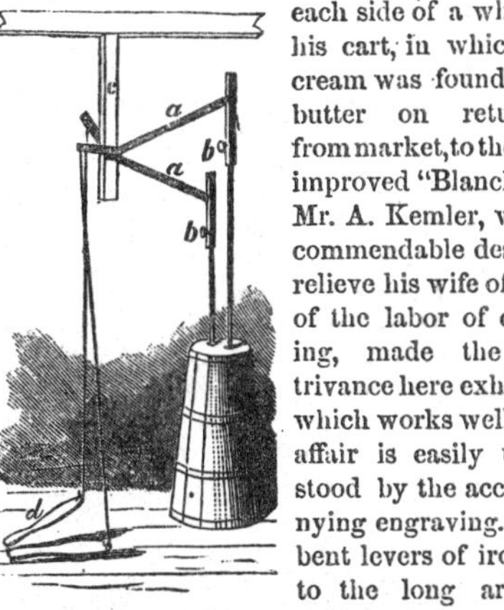

ATTACHMENT TO CHURN.

A Farmer's Barn.

Gentlemen of wealth and city merchants who have $15,000 to $40,000 to invest in a barn, have plans enough offered to their inspection, and they are the very ones who need none, because they employ a city architect for $1,000 or more to draw the plans, make the corrections, alterations, and estimates, and superintend the erection. If such barns are convenient it is very well, and quite wonderful withal; and if they are as picturesque as a Swiss cottage, and as inconvenient too, it matters little. Plain farmer folks, however want plain barns, and they want to do about two-thirds of the work themselves. They will dig and lay the foundation, cut the timber, and haul it to and from the mill; with the aid of a carpenter, plan the frame, and see that they have the right stuff. They will spend winter evenings hewing treenails, and planning conveniences, and when spring comes are ready to go-a-head, have a raising and shortly a barn.

Fig. 1.—ELEVATION OF BARN.

There are many good things about the old-fashioned barn, with its stalls for cattle and horses on each side of the barn floor—with its barn yards and sheds, and big lofts, and deep bays.—Barn cellars are good too—and it is best to have them, if possible. The horse-fork makes the deep bays less attractive, and so modern innovations modify old ways, usually for the better. A western farmer situated where he can not dig a cellar for his barn, asks for a plan for one, all upon one level. We give one which is of the old-fashioned type, but more convenient, and planned so that it may be built very cheaply—and either smaller or larger than designed. Figure 1 shows the front elevation, the main barn having 16-foot, and the wings 14-foot posts. The plan as engraved in figure 2, makes the main barn 36 feet wide and 40 feet long. This gives a barn floor of 12 feet in width, a horse stable 13 feet wide, and a cow stable 11 feet. This is not as wide as we would be glad to have either cow or horse stables, and is as narrow as will do under any circumstances. We allow 4 feet in width for cow stalls, $2^1/_2$ feet for mangers, $5^1/_2$ feet for the cows to stand upon, and 3 feet for a depressed cement walk behind the stalls to answer also as gutter and manure holder. The horses, according to this plan, will have 9 feet standing room, and a 4-foot walk at the rear, while the stalls are all 5 feet wide. A passage crosses the barn, near the rear, taking the space of one horse stall on one side, and one cow stall on the other; at the rear of the passage is a box-stall for horses, in size 10 × 14 feet. The continuation of this passage through a door at the left, crosses to the carriage, harness, and tool house, through a shed 16 × 25, open in front, having double doors at the rear, and being paved across the rear end. This shed will be found a comfortable place to groom horses; for them to stand, out of a draft to cool off after hard labor; for them to be harnessed and unharnessed in; and for sheltering farm wagons or other vehicles or implements at certain seasons. Here also should horse manure be deposited daily; and a liquid manure pit should be made, so that the heap of dung and litter, or a more bulky compost may be kept wet down and fermenting. On the right-hand side of the barn a wide shed, which may be still wider if desired, is in the same way connected with the passage across the barn floor. At the rear of this shed is an 8 × 10 box for a lying-in stall for cows, and two stalls for young stock, which are 6 × 10 feet each in size, and capable of accommodating four head, or of being converted into two, small loose boxes. The passage across the shed may be simply a paved or cemented walk. It meets a platform walk one or two steps up at the right, which passes in front of the pig-pens. A projecting roof affords shelter outside the shed.

The hog-pens are on the extreme right. They are planned to be in size each 8 feet by 10. Three have yards only 8 feet wide, but two, much larger. This side of the barn would of necessity be the great manure factory. A liquid manure pit is provided for in the barn-yard. There is room under the shed for the manure likely to accumulate in summer from the cows, but during the winter it should be regularly laid up around the pump, and kept fermenting.

If the raising of roots is followed extensively on any farm where a cellar can not be dug—they must of course be stored in heaps in the field—but enough to last through one or two months perhaps, might be stored, either in the loose box at the rear of the horse-stalls, or in the opposite corner; and if the walls are well lined with hay, the frost will not easily get in.

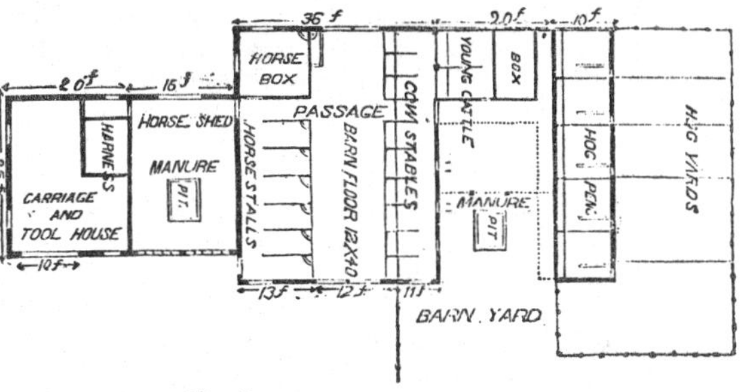

Fig. 2.—GROUND PLAN OF BARN.

The grain-room should be in one of the rear corners above the loose box or the cow stables. If water can be brought in pipes, or obtained by driven wells, the hydrants or pumps may be located wherever convenience dictates. One should certainly be at or near the rear doors, and it would be well to have one at the carriage house.

Foundations for Corn Houses, etc

We are prone to devote too much labor and expense to the superstructure, and not enough to the foundation. We are satisfied, if our corn houses are set upon eight-inch posts, which are two and a half feet in the ground. If the ground is such that posts are heaved by the frost, we make the holes rather large and fill in stones around the post so that water shall not stand near the surface. The posts rot and are renewed with some trouble. Other foundations, upon which buildings rest which have no cellars, are usually surface structures of stones, or have only a single course of stones lower than the surface. Mr. M. C. Grout, of Poland, N. Y., describes to us the manner in which he lays foundations for such buildings, particularly corn cribs. He says: "It is as cheap and easy to make a permanent foundation as any. My corn house foundation is of cobble stones, two and a half feet through, coming up to the surface of the ground. Upon this I have a flat stone about four or five inches thick, and a little smaller across than the posts. The ends of the posts rest upon these flat stones. The posts are of varying lengths, according to the uneven surface of the ground, and the sills are framed into them. Below the sills the posts are cased with inch stuff, painted like the building, and tin is put around just below the sills. My corn house is 16 × 20 feet, and 10 feet between sills and plates. Wooden posts

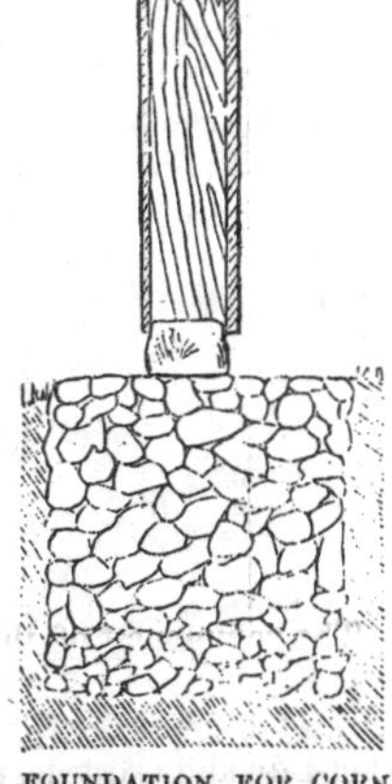
FOUNDATION FOR CORN HOUSE.

soon decay, and stone ones are moved by frost."—The arrangement described by our correspondent we have had engraved, and it strikes us that one very important point is not touched in his description, which is this: If the posts are properly cased, the stones on which the posts rest being small enough, the casing will extend a very little below the top of the stones and so effectually prevent any water from rains, ever getting to the bottom of the post. One thing must be guarded against; namely, using round posts, or those not well squared, at least at the bottom; for in the corners, there would be room for mice to run up between the posts and the casing, and thus gain access to the crib.

The Swivel Clevis for Whiffletrees.

C. H. Polhemus, of Middlebush, N. J., some time ago sent the following note with the drawing of the simple implement engraved.—Another correspondent sent a similar drawing, but the hook had no spring. We made the trial to ascertain if the hook without a spring would do equally well. It does not, because more liable to unhook, especially when the hook is fastened to another clevis and lies flat. Mr. P. writes: "It is these small and simple things that go a great way towards making farm work slip along smoothly. This drawing represents a clevis which has a swivel hook; the bolt passes through the whiffletree, or evener, and in changing from one plow to another, or to the harrow, simply place your thumb on the spring in the hook and snap it into the ring of the plow or harrow, as the case may be. Any one who knows the annoyance of backing up a team while you are under their heels so as to pull out and replace the bolt as in the old way, will, I am quite sure, appreciate this much simpler way of doing the work, particularly in fly time. Any blacksmith can make one from the drawing."

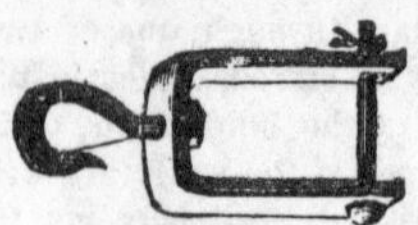

SWIVEL CLEVIS.

A Broad-wheeled Wheelbarrow.

We are often forced to take half-loads or less upon wheelbarrows because the narrowness of the single wheel will not admit of heavier ones being wheeled over soft ground. It is not only upon plowed ground or that in which crops are growing that the wheelbarrow is nearly useless, and the cumbrous wagon, stone-drag or tool-boat, of necessity, are substituted, but sward presents an almost equally soft footing in wet weather. A neighbor was in perplexity;—purslane was rampant in the mangel patch; it was worth a good deal for the hogs; to take a wagon in would sacrifice too many of the roots; the wheel of the wheelbarrow sunk nearly to the hub in the dry, light soil, with half a load, and it was a great waste of time to bring the weeds out by armfuls or forkfuls. We suggested the use of a very broad wheel, and for fear such a one could not be found or easily made, the idea of using a small lagerbier keg occurred to us. The suggestion was followed with entire success. An old keg was bought at half-price, the hoops were tightened and fastened in place by punching a few *teeth*, so to speak, in their outer edges. A hole three-quarters of an inch square was cut in each head, a square iron rod inserted, having its ends filed round and cut the right length. This rod was wedged in place, and a barrow was soon improvised which would carry a heavy load over soft ground. The axle must be secured against working either to one side or to the other, as the wheel has no *hub*, which generally performs this office. Should any one object to this spoiling of a good keg, which would be worth perhaps a dollar and a quarter, he may get his blacksmith to weld a cross-head upon each axle end. These should have two screw holes in each arm of the cross, and it would probably be well besides, to put upon each head over the iron cross-head, a piece of inch board sawed round to fit snugly, and screwed on. When bunging up the keg it would be well to put in a little water, to keep the interior moist, so that it will not shrink.

A BROAD-WHEELED BARROW.

Stallions for Common Labor.

There are very few geldings in France. The reason is, the stallions are not unmanageable, vicious, and dangerous as work-horses, but docile, obedient, easily managed, and intelligent. There is nothing in the nature of things to prevent our having the advantage of the greater toughness, strength, spirit, fearlessness, safety, (in being less liable to take fright), freedom from disease, and longer serviceableness of the stallion over the gelding, were it not that we and our ancestors have so abused the temper of the horse, that his progeny exhibit, among the unaltered males, vicious and treacherous tempers, such as make them unsafe and unreliable as work-horses, even under the kindest and most uniform treatment.

The English thorough-breds, unexcelled for spirit, endurance, fleetness, and wind, are the most vicious of all horses. They came from the gentle, docile, affectionate Arab, and it is only the training and abuse of the English stable-boys and grooms, we verily believe, which have thus, in the course of generations, ruined the temper of the most noble of the breeds of horses. Its blood is infused through all our common stock, and to it we owe most of the characteristics for which we value our horses. Where thorough-breds have been bred for generations under different treatment, as under the handling of the negro grooms and riders of the Southern States, their tempers improve, and extraordinary exhibitions of vice are rare, even among stallions. The habit of using stallions is followed a good deal by French Canadians, who send to this country so many of the so-called "Kanuck" horses. These horses are small, close-knit, and powerful, and when entire, tough beyond comparison. Wherever we meet with them, they are praised for easy keeping qualities, great endurance, and freedom from ordinary ills, and are seldom complained of as vicious. Do we not, in our ordinary treatment, sacrifice a great part of the usefulness and serviceableness of the horse, in rendering him more tractable, more liable to disease, and less intelligent and spirited? Is it not worth while to make the experiment oftener of rearing stallions for labor, though it require more patience, gentleness, and kindness, on the part of those who handle them, and repeated floggings, administered with a will, for any stable boys who dare to pinch or tickle, or to ruffle their tempers?

THE HOUSEHOLD.

☞ *For other Household Items, see "Basket"*

A Farmer's Holiday.

BY CONNECTICUT.

We do not mean the fair when we speak of a farmer's holiday. They are good, and we advise every farmer and his wife and children to go once a year and take their best products with them. But if you take up your herd of six best Ayrshires, and Joe takes the working cattle, and Ben takes the stallion, and your wife takes the butter, and Susan takes the bread and the bedquilt, and the flowers, you will have a pretty busy time of it, and will feel when night comes that the fair is so much like work, that you cannot call it play. A picnic will do. A Sabbath-school picnic if you please—if nobody has to make a formal speech—or to be bored with one. But a day of absolute rest we insist upon, as one of the best things that can befall a farming community. We work hard; we are isolated; we have too little social enjoyment; and some of us have almost forgotten how it would seem to be turned into our own green fields and forests with no care upon our minds. Now, to have a good time, we want some pleasant spot as the place of gathering—a lake with wooded banks, the sea shore, a grove, a hill or mountain with berries or nuts, or some amusement for the children and young folks. It should not be so far off or so expensive to reach that any will be deterred from going. Let every family take along its own provisions and make common stock at the gathering. Then we want congenial society; men and women of like tastes and sympathies, so that there shall be no worry about dress or grammar; no stiffness or stilted propriety. Then the less management the thing has, the better. You want some one or more to take the responsibility of appointing the time and the place, and of giving the invitations; and then let the amusements be determined by the tastes of the company. If you cannot trust the company to arrange this matter, you should not be of the party. An excursion by rail or steamer is a very good thing; as then there are no horses to be fed at the place of gathering, and there is no care except for the picnic baskets. The entire change of scene, the ride or sail, is a great refreshment. But in this case the company will have to be much larger to make the thing pay, and the manager will have to take much more responsibility. We lately fell upon a party of this kind encamped for the day in a beautiful grove upon the shores of one of the lakes in which New England abounds. The railroad cars dropped the company at the grove which had been conveniently fitted up for such occasions. There was abundant shade, green grass, nice springs, flowers, and rustic seats and tables for a multitude. We hail the multiplication of these places of rural resort, away from cities and grog shops, as a good omen of our times. Let us have more of Nature and occasionally a day of rest.

Melon Mangoes.—The late, small, and smooth muskmelons are used for this pickle. Cut out a plug at the stem-end, or, as some prefer it, from the side; scrape out the contents, replace the plug and secure it with a wooden pin, and put the melons thus prepared into a strong brine. When they have been in the brine for twenty-four hours or more, they are ready for stuffing. The stuffing is made of any pickle material at hand: shredded cabbage, broken cauliflower, small onions and cucumbers, green beans, peppers, mustard seed, nasturtiums, scraped horseradish, and the like. Cabbage and the other stuffing, except the aromatics, are better for being scalded and cooled. Stuff the melons according to fancy, and then sew each plug in its place by means of a needle and coarse thread. Place the stuffed melons in a jar, add cloves, pepper, and other desired spice, and pour boiling vinegar over them. Repeat the scalding of the vinegar for three days in succession.

The Bake-kettle, or Skillet.

BY AN "OLD FOGY."

When cooking stoves came in, the bake-kettle, or covered skillet, went out, and with it went a large part of what was good in our American cookery. How many of your readers ever saw a bake-kettle? Probably only those who enjoy the blessing of a wood fire to cook by. Just send an artist down to the back woods of Maine or away "out west," and have a drawing made of this most capital kitchen utensil. [We have had the drawing made, and did not have to go so far from home for it.—ED.] "Pioneers" will know what I mean without a drawing. It is a shallow kettle with a lid, which has a turned-up edge, and upon which coals are placed; and the thing to be cooked is "between two fires." Those who are roughing it in log cabins or in camp, know what a useful thing it is. In it the bread is baked, meat roasted or fried, coffee browned, dish-water heated—in short, it is the one thing handy to have in the house. I say, much good cooking disappeared with the bake-kettle. It allowed food to be cooked as it seldom is by the stove—long-continued, slow cooking, with all the juices and flavors kept in. Were there ever such chicken and veal pies as our mothers used to make in the bake-kettle? We have nowadays what is called roast veal—half burned and wholly spoiled in the stove oven. But stuff a knuckle of veal and put it in the bake-kettle and let it "sizzle" with fire above and below for three or four hours. It cooks quietly and slowly, all the moisture is retained, and comes out not only a delicious, but a digestible morsel. The French who, say what you will, bring more skill and common sense to the subject of cooking than all the rest of the world, *bruise* a certain class of dishes, and they have for the purpose a *braisier*, or braising-pan, which is really only the old bake-kettle Frenchified. Ducks, pigeons, and fowls, even if tough, may be admirably cooked in the bake-kettle. Our people have much to learn about the advantages of slow and long-continued cooking—a treatment which is hardly practicable in a stove oven, as that dries as well as cooks. The "modern improvements" in the way of stoves and ranges have resulted in a deteriorated cookery. Let us go back to the days of our grandmothers so far as to restore the bake-kettle to the kitchen—or, if it will be fancied better, we will call it a "*braisier*."

BAKE-KETTLE.

About the Egg-Plant.

C. Mathers writes that we speak in praise of the Egg-Plant, but do not give directions for cultivating or cooking it. It is rather late now to talk about its cultivation, but if our correspondent had looked under the head of Kitchen Garden in Hints about Work, he would have found sufficient said upon the subject. In brief, they require the same culture as tomatoes. The plants must be started in the hot-bed or in the house, and given a rich and sheltered spot in the garden. As an article of food, the fruit is much liked by those who are accustomed to it, while others positively dislike it. In this respect it is much like tomatoes. The fruit is sliced and fried, being served as a dinner or breakfast dish. It has no positive flavor, being only a rich, marrowy substance. The best way, according to our notion, is to pare the fruit, cut it into slices half an inch thick, and then salt the pieces and stack them upon a plate. In an hour or two they will have parted with considerable water. They are then to be dipped in egg and sprinkled with cracker crumbs and fried. Serve very hot. Cooked in this way they are, to our taste, very acceptable. Fried, without the covering of egg and crumbs, they take up much fat and are very indigestible. The fruit is used at any time before the seeds become hard.

Drying Fruit.—When much fruit is dried, it is necessary to have a house for the purpose. Small quantities should be so arranged as to be placed near the kitchen fire when taken in at night or during stormy days. Those who have hot-bed sash, can easily arrange a drying apparatus which will dry rapidly and at the same time keep off insects. A hot-bed frame with a bottom to it, and raised above the ground, makes a capital drying box. The sash should be elevated at one end to allow the moisture to pass off, covering the opening with netting

Something about Pickles.

It is a little singular that a large share of the questions put to the Household Department at this season should be about an article of food which is not nourishing—pickles, which to the best stomachs are only appetizing and to the weakest positively injurious. Still people will eat pickles, and whatever our "physiological" friends may say, we do not doubt that things so generally craved have some use in the animal economy. When our boys in the army had the chronic diarrhœa, our army surgeons usually allowed them to eat pickles and other things, that under ordinary circumstances would be considered fatal, and to the surprise of everybody the hopeless patients often recovered. So without discussing the dietetics of the matter, we accept pickles as a fact. To look at the matter philosophically, a pickle is a mere vegetable sponge to hold vinegar. Any vegetable tissue that is not so fibrous or tough as to be unpleasant to masticate, and which has no disagreeable flavor of its own, will answer for pickling. If the article pickled has an acceptable flavor of its own, all the better. It is the possession of this that makes the cucumber the most popular of all pickles. Vegetables which have no marked taste, such as green muskmelons, are made flavorous by the free use of spices—like the sailor's wonderful stone, which would make a nice soup when the farmer's wife allowed him to gather the odds and ends of her kitchen and garden to boil with it. It is customary to salt pickles before putting them into vinegar. Why do we?—It is not for the purpose of flavoring them with salt, for this can be added to the vinegar. This matter of salting pickles brings up the question of *osmose*, which we cannot find space to discuss. Briefly, when a fresh vegetable is placed in salt and water, an interchange takes place between the juices contained in the tissues of the vegetable and the brine by which it is surrounded. The natural juices pass out and the brine passes into the vegetable; the brine being denser, it, according to a well-known law, passes in more slowly than the juices of the vegetable pass out, and the salted things shrivel. When salted pickles are placed in water the case is reversed, their shriveled tissues are full of brine, much heavier than the water by which they are surrounded, the brine passes out, and the water goes in and restores the plumpness. Soaked pickles with their tissues full of water being put into vinegar readily become penetrated by that liquid. The question of salting pickles has nothing to do with flavor, as the finest pickles are those from which the salt is most completely soaked. One of the most frequent questions is, "How can I make pickles like those put up at the factories?"—We have answered this more than once, but will repeat, that the pickles referred to are put up in colorless vinegar made from whiskey. Diluted whiskey will make a vinegar which is almost colorless and of a pure sour taste. Cider makes a vinegar which has a color, it is true, but a most agreeable flavor. Home-made pickles should be prepared with regard to flavor rather than appearance. As a general rule, vegetables to be pickled are first put into brine, then soaked to freshen them, and then placed in vinegar, which may be spiced or not, according to taste. One point is to be noticed: when freshened pickles are put into not very strong vinegar, the water with which their tissues are filled, so weakens the vinegar that the pickles are not only not sour enough to the taste, but not enough so to keep well. It is not necessary to enumerate the things that may be pickled, as there are but few fruits or vegetables that may not be so treated—pickled peaches are delicious and pickled purslane is not to be despised—a wide range surely. Some good housekeepers have, besides the regular cucumber and other standard pickles, a jar of

MIXED, OR INDIAN PICKLE.—The basis of this is usually sliced cabbage, and cauliflower broken into bits and put into brine. After these are ready, they are covered with spiced vinegar; and then such pickle materials, fruits, or vegetables as occur during the season, are added from time to time, taking care that the newly added things are covered by the vinegar. At the close of the season the vinegar is drained off, heated to the boiling point, and poured over the pickles; this is repeated two or three times, when the pickles are stored away for use, and are usually better the second year than the first. In the making of the spiced vinegar, probably no two will agree. As a suggestion we give two recipes. The various directions differ greatly, the chief object seems to be to get in enough spice. In looking them over, we are reminded of the toper's directions for making punch, "too much of lemons, sugar and whiskey, and not enough water."—One recipe gives: Vinegar, 6 pints; salt, ½ lb.; bruised ginger root and whole mustard seed, 2 oz. each; mace, 1 oz.; shallots, ½ lb.; Cayenne pepper, a dessert spoonful, and some sliced horse-radish. Simmer together for a few minutes, then put into a jar and cover close. Another, claimed to be "very superior," directs for each gallon of vinegar 6 cloves of garlic, 12 shallots, 2 sticks of sliced horseradish, 4 oz. bruised ginger, 2 oz. whole black pepper, 1 oz. allspice, 12 cloves, ¼ oz. Cayenne pepper, 2 oz. mustard seed, ¼ lb. mustard (ground) and 1 oz. turmeric. All the above, except the mustard and turmeric, are put into the jar with cabbage, cauliflower, and other pickle vegetables, and the vinegar boiled and poured over them. The ground mustard and turmeric are to be made into a paste, with cold vinegar added.

A Draped Center-Table or Stand.

A friend who contrives to have things make a handsome appearance with a moderate outlay, gives us a drawing of the manner in which she arranged a center-table. The standard is such as are used for marble-topped tables, and was procured of a cabinet-maker. Instead of a marble top, one of pine was used, which was smoothly covered with green furniture rep, neatly tacked on. A curtain

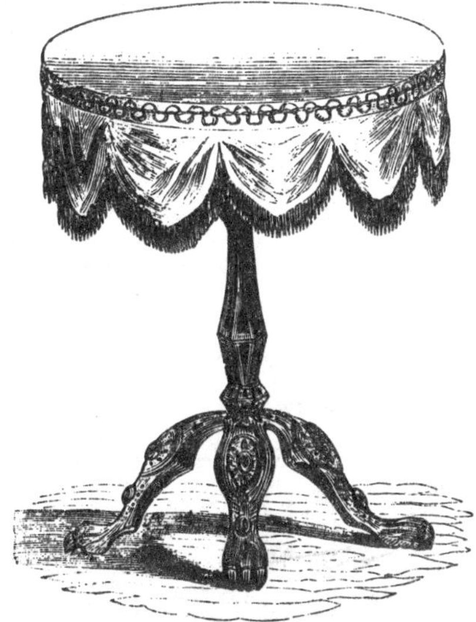

DRAPED CENTER-TABLE.

of the same material, a quarter of a yard deep, and made a little full, has a fringe at its lower edge, and is tacked by its upper edge to the table top. Gimp of a color to match is used to cover and hide the edges and tacks. The curtain is caught up in plaited festoons every quarter of a yard. Other materials may be used, and other colors to harmonize with the surroundings. By the exercise of proper taste and skill, one may make very pretty chamber sets by draping very rough kinds of furniture.

Domestic Wines.

Here is quite a file of letters asking how to make wine of blackberries, elderberries, strawberries, and —of all things in the world, tomatoes. We think that, taken as a whole, these compounds are more productive of evil than of good. All fruit juices, some with and some without the addition of sugar, will make alcoholic liquids, which are by some used "medicinally," and by others in the exercise of hospitality. Many of our readers can recollect having swallowed vile compounds of this kind rather than offend the kind hostess who presented it saying, "it is only some blackberry wine that I made myself."—Of the medicinal value of these "wines" we have strong doubts. If alcohol is needed as a remedy, which it sometimes is, but less frequently than many suppose, a physician will prescribe pure whiskey in preference to any of these domestic productions. Alcohol is the same thing, no matter how produced, and whiskey contains a nearly constant proportion of this, while the home-made "wines" may be stronger or weaker, and be more or less qualified by a greater or less amount of sugar, and aromatic or other principles contained in the fruit from which it is made. There is a great deal of nonsense about the medicinal effects of blackberry, elderberry, and other "wines." We have noticed that when these things were known to be in the house, the complaints which they are supposed to cure were of more than ordinary occurrence. That blackberry cordial, for we do not consider it a wine, may be sometimes useful we do not doubt; but put the blackberries in a jug, and cover them with whiskey and call it "tincture of blackberry," and keep it as a medicine; the sole virtue of which over and above the stimulating quality of the whiskey is due to a little astringency derived from the berries, and in this particular a handful of the blackberry root is worth a bushel of berries. So much for the medical view. As to the question of hospitality—at the present time it is the exception where wines are offered to visitors. We believe, with our present knowledge, that it is better not to offer them at all. We know that there are many who think differently. There is scarcely an Agricultural Fair but that offers premiums for Currant, Strawberry, Blackberry, Rhubarb, and other "wines." We never tasted one of these compounds that was not an abomination, and a desecration of the fruit from which it was made. If the managers of our fairs would exclude all these things they would do well. They should not be made, encouraged, nor offered to one's friends. We think that hospitality can be manifested without presenting wines or liquors to one's guests. Did we hold the opposite view, we should say that the only presentable things were pure whiskey, pure grape wine, and pure cider. Not that we advocate them, but one offered these, knows exactly what he is taking.

Hail.—"J. M.," Goshen, Ind. Hail storms come so suddenly that we know of no precautions that can be taken to save your glass from breakage.

Tomato Pie, by Mrs. K. T. H., Sevierville, Tenn.—Take two large ripe Feejee or other Tomatoes of the same size, drop them into boiling water to remove the skin, then, with a sharp knife, cut them into thin slices, put the crust in an ordinary pie-pan, as for berry pie; cover the bottom with a layer of the Tomatoes, then a layer of sugar and butter, then of tomatoes, then of sugar and butter as before; flavor with either lemon, orange peel, or nutmeg, to the taste. Cover with the top crust, bake, and bring to the table hot—(cold tomato-pie is not good). Two very large tomatoes, two tablespoonfuls of sugar, and one of butter, are enough for one pie, baked in a pan ten inches across. If there are too much tomato, sugar, and butter put into one pie, there will be too much juice; a little practice in making will make all right.

containing a great variety of Items, including many good Hints and Suggestions which we throw into smaller type and condensed form, for want of space elsewhere.

Pear-Trees and Hot Weather.—S. Wood, Jr., writes that during the heats of July, two of Manning's Elizabeth pear-trees, out of a collection of thirty varieties, had their leaves crisped as if burned by fire, and asks the cause. If some of his trees had been winter-killed, he would have said they were "tender," meaning not able to withstand excessive cold. Trees are also tender to excessive heat, hence many varieties can not be grown in the Southern States, but we hear much less of tenderness in this direction than in the other.

Peach-trees in Baskets.—Mr. A. C. Chamberlain, of Brooklyn, presented us with a four-year old peach-tree grown in a wire basket a foot across. It had a half-dozen, well-grown, ripe peaches upon it. The basket is filled with moss, through which is diffused some fertilizer, the composition of which Mr. C. keeps secret. As a novelty, it will please the curious, but we cannot see any advantage over the well-known way of growing the peaches in pots.

Pineapple Fibre.—Nora N. This is a fabric made from the fibres of the leaves of pineapples and related plants of the same family.

Wild Wistaria.—Nora N. There is a wild Wistaria, but it has blue and not maroon-colored flowers. You probably have in view the *Apios tuberosa* called Wild Bean and Dacotah Potato

Coleuses and Begonias.—"J. G. R.," Griffin, Ga. Both these need a greenhouse temperature of not lower than 60° during the winter. The Coleuses are particularly difficult to keep unless there is the proper heat.

Castor Pomace.—Mrs. W. A. B., Windham Co., Conn., says: Those who use Castor Pomace as a fertilizer should observe caution in handling and using it; if it comes in contact with the eyes, either by rubbing the eyes with the soiled hands, or if the wind blows it into them, it will cause inflammation. It should be kept away from cattle. A valuable cow came near losing her life a short time ago by getting to compost where Castor Pomace was used, repeated doses of melted lard saved her life, but she is permanently injured."

Poisoned Dogs.—"W. B. M.," Lewiston, Idaho, writes that, when living in Texas, he often had hunting dogs poisoned by strychnine that was set for wolves. He always cured the dogs by pouring down their throats as much milk-warm oil, or grease, as their stomachs would hold.

Root Pruning.—"W. T. W." The "practical" part of the operation is very simple. A circular trench is opened to expose the roots, and they are then severed by means of a very sharp spade. The point where there is the most difficulty is, to judge how much to prune. The object is to check the growth of the tree and induce it to form fruit; and while removing too large a share of the roots will cut off too great a supply of nourishment, the removal of too little will defeat the object sought. For young trees, Rivers advises that the trench should be a foot from the tree for every inch of the diameter of the trunk. Thus a tree 3 inches through would have the trench 3 feet from the tree on all sides, making a circle 6 feet in diameter, of which the tree is the center. All the roots outside of this circle are to be cut off. But no fixed rule can be applicable to trees in all conditions, and much must depend upon the judgment of the operator. The work may be done in autumn or in early spring.

Sparrows.—"D. G. H.," East Bridgewater, Mass. Sparrows are usually for sale in the spring. There are no "means to secure their staying," yet if boxes are put up for them, they are likely to take possession of them when set at liberty; but they may fly off and reject your hospitality. We cannot tell what the prices will be next season.

"Black Bug" on Cherry-Trees.—A correspondent asks, "What is the occasion of the black bug on cherry-trees? Are they destructive? and can they be got rid of?"—The "occasion" is that the insects are there to get a living. We suppose the "black bug" to be a plant louse. It lives by sucking the juices from the young shoots, and is injurious. It can be killed by the use of tobacco water, applied as most convenient.

Oiled Paper.—In these days, when so many plants are sent by mail, we find a frequent use for oiled paper as a wrapping material. We have used that prepared by Mr. F. Trowbridge, South Milford, Conn., and found it very serviceable. Plants put up in slightly dampened moss, rolled up hard in oiled paper, and then covered with common brown paper, will go safely a long distance. The oiled paper is much lighter than India-rubber cloth, and for the purpose, quite as useful.

Fish Ponds and Muck.—"M. Q. E.," Centreville, Ind. It is quite practicable to make fish ponds in muck beds. There are very frequently good gravel and sand, and springs at the bottom. In this case you can have trout. Make the ponds long and narrow, and screen the outlet. If the water proves too warm for trout you can make the ponds of any shape that pleases, and put in black bass. The muck will generally pay for the digging, with a large surplus. If the sides of the pond are left with a slope of 45 degrees, they will not need stones nor plank.

Soaking Grain for Pigs.—R. H. Dixon writes that the plan of soaking and cooking whole grain alluded to in "Harris on the Pig," does not prove successful with him. "The pigs will not eat more than half as much as they will if the grain is ground before cooking." His plan was to "put one bushel of corn to a barrel of water in a common cauldron holding forty-five gallons. Get the water to boiling heat. Then shut up tight to keep the heat under the boiler, and leave it for twelve or sixteen hours, when the corn will fill the vessel. This is cheaper than boiling the grain until it is cooked. It will be seen that the corn swells to three times its original size. The pigs had been used to cooked corn-meal, and would not eat as much of the cooked whole grain." We have no doubt that cooked meal is better than soaked or cooked whole grain, but where it is not convenient to get the corn ground, soaking or cooking it whole is better than feeding it dry.

The Humming-bird Moth, also called Bee-moth, and Clear-wing, is the insect sent us by D. H. Horning for a name. It is *Sesia Thysbe*. It is a day-flying moth, and is often seen hovering about flowers with movements much like those of a humming-bird, and is readily distinguished by the large transparent spots upon the middle of the wings. We do not know the larva, but no doubt it is destructive to vegetation. Neither Harris nor Packard, the only works we have at hand, say anything about the larva or what it feeds upon. We wish Mr. Riley would write a work upon insects—then we should have one that would tell just the things that persons not entomologists wish to know.

Wild Animals in Maryland.—In looking over one of our exchanges, the Civilian and Telegraph, of Cumberland, Md., we were struck with the long list of bounties paid for the "scalps" of panthers, wolves, foxes, etc. A hurried footing up and classification of the list shows that during the past year there were killed in that county, 1 panther, 7 wolves, 22 wild cats, 337 foxes, 7 owls, and 3 hawks, upon which bounty was paid at the rate of $5 for the panther, $20 each for the wolves, $1.50 each for cats, $1 each for foxes, hawks, and owls. This may not be extraordinary, but it strikes us as remarkable that so many of the "varmints" should be killed in a single county, within a few miles of the National Capitol.

Purification of Cisterns.—D. L. Williams, E. Hartford, Conn., and several others, have written to say that they keep the water in their cisterns pure by continuing the conductor, or inlet pipe, nearly to the bottom of the cistern. By this arrangement the fresher water is at the bottom and that which has been longest in the cistern passes out at the overflow.

Catgut.—"J. M. E.," Fairview, Pa., says that there is a dispute as to whether violin strings are made of "pussy's inwards," or whale's sinews, and the question is referred to us. We decide it easily—neither. Catgut is made from the intestines of the sheep, which undergo various processes of soaking in ley, scraping, drawing, twisting, and sulphuring. The best are made in Italy, in part, because the workmen are more skilled, and in part because the sheep are so poor and lean that their intestines are tougher. Very poor strings are made in England, and it is said to be because the sheep are fat. The name, catgut, is said to be a corruption of *gut-cord*.

AMERICAN AGRICULTURIST

FOR THE

Farm, Garden, and Household.

"AGRICULTURE IS THE MOST HEALTHFUL, MOST USEFUL, AND MOST NOBLE EMPLOYMENT OF MAN."—WASHINGTON.

VOLUME XXIX.—No. 10. NEW YORK, OCTOBER, 1870. NEW SERIES—No. 285.

FRESH MILK.—*Drawn and Engraved for the American Agriculturist.*

Those who live in cities and large towns manage to pass a portion of the summer in the country. The more wealthy have their country residences, while those of limited means content themselves with a visit to a farm-house for a few weeks. With all classes one of the strongest inducements for this change of residence is the ability to procure an abundance of fresh milk for the children. The change from heated streets to open fields is not greater, than that from the liquid dispensed by the milkmen, to the pure milk, fresh from the cow. How the little ones thrive on it, and with what eagerness they watch for milking time! The artist has represented a happy group of these city children in the full enjoyment of their healthful country fare. Milk is the natural food for all young animals, children included, and contains all the elements necessary to growth. The milk of the cow differs from human milk in containing much more caseine, or curd, considerably less of sugar of milk, and more of the mineral constituents; the proportion of all solid substances in the two is nearly the same. The milk of the goat is more nearly identical in composition with human milk, than is that of the cow, though that contains a larger proportion of caseine. While milk forms so important a diet for young children it is liable to become a source of disease from the readiness with which its composition is affected by the health of the animal furnishing it, and the rapidity with which it undergoes change after it is drawn. Nature has indicated in the most positive manner, that the food should be transferred from the mother to the young without change. Whenever we depart from this in any particular, unpleasant effects are likely to follow. Let the milk for children really be Fresh Milk.

UNIVERSAL CLOTHES WRINGER.—

But one invention has held its own in the household, and that is the Clothes Wringer. We have used one of those whose name heads this article, for *ten years*, and it has done good service during that time, although in weekly use. We consider the fact that the frame and all parts of the machine are made of wood to be in its favor. There can be no possibility of injury to the clothes by rust. Another advantage of this Wringer is that of a patent stop in the form of a screw, placed over the wheels, preventing them from getting out of gear. But the principal advantage of this Wringer over others is the patent double gear. This is the invention of the late Dr. Warren Rowell, and one of the best devices in mechanical movements that has come under our observation for a long time.—*N. Y. Mechanic, Dec.* 1, 1869.

The Universal Wringer has been in use in our family for years, giving entire satisfaction. We speak whereof we know when we say it is one of the best labor-saving machines ever invented, having several points of superiority over any Wringer we have examined.—*New-York Liberal Christian, April* 2d, 1870.

You cannot do a better thing for your wife on a washing day than provide her a Doty Washer and a Universal Wringer. It will keep aches from her back and arms, wrinkles from her forehead, and roughness from her hands. It will do the work of a hired woman, and save your linen from being scrubbed out and her temper from being chafed out.—*New-York Weekly Tribune, March* 22, 1870.

A Remarkable Statement.

TROY, N. Y., August 22d, 1870.

MR. CHARLES PRATT, 108 Fulton St., New York.

Sir:—We have a short story to tell you about "Astral Oil." In the rear of our store, about three years since, we had a warehouse built, with gravel roof, for storing Paints, Oils, Glass, Acids, Kerosene Oil, etc., etc. Unfortunately for us, it was set on fire early Saturday morning and entirely consumed. The 100 cases of Astral Oil bought from you a few days since were in the warehouse, and strange to say, came out of the fire almost unharmed; they were piled up two and three tiers high; the wood cases, of course, were burned off, and several of the top cans unsoldered on top; but there was *no explosion*, and the Oil in the open cans did *not burn*. The rear of our store was very badly damaged. The "Astral Oil" was within six or eight feet from the store, and, of course, in a hot place. Had the explosion taken place, nothing could have saved the store. We had several barrels of *Kerosene* in the warehouse, which burned lively. To say the least, the Oil has stood a very severe test, and we think it is all you claim it to be. This morning before anything was disturbed, we had a picture taken of the ruins, showing the Oil as it remained after the fire was put out. Yours respectfully, STODDARD & BURTON.

☞ The picture above mentioned and original letter may be seen at our office, also several of the damaged cans, as they came out of the fire.

The above is but one of many testimonials as to the *perfect safety*, under all circumstances, of PRATT'S ASTRAL OIL. None can afford to be without it who have a regard for the safety of their families.

For sale by all respectable dealers, and wholesale and retail by OIL HOUSE OF CHARLES PRATT, Established 1770, Manufacturers, Packers, and Dealers in strictly pure Oils, No. 108 Fulton Street, P. O. Box 3050, New York.

Newspaper Advertising.

A Book of 125 pages, contains a list of the best American Advertising Mediums, giving the names, circulations, and full particulars concerning the leading Daily and Weekly Political and Family Newspapers, together with all those having large circulations, published in the interest of Religion, Agriculture, Literature, &c., &c. Every advertiser, and every person who contemplates becoming such, will find this book of great value. Mailed free to any address on receipt of fifteen cents. GEO. P. ROWELL & CO., Publishers, 40 Park Row, New York.

The Pittsburgh (Pa.) *Leader*, in its issue of May 29, 1870, says: "The firm which issues this interesting and valuable book, is the largest and best Advertising Agency in the United States, and we can cheerfully recommend it to the attention of those who desire to advertise their business *scientifically* and *systematically* in such a way: that is, so to secure the largest amount of publicity for the least expenditure of money."

45,000 **Now in Use. GEO. A. PRINCE & CO.'S Organs and Melodeons will be delivered in any part of the United States reached by Express (where they have no Agent), free of charge, on receipt of list price.** Send for price-list and circulars. Address GEORGE A. PRINCE & CO., Buffalo, N. Y. GEORGE A. PRINCE & CO., Chicago, Ill.

One Pound of Crampton's Imperial Laundry Soap will make twelve quarts of Handsome Soft Soap. Ask your Grocer for it and Try it. CRAMPTON BROTHERS, 84 Front St., New York.

REVOLUTION

In Tin-Lined Lead Pipe.—Price per pound greatly reduced. To satisfy the public beyond the possibility of doubt as regards both the strength and durability of our TIN-LINED LEAD PIPE we have recently increased the thickness and weights of the several sizes and letters, making them in all cases, to correspond *exactly* with *like sizes* and *letters* of the ordinary LEAD PIPE. This establishes OUR TIN-LINED LEAD PIPE as the strongest flexible pipe in use. With the increase in weight we have also reduced the price per pound *nearly one-half*, making the cost per foot on the heavy pipe, adapted to plumbing, about the same as before; whilst the lighter pipes, used extensively for conveying water from springs and ponds, are *reduced in cost fully one-third*. Architects and Plumbers cannot fail to be satisfied, that with weight for weight, and thickness for thickness, a pipe *so constructed* is decidedly stronger than the ordinary lead pipe. The increased thickness of the lead coating allows the joints and connections to be made with *ordinary plumbers' solder*, without effecting the strength of the pipe. In ordering or making inquiries for price, send bore of pipe required, and head or pressure of water. Circulars and sample of pipe sent by mail free. Warranted to be stronger and more durable than Lead Pipe of corresponding sizes and letters.

Address the COLWELLS, SHAW & WILLARD M'F'G CO., No. 213 Centre Street, New York. Also, manufacturers of Block Tin Pipe, Sheet Lead, Lead Pipe, Solder, &c.

COMPOSITE Iron Works Co.

IRA HUTCHINSON, Pres't. IRAH CHASE, Vice-Pres't.

(Formerly CHASE & CO.)

EXCLUSIVE MANUFACTURERS OF

Patent Composite Iron Work,

GATES, RAILINGS, GUARDS, BEDSTEADS, ETC.

IMPROVED WIRE WORK

FOR BANK, COUNTER, AND OFFICE RAILINGS, LAWN, COTTAGE, AND FARM FENCES.

Also, VASES, FOUNTAINS, STATUARY, STABLE FITTINGS, Etc.

109 MERCER ST., near PRINCE ST., NEW YORK.

Amusette, or Portable Family Billiards.

Can be placed upon any Table. It takes but a few minutes to put it up or take it down.

☞ Every family should have it. ☜

COMPLETE SET, including BALLS & CUES.

Sent by Express to any address on receipt of price.

Price, $6.00.—Send for Circular. E. I. HORSMAN, Sole Agent for the U. S. & Canadas, 100 William St., N. Y.

COLLEGE.

Young men thoroughly and practically qualified for the various duties of Commercial Life. The most successful and popular BUSINESS TRAINING SCHOOL in the world.

The "Business Review," giving full information of terms, etc., will be sent free on application. Address

C. R. WELLS, New Haven, Conn.

New Haven Family Knitter.

Latest, cheapest, and best for family use. Knits every thing, and gives universal satisfaction. Company invite any test or comparison. Price, $30 plain; $35 silver-plated; $40 gold-plated. Agents wanted. Address NEW-HAVEN FAMILY KNITTING MACHINE CO., New Haven, Conn.

THE FLORENCE SEWING MACHINE.

The following Statement made by the FIRST purchasers of Florence Sewing Machines on the Pacific Coast, including ALL who bought in 1863, whose present residence is known, is conclusive evidence of the great superiority and unrivaled excellence of the Florence, and is the best testimonial ever given in favor of any Sewing Machine.

"*Our Machines sew as well now, and are as good for use in every respect, as they were when purchased nearly* SEVEN *years ago, and they* HAVE NEVER COST US ANYTHING FOR REPAIRS. *We recommend the Florence as being* WITHOUT EXCEPTION *the* VERY BEST, MOST RELIABLE *and* LEAST COMPLICATED *Sewing Machine in use.*"

May, 1870. (*Signed*)

L. H. BAILEY, Portsmouth House, San Francisco.
Miss ANNE BRADLEY, 534 Howard St., San Francisco.
Mrs. CLARA J. BAUM, 236 Sixth St., San Francisco.
Mrs. EDWARD BABSON, San Francisco.
Mrs. AGNES BRODIE, 82 Everett St., San Francisco.
Mrs. Dr. R. P. CHASE, 714 Howard St., San Francisco.
Mrs. E. J. CRANE, St. James House, San Francisco.
Mrs. D. B. COFFIN, 12 Perry St., San Francisco
Mrs. C. HUCKS, 708 Lombard St., San Francisco.
Mrs. J. B. LARCOMBE, 704 Howard St., San Francisco.
Mrs. M. A. MERCHANT, 510 Mason St., San Francisco.
Mrs. HENRY MILLER, 701 Post St., San Francisco.
Mrs. OHM, 217 Geary St., San Francisco.
Mrs. DELIA PARKER, 517 Pine St., San Francisco.
Mrs. SAMUEL SOULE, 323 First St., San Francisco.
Mrs. EMILY SLOSS, 7 Garden St., San Francisco.
Mrs. C. L. TAYLOR, 709 Bush St., San Francisco.
Mrs. A. J. TURNER, 320 Beale St., San Francisco.
Mrs. S. P. WISWELL, 339 Third St., San Francisco.
Mrs. B. WELLER, 129 Second St., San Francisco.
Mrs. E. M. HINCKLEY, 521 Folsom St., San Francisco.
LUIS A. ARGUELLO, Cor. Santa Clara and Main Sts., Santa Clara.
MARIANO MALARIN, Santa Clara St., Santa Clara.
D. W. HERRINGTON, Cor. Jefferson and Lewis Sts., Santa Clara.
Mrs. BELLE FITTS, Santa Clara.
E. LAMORY, Cor. Main and Franklin Sts., Santa Clara.
H. STEGEMANN, Franklin St., Santa Clara.
SISTERS NOTRE DAME, 93 Santa Clara St., San Jose.
Mrs. M. L. HENNING, 414 Third St., San Jose.
Mrs. SHERMAN DAY, Seventh St., Oakland.
Mrs. O. S. DUNHAM, Franklin St., Oakland.
Mrs. WM. S. SNOOK, Oakland Point, Oakland.
WM. B. MCGUIRE, Bath, Placer Co.
Mrs. LUCY CLEVELAND, Ukiah City.
Mrs. LEONORA HARD, Antioch.
Mrs. DANIEL DUNN, Trinity St., Vallejo.
WILLIAM MELLOR, 102 D St., Marysville.
Mrs. M. E. BUCKHALTER, Main St., Dutch Flat.
HENRY JACKSON, Watsonville.
Mrs. E. C. COMSTOCK, 104 Main St., Napa City.
O. D. METCALF, 2 Main St., Hydesville, Humboldt Co.
Mrs. JENNIE A. EGAN, Spadra, Los Angelos Co.
JOHN F. PINKHAM, Placerville.
Mrs. J. CRAWFORD, Copperopolis.
Mrs. J. C. FISCHER, Rocklin, Placer Co.
A. B. BEAUVAIS, Columbia.
Mrs. Dr. S. L. ROBERTSON, Soquel, Santa Cruz Co.
JAMES BELL, near Sonora, Tuolumne Co.
Mrs. SARAH E. MILLER, 344 G St., Sacramento.
Mrs. SARAH LONGTON, Cor. J and Seventeenth Sts., Sacramento.
Mrs. P. G. TEFT, Aurora St., Stockton.
Mrs. ELIZABETH KEEFER, Woodland, Yolo Co.
Mrs. M. A. NOYES, Pt. Gamble, Washington Territory.
O. C. SHOREY, Seattle, Washington Territory.
W. L. BARLOW, Salinas, Monterey Co.
Mrs. MARIE DAEGENER, Columbia.
Mrs. JOSEPHINE BADER, San Andreas.
Mrs. LUCY COOKE, Dutch Flat.
Mrs. JOHN O'NEILL, near Jamestown.
Mrs. MARY GUNDRY, Amador City.
Mrs. ELIZABETH MILES, Victoria, Vancouver Island.

Agencies—505 Broadway, New York, and in all other principal Cities and Towns.

VICK'S ILLUSTRATED CATALOGUE

OF

HYACINTHS, TULIPS, LILIES,

AND

OTHER HARDY BULBS FOR FALL PLANTING,

Is now published, and will be sent FREE to all who apply.

Address,

JAMES VICK,
ROCHESTER, N. Y.

40,000 Camellias. 10,000 Chinese Azaleas.

PARSONS & CO. offer well-grown specimens of these plants in the quantities mentioned.

The smallest order will be supplied, and dealers who purchase by the thousand are invited to examine their stock.

The Camellias include the best varieties, and a very large proportion of Double White. The Azaleas include the new and fine sorts. For Catalogues apply by mail to

PARSONS & CO., Flushing, N. Y.

INVENTORS and others interested in *Patent Business* should address EDSON BROS., Patent Lawyers and Solicitors, 409 NINTH ST., WASHINGTON, D. C., for ADVICE and CIRCULAR.

AMERICAN AGRICULTURIST.

NEW YORK, OCTOBER, 1870.

We have passed through a season of terrific heat, long continued, and little relieved by rains. The storms of the summer and early autumn have borne great blessings to the thirsty land, but they have been accompanied by violent winds and wonderful exhibitions of electricity, which have wrought local damage to the great distress of many farmers. These violent storms are usually quite circumscribed in their destructive effects, and around the outer sides of the tempestuous centers the blessed rains have fallen, moistening the parched earth and reviving drooping vegetation. Parts of New England, judging from our correspondence, suffered more than other sections. The cattle have been fed on hay; young stock on weeds and the undergrowth of the woods, and sheep on bushes and swamp grass. Late crops have suffered; corn, in some places, potatoes, almost everywhere, turnips, generally; nevertheless we are not likely to want. The great granaries of the West will feed the East, and there are many things we can do without. The modern hay presses put hay into such small compass that it can be as economically transported as corn. Bulk makes little difference, provided a railroad car can receive its full load of 10 or 15 tons, and this can be accomplished. While, therefore, the canal still competes with the railroads, it might be well to secure the transportation of hay to Eastern markets where there is likely to be a dearth. Cottonseed cake, linseed cake, and other forms of concentrated provender will be used extensively this winter, and contracts should be early entered into for them, or a stock laid in.

The Fairs this year are held so that there seems to be less collision than usual, and persons can go from one to another, which, with exhibitors as well as Agricultural Editors, is very desirable.

Now is just the time of the year to start Farmers' Clubs. There is little enough social enjoyment in most rural neighborhoods, and a well-managed Farmers' Club exerts a civilizing and refining influence among the farmers' families. It makes people think of something besides the humdrum of the daily routine of work, eating, and sleeping. It promotes a taste for reading and thought about the daily pursuits of life, and in every way is improving and pleasurable.

Hints about Work.

Farm Buildings.—Animals will soon need shelter, if they do not already, and should be regularly stabled at night, as in winter. In putting the stables in order, look well to the floors, and to the timbers in contact with manure, or the liquids of the stable; renew those which are in any degree weak. Tighten weather-boards, patch roofs if leaky. See that all buildings around the barn-yard have good eavestroughs and gutters, so that no more water than necessary enters the barn-yard

Road mending at this season must be done with clean gravel and sand, otherwise the mended spots will be very soft, and will cut into ruts in fall and spring. This material, with small, broken stones in some places, is the only fit substance for mending roads, where it can be obtained. Make good, hard turnouts for water, on roads upon descending grades, distributing the road wash upon meadows or pastures, if possible. Bridges and embankments should be well looked to, and put in order for winter before cold weather sets in.

Ice-Houses.—Experiment has proved that though it is really cooler under ground than above, during most of the year, yet, that if the wooden walls of ice-houses be properly constructed they will prove a better protection to the ice above ground than stone walls, with a lining of plank and non-conducting filling below ground. Besides it is decidedly cheaper to build above than below the surface. We give a plan for an ice-house and dairy on page 100. The manner of constructing the ice-house portion is applicable to one used simply for keeping ice. The points to be borne in mind are simply these. 1. Perfect drainage. 2. Close, strong, non-conducting, double walls of plank on the inside, having the space between them not less than eight inches wide, and this filled with some dry, porous material, like saw-dust or spent tan. 3. An opportunity for a change of air above the ice—entirely avoiding drafts. A house 12 x 12 feet square on the inside, is the best size for a private family. 10 x 10 feet square will generally keep the ice well, but it will waste badly, and soft ice, such as we had last winter, will not last through the summer.

Orchard and Nursery.

Apples.—The abundance of the crop in most localities renders it especially necessary to send only selected fruit to market, if satisfactory prices are looked for. Hand-pick and barrel carefully. Late varieties to be stored should be kept as cool as possible. Make poor fruit into

Cider and Vinegar, as this is the most profitable way of disposing of it. Cider-making is usually deferred until cooler weather, as it is better when the fermentation goes on slowly. Perishable fruit may be ground up and pressed for vinegar at once.

Pears.—The unusually warm and dry summer has, in many cases, caused the late varieties to mature earlier than usual. This may affect their keeping qualities. They must be kept at as low a temperature as possible. Do not mix varieties; and store all in such a manner that they may be readily inspected from time to time.

Planting of all except stone fruits, is advisable in those localities where the autumns are mild. The soil is in much better condition than in spring, the trees are usually more promptly procured from the nurseries, and the work is not so apt to be hurried. Prepare the land and stake the places for the trees, so as to be ready to put them out as soon as they arrive. In planting an orchard, put trees of the same variety together, for greater convenience in gathering the fruit. In planting, see that the

Labels attached by the nurseryman are not so tightly wired as to injure the tree. Do not trust to labels, but make a record of the position of the trees at planting time.

Nursery Stock may be trimmed into shape. Good cultivators manure between the rows of young trees.

Fruit Garden.

Planting may be done in all localities where the season is mild; we except strawberries, which are now better left until spring.

Grapes, except for wine, will have been mostly gathered. Fruit to be kept until winter should be in a cool, dry room, packed in small boxes. Isabella, Catawba, Diana, and Iona, are good keepers. Concord very poor, and Delaware nearly as bad, though this will keep for weeks if carefully handled.

Strawberries.—If plants have been struck in pots they may be set now, but otherwise spring planting is preferable.

Blackberries and Raspberries may be set out, cutting the stems off to the ground. The nurserymen send out canes with the roots, but these are of no use save to serve as a handle in planting. These are readily propagated from pieces of the root, 2 or 3 inches long, which are to be packed in a box with alternate layers of earth; bury the box where water will not stand, and out of the reach of frost.

Currants and Gooseberries.—Prune, removing old stems if the bush is crowded, and shortening the new growth one-half or more. Make cuttings of the new wood, about 6 inches long, and set 4 inches apart in trenches, with one bud above the surface. Press the earth well against the cuttings and mulch with leaves or litter when heavy frosts come.

Kitchen Garden.

Beets should not receive heavy frosts as it detracts from their sweetness.

Cabbages and Cauliflowers.—The plants from seed sown last month are to be set in cold frames. Put them 2½ inches apart each way and set them down to the leaves, covering the stems with earth. The sashes should not be put on until freezing weather.

Celery.—The drouth killed the planting in many places, and celery is likely to be scarce. Even that which stood the drouth is likely to be small. Keep it growing as long as possible, but bank up before severe frosts.

Lettuce.—Plants intended for forcing or for early spring planting are to be put into cold frames, the same as directed for cabbages. In mild localities the plants will endure the winter if slightly covered.

Spinach will need thinning; use for the table the plants that are removed, or market them. Stir the soil between the rows.

Squashes.—Gather whenever there is danger of frost, place in piles and cover with vines at night; give them a few days' sunning before storing. Handle as carefully as if they were eggs, as the least bruise will be followed by decay. Store where they will be cool and dry, and not freeze. Large growers have a house devoted to storing them, in which a fire can be made in very cold weather.

Sweet Potatoes.—Do not wait for a severe frost before digging, or the potatoes will not keep. Take up the crop as soon as the leaves show that they have been touched; selecting, if possible, a warm day for digging, and allow the potatoes to dry in the sun. Pack in perfectly dry leaves or cut straw. The essential point in keeping them is to not allow them to become cooler than 60 degrees.

Flower Garden and Lawn.

This is the most favorable season for grading, draining, road-making, transplanting deciduous trees and shrubs, and doing much of the work that is usually left until spring.

Pæonies.—Remove this month, as they seldom flower if disturbed in spring.

Herbaceous Perennials are also better transplanted at this season. When they have been in place for three or four years they should be lifted, the large clumps divided and reset in fresh soil. At the time of doing this there will be an abundance to give to less fortunate neighbors. Thin out seedlings sown earlier, and at the approach of severe weather give them a covering of leaves.

Dahlias.—When the frosts have disposed of the portion above ground, choose a sunny day and dig the roots. Let them dry off and store them in any dry cellar that will keep potatoes in good condition. Let the labels be legible and properly attached.

Winter-covering.—Many make a mistake in going into winter quarters too soon. When the ground begins to freeze, is soon enough. Materials should be collected. Leaves are among the most valuable; these and salt and bog hay and cedar or other evergreen boughs are the materials generally used.

Greenhouse and Window Plants.

Bulbs should be potted for winter blooming as early as they can be had. Use rich soil, with some sand to keep it light, and place the pots in a cool, dark cellar, or in a frame, and cover them with several inches of coal ashes or tan. When pots are covered in this way it is better to invert a thumb pot directly over the bulb to protect the shoot in case it should start before they are removed.

Insects.—Take none into the house with the plants. If any plants are infested keep them apart from the rest until they are thoroughly fumigated.

Forcing Plants.—A number of hardy shrubs and herbaceous plants are forced for winter decoration, and near cities, to sell to people who do not know any better than to buy them. Astilbe (Spiræa) Japonica, is one of the best, and makes a charming plant for winter blooming, and for cut flowers. Lily of the Valley, Dicentra, Deutzia gracilis, and others, are used. The plants are to be potted and kept in a cold frame or cool cellar, where they must not get "killing dry," and in February they are brought into heat.

Annuals.—Sow seeds in pots of such as may be wanted for winter blooming.

Appliances.—Get soil, sand, moss, pots, and whatever may be required, under cover in good season.

The Virginia Itea.

The Itea is one of several, pretty, native shrubs that have failed to find a place in our ornamental collections, probably for the reason that no dealer has seen fit to praise them in his cata-

VIRGINIA ITEA.—(*Itea Virginica.*)

logue. This spring we flowered the Itea for the first time, and though we had frequently seen the wild plant, were not prepared to find it such a pleasing addition to our collection. The shrub grows from four to six feet high, is much disposed to branch, and at the end of each small branch, bears a spike-like raceme about three inches long, of rather crowded white flowers. The flowers appear in June, and continue for a considerable time. The Itea is found along the coast from New Jersey southward, and naturally grows in wet places, though it flourishes well in almost any soil. It is readily increased by suckers and by seeds. The shrub does not seem to be well enough known to have received a common name; its botanical one, Itea, is the Greek name for the Willow, and is said to have been applied to this plant on account of the willow-like rapidity of its growth.

The Pyracanth Thorn for Hedges.

The Pyracanth Thorn, *Cratægus Pyracantha*, is a small evergreen shrub from the south of Europe. It is abundantly furnished with thorns and its small, horizontal branches terminate in sharp points. The foliage is of a rich green; its small white flowers are produced in great abundance, and are followed by small, red berries which are in such profusion and give such a glow to the plant during winter that the French call it *Buisson ardent*, or Burning-bush. Its berries are not much eaten by birds, and as they, as well as the foliage, remain on all winter, the plant is a very ornamental one. It is used in Europe for covering walls, and is sometimes grafted standard high upon the Hawthorn to make an ornamental tree. We are not aware that it is there employed as a hedge plant. In the southern portions of this country it has been thoroughly tested for hedging, and has been found to answer admirably; this and the McCartney Rose being the two favorite plants for live fences. The late Thomas Affleck, writing for Texas, considers the Pyracanth the most valuable for hedges for that country, and that it requires only one-fourth the care that the Osage Orange does. Unfortunately the plant is not hardy at the North, though it endures the winters near Washington. Some years ago a seedling Pyracanth sprung up in an English nursery, which bore orange-colored fruit, and plants of this were propagated under the name of White-fruited, as affording a pleasing contrast with the ordinary red-fruited sort. Frequently these varieties are not as hardy as the type; but Messrs. Parsons & Co., of Flushing, N. Y., have found the White-fruited Pyracanth to be more hardy than the red, it having in their grounds stood unharmed through the severest winters. It grows rapidly, and makes a most compact and impenetrable hedge. The Pyracanth is propagated by cuttings about six inches long, made in the fall, and set two-thirds of their length in the ground. The nursery bed should be mulched in summer, and the plants should remain in it for a year. The proper time to set the hedge is the fall, as the plants start very early and are impatient of removal in spring. In the engraving we give a small twig of the natural size to show its thorny character and the shape of the leaves; also, a cluster of flowers and one of fruit, both of which are smaller than in our common thorns.

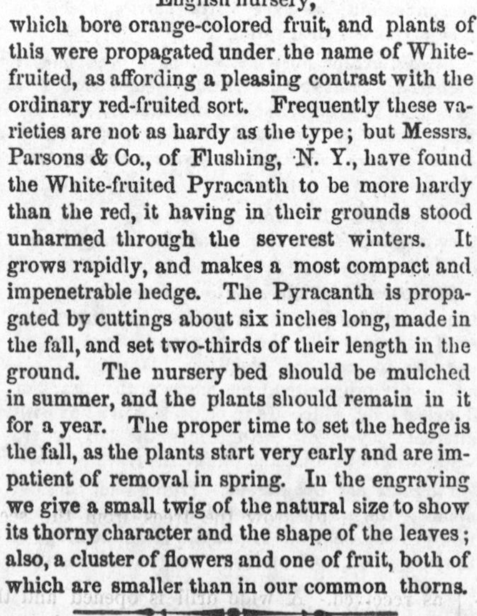

Forcing Rhubarb.

BY PETER HENDERSON.

Forcing Rhubarb is one of the simplest of all our forcing or forwarding operations. The roots are dug from the open ground in fall, put close together in a box or barrel, and soil sifted in to fill the interstices between the roots; they are then placed in a situation where the temperature will range from 55° to 75°, with a moderate amount of moisture. By this treatment, Rhubarb may be had from January to April. The roots may be placed wherever there is the necessary temperature; light is not neces-

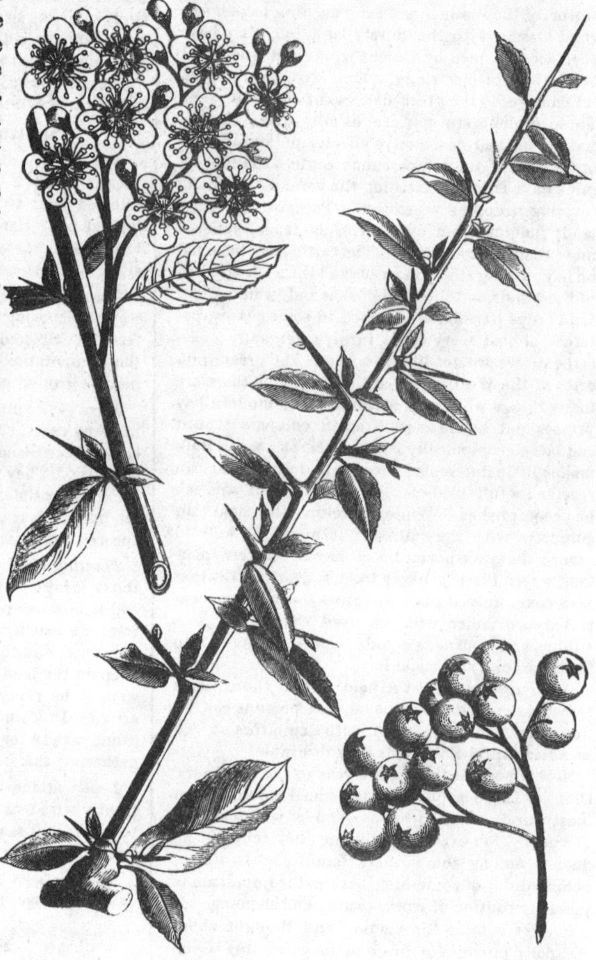

WHITE-FRUITED PYRACANTH THORN.

sary at all, in fact the stalks of Rhubarb are much more crisp and tender when forced without exposure to direct light; hence the roots may be placed in the furnace room of a cellar, under the staging of a greenhouse, or in an early forcing grapery. A florist in Boston, a few years ago, told me that he had sold enough Rhubarb grown under his greenhouse stages, to pay his coal bill, (over $100), besides having all he wanted for his family use.

Chestnuting—The Chestnut-Tree.

Many a reader will stop as he sees the fine engraving of a nutting party, which Mr. Perkins enables us to give on the following page, and recall the time in which he participated in such a scene. How vividly all presents itself in memory; the fine air of the clear autumn morning; the golden burs of the chestnuts shining against the dark green leaves; the pleasant "woodsy" smell; the climbing and beating the trees; the gathering the nuts; the pricking with the burs; the young companions—and all that go to make up a happy day at chestnuting. Leaving the picture to tell its story of a nutting frolic, we say a word in favor of the tree itself. The chestnut is neither the best of timber, nor the best of fuel, but it is valuable for both these uses. For fencing it answers an excellent purpose, and its lumber is now in great request for interior finishing. Its wood makes a valuable charcoal; its fruit is always in demand, and brings a good

CHESTNUTING.—DRAWN BY GRANVILLE PERKINS.—*Engraved for the American Agriculturist.*

price. Two great points in favor of the chestnut are the rapidity of its growth, and the readiness with which it renews itself after being cut down. It grows freely in any soils that are not too wet, and is hardy over a wide extent of country, it being found native from the 43d parallel to as far south as Florida. The chestnut presents many claims to those who are interested in tree planting, not only as valuable for the treeless portions of the West, but for those lands of the East which can only be made remunerative by covering them with forests: the great obstacle to its introduction has been the difficulty of procuring the seed or the plants. The seeds are not usually kept by seedsmen, as they soon become worthless, and unless one has friends in a locality where chestnuts grow, they are not easy to obtain. As to the matter of plants, we notice that Messrs. Storrs, Harrison & Co., of Painesville, O., make a specialty of them, and offer them at moderate prices. Those who wish to grow the trees from the seed, should make arrangements to procure the nuts as soon as ripe, and they may be sown as soon as received. A wide drill is opened and the nuts scattered rather thickly, the spaces between them are to be filled by sprinkling soil among them, and then covering with two or three inches of leaves. In the spring the greater portion of the leaves is to be raked off, and when the young plants have grown an inch or two, fine soil is drawn up to them. Squirrels, gophers, and mice, will destroy the nuts if they discover them, and in localities where these pests abound, the seeds must be kept until spring. They should be mixed with three times their bulk of dry sand. They are said to retain their vitality if packed in perfectly dry moss. The chestnut can be readily transplanted if removed while young, and nursery trees that have been transplanted twice, the tap-root being removed, are quite as likely to live as most other deciduous trees. Our native chestnut is a variety of the European, differing in the size of its fruit. The fruit of ours is much smaller, but at the same time superior in sweetness and flavor. There is a great difference in the product of our wild trees, some of them yielding fruit twice the size of the average. There is no doubt that by selection and cultivation, the size of the nuts of the American variety could be greatly improved.

Ice-house and Summer Dairy Combined.

Perfect control of the temperature of the dairy is a great step gained towards making the best butter. It is only by means of ice, or very cold spring water that we can keep the most desirable temperature in very warm weather.—During most of the year there is little difficulty in maintaining sufficient coolness. In winter the problem is how to keep a dairy warm enough and not get it too hot. This is the battle with the weather that we wage almost the year round. In former years we have given numerous plans for ice-houses, both large and small, with cool rooms or refrigerators attached. We have lately had our attention called to the desirableness of a combination of the dairy and ice-house, and present the following plan which we deem entirely practical.

Fig. 1.—ELEVATION OF ICE-HOUSE AND DAIRY.

The plan proposes an ice-house above ground and a dairy half below; the ice room half covering the dairy, and the rest of the dairy being

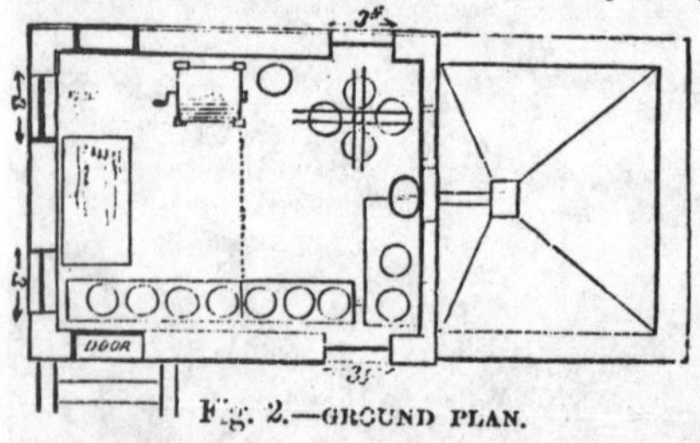

Fig. 2.—GROUND PLAN.

covered by a cool room, which forms the entrance to the ice-house. The exterior walls of the ice-house are of wood, those of the dairy are of stone. The floor of each room is laid in cement with a slope sufficient to carry off the water. The drainage of the ice-house is collected and made to pass by a pipe, into a vessel in the dairy, where the end of the pipe is always cover-

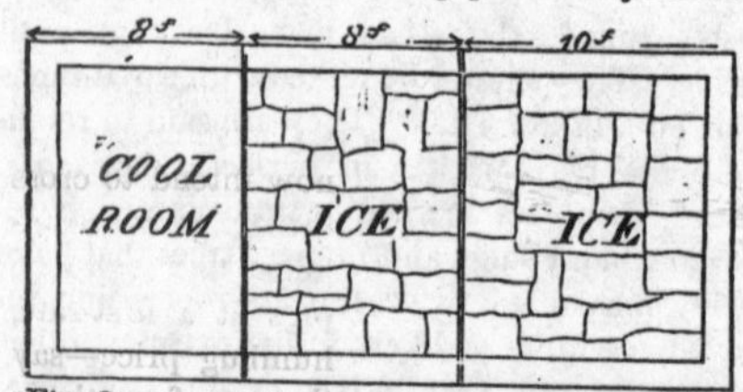

Fig. 3.—PLAN OF UPPER PART OF ICE-HOUSE.

ed with water. The water is allowed to flow through shallow troughs in which milk pans may be set. The amount of water would not be large, but it will be cold and ought not to be wasted.

The building represented in the perspective elevation, fig. 1, is 28 feet long by 14 wide. The ice-room seen in figs. 2 and 3, is 10 × 12 feet on the ground, and about 12 × 16 feet, including the space above the dairy. The sides of the building are 9 feet above the ground, and the hight of the dairy 7 feet in the clear. The outside walls of the ice-house are made of 2-inch plank, 10 inches wide, set upright, having inch-and-a-half planks nailed on the inside, weather-boarded neatly on the outside, and filled with spent tan-bark or other dry, non-conducting substance. The partition wall between the dairy and the ice-house, and between the cool room and the ice-house, is half the thickness, and not filled. Thus forming closed air spaces between the studs. These spaces communicate with the dairy, by little doors near the floor and so currents of cold air may be established and perfectly regulated, entering the dairy on the side towards the ice-house. These, with a ventilator V, at the top of the room for carrying off

Fig. 4.—SECTION OF ICE-HOUSE AND DAIRY.

the warmest air, will surely cause the temperature to be easily governed. This description, with the engravings, sufficiently illustrate the idea to enable any good builder to carry it out.

THE LOCUST AND AILANTHUS. — A New Jersey correspondent writes that he visited a Locust grove through which were distributed a number of Ailanthus-trees. The Locusts were 50 or 60 feet high, and had never been troubled by the borer, while trees not over a fourth of a mile off were badly injured. He asks if we or any of our correspondents have ever known the Ailanthus to protect the Locust from the borer. We have not, and it would be a most fortunate discovery if it were found to be the case, as it would, as our correspondent states, "add millions to the wealth of the country." He asks if we ever knew any insects to attack the Ailanthus. The Ailanthus silk-worm which was introduced into the country several years ago, as an experiment, has become thoroughly naturalized in New York and Brooklyn.

How to start or back a Heavy Load.

A team will usually draw a much heavier load than it can back, and the power of a man applied upon the wheel is an assistance constantly made use of both in starting and backing heavy loads. We have been observing with interest the six-horse and eight-horse teams drawing immense blocks of granite for the foundation of the new post-office in New York, and have been surprised to see with what ease a single pair of horses would back the load which eight were required to draw. It was done in the way we indicate in the accompanying engraving; namely, by attaching the horses to the rim of the wheel near the top. Thus their power to move the load was doubled by the leverage

BACKING A HEAVY LOAD.

Saving Fodder.

In many localities the drouth has made a short crop of hay, and though of excellent quality, it will not suffice to carry the ordinary stock of the farm through the coming winter. All the substitutes ever used for good hay will be wanted, and it becomes farmers to make the most of them. Oat and wheat straw that are often allowed to rot upon the ground, or are used for bedding, make a very good fodder, and should be carefully stacked or stored in the barn at the time of threshing. If cut and mixed with corn meal or with wheat bran, cattle will thrive upon the feed quite as well as upon good hay. Those who have sowed corn fodder abundantly, have a good substitute for hay. Save what is left from the fall feeding. Make the most of the fodder from the cornfield, which is often damaged from careless stacking. It is not yet too late to cut salt grass and bog meadows, that are often neglected from the abundance of other feed. To be sure there is not much nourishment in these grasses, but there is some, and when run through the cutter, and mixed with meal or roots they serve a good purpose. Carefully store the turnips that have been sown between the corn rows, or by themselves. It frequently happens that a dry summer is followed by abundant fall rains, which come in season to allow us to make a large turnip crop.

FROSTS.—It is often the case that after a few frosty nights, we have many days of fine weather, and if plants can be protected during the early frosts, their season may be prolonged. Straw mats, a sheet, or even newspapers, supported above, but not touching, the plants, will protect them from the usual early frosts.

The Wolverine.—(*Gulo luscus.*)

The Wolverine, now rare in the United States, but abundant further North, is probably identical with the Glutton of Northern Europe and Asia. Some naturalists have even endeavored to establish two species on this Continent, but as the differences depend almost solely upon color, the new species could not be accepted. The Wolverine is mercilessly slaughtered by hunters whenever found, and hunted relentlessly where its presence is suspected, not because of its value, for its fur is of little account, but for its destruction, as it causes the trappers great damage by destroying the lines of Sable traps for the sake of the bait. The animal is about three feet long, with a tail six inches in length, covered with long hairs. The head is broad and thick, but the muzzle sharp. The feet are five-clawed, and very large, giving the animal the ability to walk upon the snow. So large are they that its tracks are occasionally mistaken for those of the black bear. In color, different individuals vary greatly. The muzzle is dark to the eyebrows, across which a lighter band passes, which extends down upon the side and passes over the rump. The rest of the body is dark brown above, and the legs, tail, and under parts are nearly black. In some of its characteristics the Wolverine closely resembles the bear, and it was regarded as a bear by Linnæus; now, however, our best naturalists classify it with the otters and weasels, with which it has apparently less affinity. It may, indeed, be regarded as forming with badgers and raccoons, a chain of associating links between the more positively marked animals of these families. The female has usually but two young ones at a litter, hence the species does not multiply rapidly, and has gradually of late years become extinct over a large part of the Union where it once abounded. It is said to have been common in the mountains of Tennessee and Carolina, but is now rare, even in Michigan. Most of the marvelous tales told of the Glutton and Wolverine, are purely fiction, or have a basis only in the fact of the Lynx or some other animal having been mistaken for this one. It has a very acute sense of smell, and will scent out the *cachés*, or places where arctic hunters deposit their provisions, and its strength is so great that it will often dig under or uncover them, though loaded down with logs and stones. It climbs trees, but is not at home in them. On the whole, it is one of those few animals which are of no use to man, and with so many hurtful traits that its extermination will not be regretted.

THE WOLVERINE.—(*Gulo luscus.*)

Swine of the South Sea Islands.

We have been much interested lately in seeing and learning the peculiarities and merits of some swine from the Southern Pacific Islands, imported by Mr. Jas. P. Swain, of Bronxville, N. Y., who is known to the readers of the *Agriculturist* as a breeder of discrimination, and on the lookout for valuable neglected breeds, crosses, and points, in our domestic animals. The pigs, shown in the engraving, are young, of breeding age, perhaps a year old, and weigh about 150 pounds. The request for further information elicits the following from Mr. Swain:

SOUTH SEA ISLAND SWINE.

"You ask for information in regard to South Sea Island Pigs. Of one thing I can answer you, your artist has made a most truthful drawing of them, even to the peculiar pigish expression of their faces. As to the origin of these pigs, I must refer you to some one more learned in the science of pigology. The Hon. S. N. Mason says that these pigs were plenty in all the islands fifty years ago, so plenty that four yards of calico would purchase a pig weighing fifty pounds, and three yards one of seventy-five pounds, and two yards one weighing over one hundred pounds. A pig of fifty pounds was large enough for a feast, and a larger one was more trouble to kill and cook and not so tender. Mason found them on the newly discovered islands, under and south of the Equator, and plenty on all those islands, and does not think there was time after the voyages of Capt. Cook to so thoroughly disseminate them. They have been left by our whale ships at the island of Barbadoes, and now they are quite plenty there. Their meat is unquestionably the sweetest of any pork known. I imported some about twenty years ago, gave them away freely, crossed them with others, and was better pleased with them than any other breed I ever tried. My neighbors liked them, but as they did not cost any thing, they took no care of them, and they became extinct in a few years, or so mingled with others, that they are no longer known. You will see that they are, unlike the wild pigs of Europe, heavy in the hind-quarters, while the wild boar is heavy in the fore-quarters. I think there is a slight resemblance between these and the China and Sandwich Island pigs. I now intend to cross these pigs with Suffolks, and sell the pigs at a first-rate, humbug price—say three or four times as much as they are worth, and see if they will not be appreciated. I should judge that every thing in the Essex that is valued came from this breed of pigs. But as these pigs were never owned nor bred by H. R. H. Prince Albert, Earl Ducie, or Lord Wenlock—were never exhibited at any Royal Agricultural Show, I fear they will not be much valued."

The Muck Mines.

We shall lose one of the great blessings of the drouth if we fail to work the muck mines. On many farms these are laid bare for only a few weeks in the year, and then extra help should be employed to get out a large quantity for future use. We have tried this muck so long and so thoroughly, and derived so much benefit from it, that we shall make no apology for frequent allusion to it. We are fully persuaded that any farmer, who has one of these mines of peat or muck upon his farm, can make no better use of his capital than by working it. If he should get two or three years' stock on hand—a thousand cords or more, it will pay a good interest upon the investment. It is all the while improving by exposure to the atmosphere, and will be more valuable in the sties, stables, and yards. Well-cured muck alone is a valuable top-dressing for the meadows. It starts the grass earlier, it absorbs ammonia from the rains and snows, and helps to protect the grass-roots in the winter.—It absorbs moisture, and is one of the best safeguards against drouth for sandy and gravelly loams.

NATIVE POTATO.—(*Solanum Fendleri.*)

A Native Potato.

Chili, Peru, and the neighboring islands are believed to be the home of our cultivated potato. The writer, while exploring the mountains of New Mexico, in 1851, was delighted to find, as he supposed at the time, the potato growing wild in that region. The plant appeared like a diminutive potato, and, as small tubers were found, it seemed very likely to be *Solanum tuberosum* in its original condition. Circumstances did not allow of the preservation and transmission of the fresh tubers, but dried specimens were made. It was found upon reaching home that the plant had already been described by Doct. Gray, as *Solanum Fendleri*, in honor of Mr. A. Fendler, who had just before made a botanical journey through New Mexico. This spring we received through the kindness of a friend a few tubers of the same Solanum, which enabled us to grow it. The plant is much smaller than the common potato, from which it principally differs in the nearly uniform size of the lobes of its leaves, as will be seen by the engraving taken from a living specimen. The friend to whom we are indebted for the tubers observed that it threw up, at a considerable distance from the main plant, stems which sprung from long underground runners. We have not noticed this tendency in our own plants thus far. Several who are curious in such matters have this Solanum in cultivation; and should cultivation lead to its improvement, our readers will be apprised of the fact. There is abundant room for amelioration, as the tubers of the wild plant are hardly as large as a boy's marble.

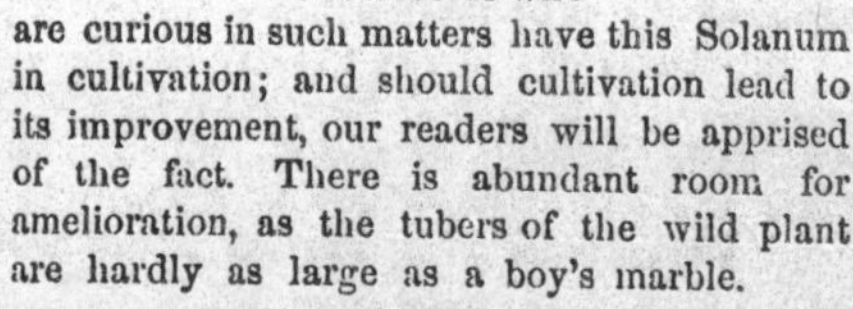

The Japanese Irises.

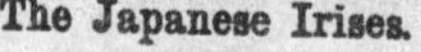

Last summer we saw in the garden of Mr. James Hogg, at Yorkville, N. Y., a bed of Irises, the roots of which had been sent from Japan, by his brother Thomas. The Iris in its various species is a favorite with us, and this one from Japan struck us as a charming novelty. The foliage is not glaucous, like that of the usually cultivated kinds, but is of a dark green, the leaves being about two feet long, and less stiff than in the common species. The flowers, instead of having the petals recurved in the usual manner, are flat, as shown in the engraving, where they are about a third less than the natural size. The colors range from white and the faintest lilac, through blue to deep purple, and present an agreeable variety in their veinings. The one on the left of the engraving was of a rich purple with golden markings in the center; and the one on the right was white, delicately veined with lilac. Many of the flowers show a strong tendency to become double. We have not been able to find a description of this species, but have the impression that we have seen it noticed somewhere as *Iris Japonica.* The plants are perfectly hardy. Some careful cultivator, by hybridizing this with other species, might produce some interesting results.

NEW JAPANESE IRISES.

Inland Water-Carriage in a small way.

Railways have thrown canals quite out of fashion the world over, yet every political economist knows that of all means of transportation of heavy goods, water-carriage is by far the cheapest. It is also the slowest. It may be used inland on rivers, brooks, canals, and lakes. We can, in fact, hardly have a steady flowing stream so small that it may not be so widened and deepened, that upon prevailingly level land it can not be used as a canal for flatboats and dug-outs. We associate with the word canal the idea of an artificial river, 30 to 100 feet wide, and capable of floating boats of several hundred tons' burden. A ditch eight feet in width, two feet deep, and half a mile long, connecting a farmer's barn-yard with his meadows at a distance, is just as much a canal as is the Erie; and there is hardly any limit to the service such a canal might be to some farmers. We have a farm in mind where such a ditch might be dug nearly a mile long; and while it would serve to drain the extensive meadows it would pass through, it would come at its lower extremity within eight rods of the farmstead, and might be filled in half a day at most seasons. This canal would probably require one lock, as the fall in the entire distance would be about six feet. Locks interpose but little difficulty when there is plenty of water. They involve, it is true, delay; but the small quantity of water necessary to float the boats ordinarily used makes it a small matter. On some of the canals in Germany, empty boats are drawn by a horse around the locks on rollers when the water is low. When locks are required, the canals are usually wider and the boats shorter and broader than otherwise. This is also necessary when very bulky loads will usually be put upon the boats; for it is easy to see that a very long and narrow boat would not carry a load of hay so well as one of grain or manure. Yet when they can be used, narrow ones are preferred, because they require so much less labor to draw, row, or pole them at a moderate rate of speed.

Not only might such canals be of great use on many inland and tide-water farms, but near our seaboard cities great quantities of vegetables might be easily transported to market, and return loads of manure obtained; two men only being required to manage a large boat, and the power of a single man on the tow-line being sufficient to draw six or eight tons at a comfortable rate of speed,—say two and a half miles an hour. One horse will draw upon a canal, fifty to seventy tons, two and a half to three miles an hour, as easily as he can draw 1,700 or 1,800 pounds upon wheels. A strong boat that will carry six or eight tons will cost not nearly so much as a common farm wagon.

Locks may be made a little wider and longer than the boats used, so that they will never bind, and should have plank sides and bottom, well caulked, and be packed around the outside of the lock with a puddling of clay uniformly moistened and well rammed beneath and around the channel. There should be a single door or gate, opening up stream, to close each end, and these gates may have a flap of rubber around their edges as a secure packing to prevent leakage; and there must be small gates to let the water in and out previous to opening the great gates. Locks are rather expensive, but need not be a serious obstacle if other things favor a canal.

Odd Washing Fluid Recipes.

Among replies to our request some time ago for recipes for washing fluids, were some compounds containing ingredients that were quite useless, and whatever efficiency they possessed was in spite of, rather than because of them. One reads thus: Crude Potash, 1 lb., or 1 box of Concentrated Lye; Sal Ammoniac, 1 oz.; Salts of Tartar, 1 oz.; Boiling Water, 2 gallons. Mix, and when cool, put into a stone jug. Soak the clothes over night, soap the most soiled parts and boil in water, to every two pailfuls of which ¾ of a teacupful of the liquid has been added. The value of this mixture depends upon the potash or concentrated lye, which are carbonate of potash. Salts-of-Tartar is nothing but a white and pure form of the same thing, and the sal-ammoniac must be quite useless.

Another is: 1 gallon of water; 1 oz. Saleratus; 1 teaspoonful Tartaric acid; 1 do. Sal-Soda; 6 oz. Gum Tragacanth, or in case of no gum, 1 lb. of good bar soap. Dissolve together. This is said to be the "great $5 recipe." Saleratus and sal-soda are so similar that there is no advantage in using both. The Tartaric acid is worse than useless, as it is neutralized by the saleratus and soda, and in turn neutralizes them, and renders the mixture the less efficient. The gum tragacanth is just so much foreign matter to be removed in rinsing.

Aunt Eliza's Cake.—One cup of sugar, one cup of sour cream, nearly two cups of flour, and one teaspoonful of soda. Flavor as preferred.

THE HOUSEHOLD.

☞ *For other Household Items, see "Basket"*

Household Ornaments—Grass Bouquets.

The introduction of a number of annual flowers, which, when dry, retain their form and color, has made winter bouquets very popular. These everlasting flowers, as they are called, when carefully dried and made up with skill, form pleasing ornaments for the household; but at their best they are not, to our taste, so desirable as bouquets of dried grasses. Many of the grass bouquets that we see are failures, for the reason that the maker of them tried to crowd too much into them. A collection of the rarest and most elegant grasses, if tied into a bunch and crowded into a heavy vase, will fail to be pleasing. The beauty of grasses depends upon their ease and freedom from restraint. Some color the grasses or incrust them with alum crystals, processes which we do not think add to their beauty, however much it may to their showiness.

A BOUQUET OF DRIED GRASSES AND FERNS.

Seeds of several kinds of "ornamental" grasses are to be had of the seedsmen, but few of these produce any thing more beautiful than may be found growing in the wild state. There are numerous wild grasses which are suitable to use in bouquets, but as they as a general thing have no common names, it is not worth while to enumerate them by their botanical ones. The best way is to collect whatever grasses seem suitable that we meet with in our rambles, tie them in small parcels and dry them in the shade in a place free from dust. Some berries of the Wax-work (*Celastrus*), and ferns dried between paper, or in a large book, will be useful to combine with the grasses. For a grass bouquet the great trouble is to find a suitable vase or receptacle. Those sold as flower vases are altogether too heavy in style to correspond with the airiness of the grasses they are to hold. Last year, in June, we figured a stand for cut flowers; something of this kind is most suitable for an ornament of dry grasses. As no water is needed, some such a stand could be very easily contrived. A glass tube or solid rod of glass, half an inch or a little more in diameter, and about two feet long, may be procured of the druggists or instrument makers. This is to be fitted into a round block of wood, heavy enough to answer for a firm base, and at its top furnished with a funnel or trumpet-shaped receptacle to hold the stems of the grasses. This receptacle may be covered with paper of some neutral tint, or may have mosses and lichens gummed upon it. For the pleasing arrangement of the grasses in such a stand as this no directions can be given; each individual can display her taste in the matter. The aim should be to avoid all appearance of crowding, and allow each kind of grass to show its natural habit. The wooden base of the stand can be concealed by the dried fern-leaves, and cones, berries, and nuts may be introduced. Those who are fortunate enough to find the delicate Climbing Fern can add much to the beauty of such an ornament by twining one of its stems around the glass rod. We give an illustration of an ornament of grasses arranged in the manner we have suggested.

Preserving Crab-Apples.

BY "L. J. R.," PHILADELPHIA.

The fruit is prepared by first cutting out all decayed portions; then wash clean, and place in a kettle with sufficient water to cover the fruit entirely. Have a tight-fitting lid to the kettle and boil, over a moderate fire, until the fruit is soft enough to pierce with a straw; drain off the water, and strain through a coarse cloth or jelly bag, and set it aside for jelly. The apples, in boiling, will have burst their skins, which are easily removed; the cores are taken out by pushing them through from the blossom end, with a goose quill or a stick of equal thickness, being careful to press the stem end against the fingers to prevent breaking the apple. The fruit is now ready to preserve whole or to make into marmalade; for either, the proportions are: 4 lbs. of fruit, 3 lbs. of sugar, and 1 pint of water. Put the sugar and water into the preserving kettle, set it over the fire until it boils, then drop in the fruit, (if it is to be done whole,) boil until clear, and remove into a jar. If there is more syrup than will be needed, boil down to the desired quantity; pour it over the fruit while hot, and cover with a cloth, cut of sufficient size to cover and tie down. This cloth I dip into a cement made of two parts of bees-wax to one of rosin, adding enough tallow or lard to keep it from cracking. While the whole is warm, I draw the cloth tightly over the top of the jar and tie down. To make marmalade, the boiled fruit must be mashed to a pulp before being added to the syrup, and then boiled and stirred until it becomes clear, which is usually in half an hour.

For making jelly, I use equal portions of the water in which the fruit was boiled, (which has been previously strained), and sifted sugar. I seldom resort to the scales. One tumbler, *even* full, of sugar to one of the juice, gives the required proportions for all jellies; boil as for other jellies, and put up in the same way. Many persons fail in making jellies by endeavoring to boil too large a quantity at one time. I have always found better success with 2 quarts or less of juice at a boiling, than when I have undertaken more.

Hints on Cooking, Etc.

Pudding without Milk or Eggs.—By Mrs. Philip. Soak dry bread in as little water as possible, and squeeze out all the water. Add sufficient sugar to sweeten, and for a small pudding, half a teacupful of chopped suet or butter, and dried fruit which has been soaked over night, or canned or fresh fruit. Mix well together, adding a little allspice. The pudding is put into a greased tin pail, a cloth placed over, and the cover put on. The pail is set in a kettle containing sufficient water to come half way up the pail; boil for two hours, or more for a large pudding. To be eaten with sauce.

Apple and Tapioca Pudding.—By Armonck. Put a teacupful of tapioca into a pint and a half of cold water, over night. In the morning set it where it will become quite warm, but not cook. In the course of the forenoon peel half a dozen sour apples and steam them until tender. Put them in the pudding dish, add a teacupful and a half of sugar, a little salt, and a teacupful of water to the soaked tapioca, and pour over the apples. Slice a lemon very thin and distribute the slices over the top of the pudding. Bake slowly three hours. At the end of this time it will be a perfect jelly. Eat warm or cold, as you like.

containing a great variety of Items, including many good Hints and Suggestions which we throw into smaller type and condensed form, for want of space elsewhere.

Weeping Trees.—R. B. Werden. We cannot tell you the cause of the pendulous character of those trees called "weeping." They are eccentric individuals, that have the peculiarity from the start, and are "born so." These and other departures from the usual habit of trees are perpetuated by budding and grafting.

The Richmond Peach.—Dr. E. Ware Sylvester, of Lyons, N. Y., sends us samples of this new variety from the original tree. He says in comparison with the Early Crawford: "It is more hardy, it is as large, as firm for marketing, as handsome, and that it is as sweet as any first-rate white peach; lacking the acidity the Early Crawford always has however ripe it may be. I grew 500 seedlings from selected seed; and the Richmond and Atlanta, a white peach not quite ripe, were the only ones I deemed better than varieties we had already." The specimens were of excellent flavor and very sweet.

"What is the Matter with these Grapes?" asks "Dobb's Ferry." Many of the grapes are discolored, have the skin broken, and are already passing into decay. Had our correspondent carefully examined the interior of the infected berries, he would have found a minute "worm," which is so near the color of the pulp of the grape as to readily pass unnoticed. This is the larva of the Grape Curculio, *Cœliodes inæqualis*, which will be found figured and described in our columns in July, 1868, p. 223. The perfect insect is $^1/_{10}$th of an inch long, and appears in July. It may be jarred from the vines early in the morning. At the West it has proved very destructive; there the fruit drops, and the larva goes into the ground to undergo its transformation late in July or early in August, while in the specimens from Dobb's Ferry (some 30 miles up the Hudson) the fruit had not yet fallen. This insect has not come under our observation before, but we have no doubt that much of the injury to grapes ascribed to bees and wasps is due to this curculio. It would be a wise precaution to gather all this infected fruit and burn it, or place it in a tight vessel and allow fermentation to take place, which would probably kill the larvæ. Riley, in his report on the insects of Missouri, 1869, says that probably a parasite is at work in destroying it, as in 1868 the vineyards in that State were almost entirely free from it.

Dahlias.—T. L. Ingell. You cannot expect Dahlias to bloom well until the heats of summer have passed. They have all they can do to hold their own through the hot months.....J. W. Bair. It is impossible to tell about Dahlias from description only. Your seedling you can name what you choose, but it is not worth while to name it unless it is really distinct. To get the finest exhibition blooms, the plant is allowed to bear but few. Good culture, in order to secure strong roots, is all that you can do to develop your seedling.

Cactus.—T. J. Ingell. The term Cactus, is a very comprehensive one, and it is impossible to tell what one you have that does not bloom. Some, of the hundreds comprised under the general head of Cactus, only bloom when of great age, and are cultivated for the singular forms of the plants. Others, again, flower when quite young. The best way to treat the majority of them is to keep them quite dry during the winter, and in spring when they begin to grow, give plenty of water.

Chestnuts in Illinois.—"D. L. M." The Chestnut will grow in any part of your State.

The Grape Culturist and the Farmers' Club.—We have heretofore spoken in commendation of the Grape Culturist, edited by Geo. Husmann, St. Louis, Mo., but we fear that we shall be obliged to retract. It quotes a talk on the Scuppernong grape at the N. Y. Farmers' Club, and then says: "It is certainly amusing—but at the same time disgusting—to see men, who have not the faintest perception of vegetable life, who know nothing about the influence of grafting, but only know that the word hybridizing is in the dictionary, and that it means some horticultural operation, persist in trying to force this grape upon Northern planters, etc."——Mr. Grape Culturist, this wont do, at all. The "Farmers' Club" is a peculiar, New York institution; it embodies wisdom, science, experience, modesty and eloquence, in a manner that no other institution ever did before, and probably never will again; and we simply demand that you show it that respect which it merits. Please keep on your own side of the Mississippi. How can you know any thing about grapes so far from New York?

Flowers in Church—A Beautiful Custom.—At the Methodist E. Church, in Middletown, Conn., (the seat of the Wesleyan University), a large vase of beautiful, fresh flowers, with small trailing vines and foliage, is always found standing on the table in front of the Clergyman's desk; and on communion days, and frequent other times, one or two extra vases are added at the right and left, with a cross of white flowers upon the pulpit, or speaker's desk. This has been kept up every Sabbath, summer and winter, for several years past. These flowers are the gift of a lady, Miss Ellen Rockwell, who raises the flowers, and wreathes them with her own hands—presenting them as a token of love for the House of God—and it seems to us, a very appropriate one. Might not this custom be appropriately imitated in every Temple of Worship? in summer at least, where greenhouse flowers cannot be secured for the winter season—though they appear most pleasing at the latter season.

A New Peach.—Messrs. Kemp & Kerr, Choptank Nurseries, Denton, Md., send us specimens of a new peach called Glendale. It originated with the Rev. R. W. Todd, of Caroline Co., Md., and is described as a vigorous grower and good bearer. The fruit is of good size and exceedingly beautiful in appearance. The quality was remarkably fine for a yellow-fleshed peach, and seemed to us the best of that class we have tasted.

Lice on Cabbages.—We have been told that salt and water will kill these, but fortunately have not had occasion to try it.

Hedge in Connecticut.—"J. W. J.," Stamford.—For you the Honey-Locust will make the best hedge to "take the place of a fence." It will turn stock in four years. Land in good condition for farm crops is all that is required. Raise seeds in seed-bed, and transplant when one year old.

Moles.—An old gardener of our acquaintance says that he drove the Moles out of his garden by the use of coal-tar. He dips a corn cob in the tar and places it in the run. He says that the mole will not travel that road again.

A Vine Lock.—Mr. Edward F. Underhill, Brocton, N. Y., sends us specimens of a vine lock, which is a peculiarly bent wire for attaching the canes of vines to the wires of the trellis, and serves as a substitute for ties. Those who have used the appliance speak highly of it as effective, and allowing the work to be done with great rapidity. The samples came too late for us to try them this season.

Plants by Mail.—Persons who send plants, seeds, etc., by mail, should recollect that no written nor printed communication can accompany them. The law is very explicit upon this point. Nothing beyond the simple label is allowable.

Tobacco Stems for Manure.—"D. L. M." Tobacco stems will decompose if laid up with stable manure, and make an excellent compost for a market garden. Even burned for their ashes they would have considerable value.

Ants.—What will kill ants? We have published all the remedies that have come to us, and still the call comes for help against the pest. The ant question is still open. Let us have experiences.

The Black, or Barn Weevil.—Harrison Y. Krauss, Bucks Co. We believe there is no feasible remedy for this pest but starvation. If no grain is stored in the barn or granaries for a full year the insects die out. If a head of rye is in the hay, the probability is the insect will be found there and the evil perpetuated. The best way is to build barracks and shelters for hay, grain, and straw, and leave the barn empty for a year—at the same time clean it out thoroughly, and make what repairs are necessary. Granaries which are isolated can be fumigated with burning sulphur sometimes, and the weevil thus exterminated.

Wood-Ducks and the Falcons.—"S. M." pleasantly writes from Bluffton, Mo.: "Your Picture of our beautiful Wood-Duck in the Sept. No. is such a complete representation of what can be witnessed out here in the wilderness, that I am tempted to give you a little account of an interesting chase with a set of the young of the above named ducks In company with another man and two boys, we went in a skiff to Hermann fourteen miles below here. About midway down, and a few hundred yards from shore, we came across a half-dozen of these little ducks, about the size of half-grown partridges. The boys were elated and at once gave chase, as they said they would make nice pets. We said nothing but well knew what would be the result. The skiff went one way and the birds another, for they would dive and reappear in every direction. In the midst of the excitement a new character appeared on the scene. Two Wandering Falcons, [We presume S. M. refers to the Duck Hawk, *Falco anatum*, a very rare bird.—Ed.] came swooping down from a lofty cliff, screaming fearfully. 'Stop,' said I, 'one enemy is enough;' and we lay on our oars and looked on. Those voracious birds would come down like a dart at the little ducks, and when we thought one had a duckling in its talons, it would appear on the surface again. After various attempts in coming down on them, they tried it horizontally, and would skim along the surface like an arrow shot from a bow, and at times I thought a bird was actually in the Falcon's claws, but quick as lightning the little thing dived under. We stayed until the robbers went back to their lofty rocks. You may be sure we had no gun along, or there would have been one or two rare birds still more rare, as our feelings were by no means pleasant toward the assailants.

White Leghorn Fowls.—R. J. Taylor, Berkshire Co., Mass., writes: "In different parts of this State there are persons who keep what they call White Leghorn Fowls. Is there any such breed of fowls, and what are their distinguishing characteristics?" A fine engraving and description of this beautiful breed was published in the *Agriculturist* for March, 1869, p. 89, which you had better send for. We are more than ever inclined to adhere to our views then expressed, that White Leghorns, to be considered pure, should be of a slender stylish figure and proud carriage; and have pure white plumage, yellow legs, thin single combs, carried erect in the cocks and drooping in the hens, and white or creamy-white ear lobes. The hens should be persistent layers of medium-sized to large, white eggs, and rarely or never sit. They are a hardy, valuable breed, especially as layers. Increased size of fowls and of eggs should be cultivated—and the points named insisted upon. There is a *Brown* variety equally well defined and valuable.

"Mark on My Paper."—We frequently have a request that we shall answer a question by some mark on the writer's paper. This it would be more difficult to do than to write a dozen letters. The mailing is done with such great system and rapidity that it would cause serious interruption to single out a particular subscriber's paper. Besides it would be contrary to the spirit, if not the letter, of the postage laws.

"W. J."—We do not give advice on purely personal matters through the paper; and do not advertise any secret compound unless we know its composition, think it proper to use, and worth what is asked for it. But why not sign such letters?

Does Fish Culture Pay?—We have never doubted that it was profitable to raise trout to sell by the thousand when an inch or two long,—or to sell the eggs of trout and salmon,—but whether it would pay to raise trout as we raise sheep, in enclosed races and ponds, where almost every particle of food they receive must be provided for them at considerable expense, is a question we have never considered as satisfactorily answered. The following statement by Mr. Furman, seems to settle the matter, in his own case. As his system of breeding is peculiar, his success may be peculiar also. He writes: "You may place fish culture among the paying pursuits of the day, as I can sell my young fish *by weight* and pay the entire expense of shanty and attention, and will undertake to deliver them on the 1st of May next to the extent of *ten tons*—many of them weighing three-quarters of a pound each."

Stable Floors, etc.—"C. B.," of Bealston, Va., writes: "In the *Agriculturist* for August, in describing a barn, you speak of a depressed walk behind cattle to convey manure. Please tell us what kind of materials you use in making this walk, and the manner of mixing them. Also, if it will support the weight of a horse or cow without cracking?"—Such a walk may be made of two thicknesses of plank, the first receiving a good coat of asphaltum, rendered fluid by coal-tar and heat, and applied nearly boiling hot, and the seams of the upper planks being filled with the same. Bricks laid in common cement mortar, or in asphaltum and coal-tar, thickened with sand, and used hot, exactly like mortar, make an excellent stable floor. Stones may be used laid in the same materials, and if covered with dry earth, make a floor easy to the cattle or horses, and at the same time impervious to water, durable, and easily taken care of.

AMERICAN AGRICULTURIST

FOR THE

Farm, Garden, and Household.

"AGRICULTURE IS THE MOST HEALTHFUL, MOST USEFUL, AND MOST NOBLE EMPLOYMENT OF MAN."—WASHINGTON.

VOLUME XXIX.—No. 11. NEW YORK, NOVEMBER, 1870. NEW SERIES—No. 286.

DEAD GAME OF THE SEASON.—DRAWN BY H. W. HERRICK.—*Engraved for the American Agriculturist.*

With the advance of the season our marshes and copses, quiet woods and plains, from which the husbandman has disappeared for the time, are more or less filled with gleaners of the harvest of nuts and seeds, which wild Nature provides for her children, or of the residue of the imperfectly garnered products of cultivated fields. Among these busy denizens of the free country and the green wood, now sere and russet, are birds and beasts, many of which are excellent as food, and all of which would multiply to their own destruction were they not held in check by birds and beasts of prey, which, in the providential ordering of nature, have a definite work to do in keeping down this excess of life. The quick shot that stops the flight and the joyous life, really cheats the fox or the hawk of his supper, while it adds not a little to the sum total of human happiness.

Great numbers of our autumn and winter game-birds are migratory and do not breed with us. The wild goose, next to the turkey and swan, is our finest game-bird. It rarely breeds within the bounds of the United States. The same is true of all the ducks but one, but none of the gallinaceous birds are properly migratory. Hence these are protected by stringent laws, which should be enforced and sustained by public opinion, or the sport and the benefit of wild game will be lost forever. After the breeding season has passed and the young have grown, in a country like ours, game of all kinds must take their chances. Though man should be their best friend during the summer, and until, by statute and usage, "the law is off," he will be found their worst enemy when the reduction of their numbers contributes so greatly to his enjoyment as a matter of sport, as well as to the gratification of his palate, to say nothing of the highest pleasure of the true sportsman, that of sending gifts of game to his friends, which are sure to be highly valued by them.

POTATO DIGGING PLOW.

(Fig. 79, Page 57 of our Large Catalogue.)
(Send for a Special Illustrated Circular.)

From the REV. WILLIAM CLIFT, "Tim Bunker."

"MYSTIC BRIDGE, CONN., January 17, 1870.
"I have used the Patent Digger upon Poquonnoc Farm the past season, with satisfaction. On gravelly and sandy loams, kept clean, it works well. The few potatoes that are left under the dirt are readily brought to the surface by the Steel-Tooth Horse Rake, which is the best implement I have tried for this purpose. The Digger is also an admirable tool for harvesting Ruta-bagas."

Order at once. Send $15 for the Iron Plow, or $25 for the Steel Plow, by bank draft, Post-Office Order, or bills per express or mail, and the plow shall be sent at once.
Address letters to P. O. Box 376.

R. H. ALLEN & CO.,
189 & 191 Water Street, P. O. Box 376,
New York.

Broom Corn Machinery

for preparing the crop for market.
Estimates for complete or partial sets of machinery furnished by

R. H. ALLEN & CO.,
189 & 191 Water St.,
P. O. Box 376. New York City.

Harder's Premium Railway Horse Power and Combined Thresher and Cleaner,

TWO GOLD MEDALS AWARDED ONE MACHINE.

At the Great National Trial, at Auburn, N. Y.
Threshers, Separators, Fanning Mills, Wood Saws, Seed Sowers and Planters, all of the best in Market. Catalogue with price, full information, and Judges Report of Auburn Trial sent free. Address
MINARD HARDER,
Cobleskill, Schoharie Co., N. Y.

H. KILLAM & CO.,

Chestnut St., New Haven, Conn.

We manufacture the finest class of carriages for city use, consisting of Landaus, Landaulettes, Clarences, Coaches, Coupes, Coupelettes, Barouches, Bretts and Phætons. Which we warrant equal in point of style, finish and durability to any built in this country.

Messrs. DEMAREST & WOODRUFF, 628 Broadway, are our Agents in New York City.

Improved Foot Lathes,

With Slide, Rest, and Fittings. Elegant, durable, cheap and portable. Just the thing for the Artisan or Amateur Turner. "Your $50 Lathes are worth $75." Good news for all! Delivered at your door.
Send for descriptive circular.
N. H. BALDWIN, Laconia, N. H.

Premium Farm Grist Mill.

Cheap, simple, and durable. Is adapted to all kinds of Horse-powers, and grinds all kinds of grain rapidly. *Send for Descriptive Circular.*
WM. L. BOYER & BRO.,
Philadelphia, Pa.

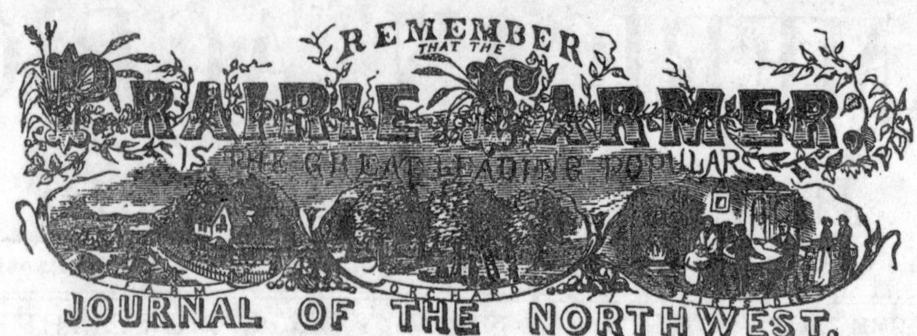

JOURNAL OF THE NORTHWEST.

For Country and Town! For Old and Young!

Published Weekly at $2.00 per Year.

Three Months on Trial for 50 Cents. Specimens Free.

NOW IS THE TIME TO SUBSCRIBE!

The Best of 1870 Gratis! The Annual a Present! The Champion Club Free!

☞ If you think of *changing* your Agricultural paper, or taking an *additional one*, give the **PRAIRIE FARMER** a TRIAL.—You will like it!

SPLENDID OFFERS TO AGENTS!

ONE HUNDRED & FIFTY PREMIUM ARTICLES!

A PRIZE FOR MERCHANTS AND CLERKS!
A PRIZE FOR POST-MASTERS!!
A PRIZE FOR SCHOOL TEACHERS!!!
A PRIZE FOR GIRLS!!!!
A PRIZE FOR BOYS!!!!!
A PRIZE FOR EVERYBODY!!!!!!

☞ IF A CASH COMMISSION is preferred to Premium Articles, Agents can deduct **Twenty per Cent** from yearly and half-yearly subscriptions.

☞ No other Weekly Journal of its class in the world, offers such splendid inducements to subscribers and agents alike, and no other is so easy and so profitable to canvass for among the industrial masses. We are determined to publish the **best** paper for the **least** money. Remember that **THE PRAIRIE FARMER** is **not a Monthly of Twelve Numbers**, but **a Weekly of Fifty-two Numbers.** It does not cost **Three Dollars**, but **Two Dollars**, per year.

Canvass for it! And Begin Now.

☞ Ask for Premium List, Specimen Numbers, and Canvassing Documents, and they will be sent FREE.

Address in all cases, **THE PRAIRIE FARMER COMPANY, Chicago, Ill.**

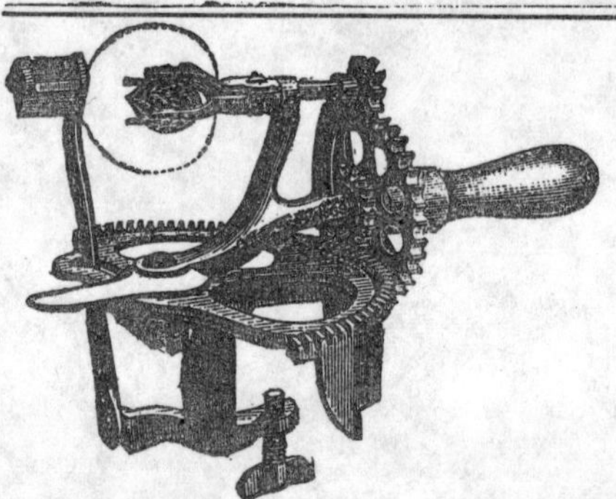

A continuous and urgent inquiry for a machine for **Paring Peaches**, has been ringing in our ears from all Peach-growing sections, for the past five years, and in response to this universal appeal, the manufacturers of the **Lightning** and **Turn-Table Apple Parers** have at last succeeded in obtaining and securing a device for **Holding** and **Paring Peaches**, which is as practical and economical as the Apple Parer, and cannot fail to come into immediate and general use. They pare **Apples** also, as well as any **Apple Parer**.

They were exhibited in the Fall of 1869, at the NEW YORK, OHIO, INDIANA, ILLINOIS, WISCONSIN, MINNESOTA, and ST. LOUIS FAIRS, and in every instance the judges were delighted with the rapidity and perfection of their work, and awarded them the highest Premium. A limited quantity of them has been made for distribution this year, in anticipation of the immense sale that must speedily follow. On receipt of $1.50 we will send a sample machine by Express.

SARGENT & CO., Sole Agents,
70 Beekman Street, New York.

P. S.—We are also Sole Agents for the **Lightning** and **Turn-Table Apple Parers.**

Silver's Meat and Vegetable Chopper,

for Farmers, Hotels, and private families, for cutting Sausage, and Mince Meat, Hash, &c. Large sizes are made for Butchers and Public Institutions. Send for Circular.
SILVER & DEMING,
Salem, Columbiana Co., Ohio.

AGENTS LOOK!—$12 a day made, selling our SCISSORS SHARPENER and other wares. Sample 25 cts. Catalogue free. T. J. HASTINGS & CO., Worcester, Mass.

INVALIDS' TRAVELING CHAIRS, from $15 to $40, for in and out-door use. Any one having use of the hands can propel and guide one. Having no use of the hands, any child of five years can push a grown person about. Invalids' Carriages to order. Pat. Carrying Chairs. State your case, and send stamp for circular.
STEPHEN W. SMITH,
No. 90 William Street,
New York.

MONEY EASILY MADE.

With our Stencil and Key Check Outfit.
Circulars Free.

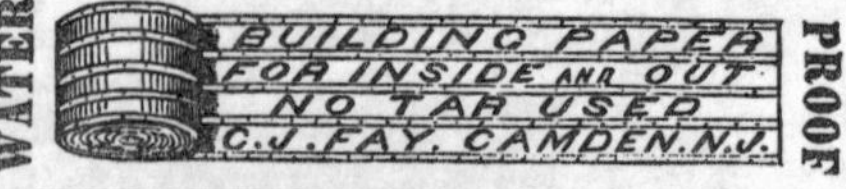

TILE-DRAINAGE.

Wanted employment by a thoroughly practical tile-drainer, competent to carry out drainage operations in the best manner. Has had extensive experience in draining in England, and in Canada. Can refer to Sheldon Stephens, Esq., Montreal, Col. Geo. E. Waring, Jr., and L. A. Chase, 245 Broadway, New York. Address JOSEPH AMBERSON, Care of Col. Geo. E. Waring, Jr., Newport, R. I.

THE UNIVERSAL SHARPENER.—An indispensable household helper. Sharpens all SHEARS and SCISSORS as well as TABLE CUTLERY. So simple, any one can use it. Never out of repair. Makes a perfectly sharp, even edge. Does not wear the blade. Lasts a lifetime. Simple, effective, convenient, and durable. Sold by Hardware and House-furnishing Stores. Price, $1. Send for Sample.
A. C. IVES,
Sole proprietor, 45 Beekman St., New York.

USE THE HILTON ALL GLASS SELF-SEALING Fruit Jar. The best in the world. For sale wholesale and retail by W. H. BARTLETT, Agent, Union Market, Boston, Mass. All orders by mail or express, promptly filled.

SEND FOR CIRCULAR.

AMERICAN AGRICULTURIST.

NEW YORK, NOVEMBER, 1870.

How humdrum and stupid the life of the farmer would be, were it not for the change of labor and thought which each season brings. We think, sometimes, almost with a longing to enjoy them ourselves, of the bland winters of half-tropical regions, and of those changeless spring like climates of some of the high valleys in the midst of the torrid zone. Yet we are happier and better off as farmers and as men for our rigorous winters, for the necessity of making provision for ourselves and for our stock, for the pinching of the frost, for the dangers we avoid, and for the losses that will surely come from neglect and carelessness during our winters. The better we are prepared for them, the more we may enjoy and profit by them. There is no part of the United States where winter is not felt with more or less severity. Even in Florida the trees, many of them, lose their leaves and now and then the "northers" clear up with frost and ice. In these warmer regions the provident farmer's faculties are exercised to provide against the drouths of summer, and it is a mistake to suppose that the seasons are not nearly as well defined in the Southern as in the Northern States.

Winter, wherever the snow does not cover the ground and where the ground does not remain frozen from day to day, is a season of pasturage for cattle, for general tillage, and for the prosecution of all sorts of field work not directly connected with growing crops. The importance of comfortable shelter for domestic animals is too little valued, far north of the line where pasturage can be relied upon; neither is the great value of the winter realized as a harvest time for *manure*. This is really only an other name for wheat, corn, and cotton; the one representing the others as truly as a "greenback"-note represents gold, or silver, or copper, or, in fact, almost any thing else we value.

Hints about Work.

The Crops still remaining in the field must be secured at once. Those stored, must be protected against frost and vermin—against wet and deterioration, and every preparation should at once be made for sudden and severe cold.

Manure.—Cart out all that can be plowed in; that which cannot be plowed under, place in heaps, either composted with vegetable matter, muck or sods, or covered with a few inches of good soil to turn the rain from the sides, and to keep the valuable products of fermentation from passing off. Make provision for plenty of material to mingle with the dung of animals during the winter; straw, buckwheat haum, and such things, of course, but do not neglect to gather

Forest Leaves.—The most convenient wagon for loading leaves we have ever used is made by taking a broad hay rigging with wide wings over the wheels, or stakes all around, and attaching hay-caps to the sides, hanging down into the box. The leaves should be gathered after a few dry days, and the most convenient way to collect them is with hay rakes, and large baskets into which the leaves can be pressed. Store them in bins or empty box-stalls, well trodden down. Leaves contain a large percentage of ash, and though not so valuable as straw for litter, are very good.

Sawdust.—As soon as the country saw-mills get water and begin to run, the neighboring farmers can secure sawdust for litter. This adds not only bulk, which is of great value, as thus the manure is divided and is more easily and evenly distributed, but it decays both in the heap and in the soil, thus affording organic matter and ash to the plant, while it retains the valuable qualities of the manure.

Orchard and Nursery.

Planting. — In advising fall planting, we have reference to the season. Do not put trees into half-frozen ground. If the weather is still mild and rains have not made the soil too wet, planting may continue; but it is only in very favorable seasons that this condition of things lasts into the present month. Where planting cannot be done properly, the trees should be

Heeled-in.—So many trees are lost by improper heeling-in, that some writers denounce the thing altogether. There is no doubt that it is better to procure trees in the fall and heel them in properly, than to run the risks of the injury they are likely to receive by hurriedly taking them up in spring and by a long transportation, at a time when the trees are excitable. In autumn a tree is thoroughly at rest, and if taken up and heeled-in, it starts much later in spring than if allowed to remain where it grew. Proper heeling-in requires that the roots shall be thoroughly covered with earth, leaving no holes for either air or water to work their destruction.

Picking and *Packing.*—Enough has been said upon these. What late fruits remain to be gathered, should, if they are intended to keep well, be treated with the care already prescribed. It is likely that this will be a poor year for keeping winter fruits, if we are to judge from the rapidity with which fall varieties have ripened up. The fruit which is usually depended upon for keeping late, will require to be retarded by as low a temperature as can be maintained without freezing.

Cider.—The best is made this month. The apples should be thoroughly ripened, and every care taken to keep out impurities. The cellar where fermentation goes on should be kept as cool as possible. When this ceases, fill up the barrels and bung up.

Fruit Garden.

Grape-Vines, as soon as the leaves have fallen, may be trimmed. Cut young vines back to their buds and draw earth up to them, or give them a covering of leaves. Old vines that have been allowed to run wild, are very difficult to prescribe for. The best general direction we can give is to cut all the past season's growth back to two buds; recollecting that every bud will make a shoot; if this is likely to produce too much wood, thin out some of the canes altogether. Even hardy varieties will do better if laid down and covered with a little earth. This may not be practicable with old vines, but it is with young ones, and as long as it can be done it will be found to pay.

Cuttings of grape wood should be prepared now. For out-door propagation, cuttings of two buds are best. Prepare them at pruning time, and tie them in bundles of 25, or of convenient size, and bury, the lower end up, in some dry place. We say lower end up, because it is desirable to keep the upper bud dormant, and allow the lower part of the cutting to be more forward.

Kitchen Garden.

Any one who has seen the advantage of preparing the soil by turning with the plow or spade at this season, will not be slow to adopt it.

Manure may be spread now and turned under, or the plowing may be left until spring. Stiff lands may have coarse manure and be plowed and left rough. If thrown into ridges without manure, frosts will have a most beneficial influence upon them.

Winter protection is in most localities necessary for spinach, sprouts, and the like. This is not to be applied until freezing weather. The object is to prevent frequent freezing and thawing, and if put on too soon it may injure the crop.

Cabbages.—After trying various plans for wintering cabbages, we think that followed by our market gardeners is the best. The cabbages, after freezing weather has set in and before the ground is so frozen that they are difficult to pull, are taken up and inverted, and three or four inches of earth thrown over the heads. If many are to be treated the earth is plowed to them, but with small crops the spade will do the work. Plants sown for the purpose are to be put in cold frames. Set them down to the leaves and do not put the glass on until cold weather. The object to be attained is to keep the plant dormant and quiet, but still alive.

Root Crops.—Roots in cellars are apt to deteriorate if not surrounded by sandy earth. They can be best kept in trenches or pits. Make pits in a dry place three or four feet wide and six feet deep, and stack the roots, beets, carrots, parsnips, salsify, etc., in sections; when a section of two feet in length is packed, leave a space of six inches and make another section, and so on. Fill the six-inch spaces with soil, and when cold weather comes on cover the tops with litter and put boards over that. This keeps the roots in contact with earth, and, while it prevents freezing, avoids the drying which so injures roots kept in cellars. If the supply is small, pack in boxes in the cellar, with sandy earth.

Celery.—The best results come from stacking the roots, which have been properly earthed up, in similar pits to those directed for root crops. The pit is to be made as deep as the celery is high, and about a foot wide. Pack in the roots closely together after cold weather comes on. It often happens that celery will make considerable growth this month. The storing should not take place until cold weather has checked the growth.

Lettuce.—Put into cold frames the same as cabbages. In the warmer parts of the country it will pass the winter in safety with a covering of litter.

Rhubarb.—It is much better to make new plantations in fall as the plants start very early in spring. Old roots may be divided. See article on forcing last month. Old roots force very well. We have had good success by placing them in a large cask in which a quantity of manure had been placed for bottom heat. Light is not necessary.

Sweet Potatoes should be harvested before any blackening frost touches the vines. Dig on a pleasant day, and let the potatoes dry in the sun. Pack in boxes or barrels with thoroughly dry straw, and put in a place where they will not cool below 60°.

Flower Garden and Lawn.

Lawns.—To judge from our own case, the summer has been particularly severe upon both old and new lawns. Bare spots occur where the new grass was actually killed by the heat, and there will be other places where weeds have obtained a foothold and must be uprooted. All bare spots should be attended to this fall. Pull up all weeds, and, if needed, put on some rich soil to restore the level, and sow an abundance of grass seed. Bone and ashes make an excellent top dressing. Use no compost that is likely to bring in weed seeds.

Plants in Pits and Cellars are not expected to grow, and they must be held in a dormant state by keeping them as cool as may be without freezing, and allowing only enough water to sustain life. Mice are fond of many plants, and injury from these troublesome pests should be guarded against.

Greenhouse and Window Plants.

Plants.—Air, light, heat, and water, are essential to vegetable growth. Air is the requisite most generally neglected. On mild days give abundant ventilation. Whether in the window or in the greenhouse, there should be a difference of 10° or 15° between the day and night temperature.

Camellias should have the foliage kept perfectly clean by syringing or by sponging.

Bulbs.—Whenever the ball of earth in the pots is well filled with roots, the plants may be started by bringing them into a warm place. It is best to keep some pots in reserve to allow of a succession.

Hanging Baskets usually dry out rapidly; give the earth a thorough soaking once or twice a week.

Annuals come in usefully for cut flowers, and all spare corners may be occupied by pots or boxes in which seed of Candytuft, Mignonette, and the like, may be sown. Thin the plants well.

Fumigation.—The frequent use of tobacco smoke will keep plants free of most insects. Use it in abundance in the greenhouse. Window plants are easily smoked by placing them in a box or under a barrel.

Remedy for Drouth.

The extreme drouth which has prevailed in many parts of the country, drying up the brooks, destroying the fish, and, in many places, making a total failure of corn and potatoes, leads us to inquire for a remedy. Is there any? Has man any power over nature? Can we add to, or diminish, the rain-fall? There are many facts in the history of the old world, which go to show, that man has much of this power, and that he may so direct his labors as to modify very essentially the climate, as well as the soil. Countries once fertile are now nearly barren, and sustain but a handful of people. Their brooks are dried up, and the rain-fall is greatly diminished. On the other hand, wells sunk in the desert make an oasis, and the spot of verdure increases with the passing years, until showers fall upon the parched sands. A remarkable instance of the effect of man's labors upon climate is now going on in the Great Salt Lake Valley, in our own country. When the Mormons first settled this region, they were entirely dependent upon irrigation for their crops. The supply of water was small, and they feared lest with the increase of their population, there might not be at last enough to irrigate all their lands, and famine must stare them in the face. But they have tilled their lands, planted trees, which are now large and completely embower their city, and their gardens are full of fruit trees and flowering shrubs. Many thousands of acres, once barren, have been made more productive than in rainy climates. Enormous sums have been spent in bringing water by artificial channels from the distant mountains to make these now fertile fields. The face of the earth has been changed, and there has been a corresponding change in the climate. They now have rains from the sky, almost enough to meet the wants of growing crops, a thing unheard of until within a few years. The effect of the increased rainfall in the Valley has had a very marked effect upon the Great Salt Lake, which is 126 miles long by 45 miles wide. It has risen 12 feet since the Mormon occupation, and the water has a smaller proportion of salt. Formerly it took three gallons of water to make one of salt, now it requires four. The change has also affected the streams that flow through the Valley, and it is estimated that the same channels carry twice as much water as formerly, for the purposes of irrigation. These facts are very encouraging, not only to the Mormons, but to the settlers along the line of the Pacific Railroad, where there is little rain. It may be expected that irrigation, and cultivation, and the planting of trees will gradually work a change in the climate, and make rainless regions productive.—In this country there is no regulation, and every man follows his own sweet will in destroying trees. We think the time is not far distant, when our government will have to look after this matter, and place some restriction upon the removal of forests, and encourage the planting of trees upon the prairies, and in the rainless regions. Wood and timber are growing very scarce and high, in some of the older parts of the country; and streams once full of water are now nearly dry for the larger part of the year. We very much want information disseminated upon this subject. The instinct of self-preservation, if it were enlightened, would lead farmers to preserve their forests upon the mountains and hills, in which our streams take their rise, and not to drain too many of the swamps in these high lands. The springs at the source of every brook should be sacredly guarded. These high lands are generally rocky, rough, and steep, and quite impracticable for the plow. They are favorable to the growth of wood.

Fencing Flooded Fields.

Many solutions have been attempted of the problem how to enclose fields liable to be flood-

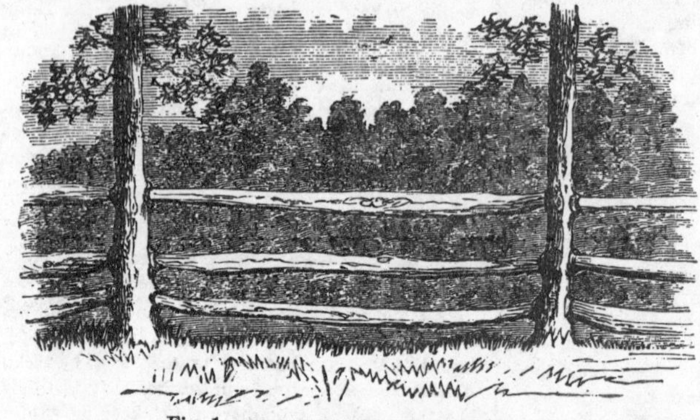

Fig. 1.—FENCE FOR FLOODED LANDS.

ed ar l washed by freshets, bearing ice, flood-wood, uprooted trees, etc., in a way to be effective, and yt not have the fences washed away as often as the water rises. Several plans have been suggested and carried out which have been more or less successful. The Connecticut River meadows are not fenced as a rule, except at certain points where the conformation of the head-lands turns the ice and drift-wood away; for no ordinary fence could stand before the flood which brings down cakes of ice a foot or two in thickness, and perhaps covering an area of one or two square rods, or it may be square acres in ex-

Fig. 2.—RAIL PROPERLY SHARPENED.

tent. The up-stream sides of the trees are denuded of bark, scarred, and bruised. There are, however, many streams, the meadows along the banks of which are not swept by such impetuous floods, where ordinary fences are very unstable property.

In these situations Soft Maple-trees will almost always grow vigorously. They afford shade, and finally fire-wood. They will bear a great deal of bruising and barking without either decaying at the heart, or being seriously damaged. This flooded land is ordinarily rich, and in four or five years from the time they are set out, they will be strong enough to mortise rails into, so as in time to make a fence like the one shown in fig. 1. The rails should be good ones, preferably of chestnut, and the ends not sharpened thin, but holding a good thickness, though sharpened somewhat like those of the common post-and-rail fence; fig. 2 represents the end of a rail properly sharpened. It is not necessary to use maples, for in all probability white willows, cotton-woods, certain kinds of poplar, all of which are very easy to transplant, would do equally well. The trees should be set 12 feet apart on the exact line, and stayed on the up-stream side by a stake driven as shown in fig. 3, to which the young tree is bound with straw at a hight of about 4 feet. This serves as a defence both against damage by wind and water, and to some extent, by cattle. The band must be removed in the course of the first summer, if the tree makes vigorous growth, and as soon as it is substantially rooted, it should be removed altogether. This fence, when well established, is a picturesque and beautiful object; and one forgets the disagreeable scars on the trunks in admiration of the beautiful rows of trees, that replace the unsightly posts, of the common post-and-rail fences, which are always rotting off, or heaving by frost; and unless newly set, are frequently becoming insecure. In cases where there is a liability to have the rails broken by flood-wood or ice, and yet where the stream is not very violent, a fence like the one shown in figure 4, has been found very good. Half the number of trees is set, and posts are placed between them. The fence panels are made of strips of pine or spruce wood of any convenient pattern, the principal top and bottom rails being of not less than 1½-inch stuff. The ends of the panel rails next the trees rest in sockets of iron, (of the form shown enlarged in fig. 4, above the fence,) which are driven into the trees. The other ends are supported upon a cross-bar nailed upon the posts and held in position by cleats nailed on, or by buttons. One of each pair of panels is chained to the tree which supports it, and the other is chained to its fellow. These chains may be made very cheaply of stout iron wire. When the freshet comes the fence is raised up and floats free from the post; and of course this allows the ends attached to the trees to draw out of the sockets, and each pair of panels floats, fastened to the tree to which it is chained, as shown in figure 5. The fence, if well put together, will withstand very hard usage from swift currents, floating logs, etc. The use of carriage bolts instead of nails at the four corners of each panel, is advisable. Breachy cattle if they learn how, may prove troublesome if placed in a lot inclosed by this fence; but it has such a substantial look and is so firm unless lifted up bodily, that no trouble will ordinarily be experienced. Should the necessity arise of having it absolutely secure, movable wooden pins may easily be placed in the crossbar of the post, and the sockets in the trees closed above so that the rails could not be lifted out. In case of a rise of water, it would be necessary

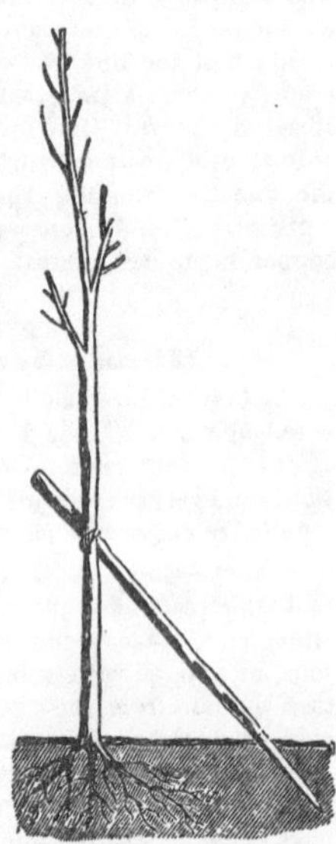

Fig. 3.—YOUNG TREE STAKED.

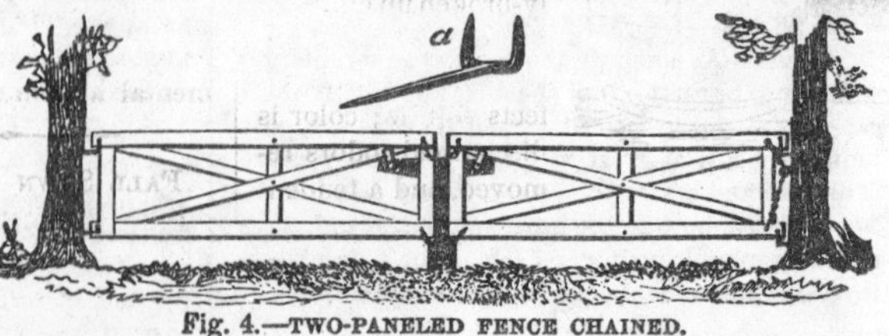

Fig. 4.—TWO-PANELED FENCE CHAINED.

to go along the fence and remove the pins before the flood had reached too high a stage. From the engraving it will be understood that the posts stand upon the up-stream side of the fence, and as every thing must have its weak-

Fig. 5.—FENCE PANELS FLOATING.

est point, it would be well to see to it that the parts which would break first in case of a sudden rise of water would be the cleats or buttons which hold the fence to the posts.

LOOK TO THE WINTER WHEAT.—It is rare to find a field of winter wheat on which there are not low spots where the water lies on the surface. Much might have been done to prevent this by "furrowing out" as soon as the grain was sown. Now it must be done with the hoe and spade. Some wheat may be destroyed by the operation, but not one-tenth of what will be "winter-killed" if the water is allowed to remain on the land. Let not a moment be lost in attending to this matter. It is by no means a substitute for underdraining, but it is far better than nothing. If the land is low, commence to dig where there is an outlet and make the water follow you up into the land. You will be astonished to find how much fall there is, even on land that is apparently on a dead level. Try it and you will save wheat enough to buy a first-class agricultural library.

A Family Filter—Home-made.

Pure water for family use is almost beyond value. The past two summers have tried wells sorely, and many have been found wanting. Some have been so low that the water became stagnant, and multitudes of farmers would willingly have used swamp water if they had known how to filter and purify it. Filtering, as generally viewed, is a purely mechanical operation, but it is *not* so of necessity. When finely-broken up charcoal is used in the filter, marked chemical effects follow; color is discharged, odors removed, and a tendency of substances in the water to decompose is arrested. For this reason charcoal is a very important ingredient in filters for drinking-water. There are many ways of arranging a filter, and the accompanying plan is suggested as of very general application. A water-tight barrel, or half-barrel, is obtained, and one head taken out uninjured. Two tinned iron pipes are fitted into the head which remains, each extending a few inches above the top of the head; one enters the barrel a few inches, and one of them goes nearly through it. Caps of tinned wire gauze are made to go over the ends of the pipes which are inside the barrel, and these are bound on with wire. The keg is then set with the open end up, and filled thus: 1st, several inches of clean gravel; 2d, 4 or 5 inches of well-washed, fine sand; 3d, about 12 inches of freshly heated and pounded charcoal, sifted, and in about as large pieces as grains of wheat; 4th, 4 or 5 inches of sand like that before used, and finally several inches of gravel—the whole well packed and settled by water, layer by layer, to fill the keg full. Then put in the head and make all tight. The exterior ends of the pipes should have screws cut upon them; then a coupling with a faucet may be attached to one, and a bigger piece of pipe, in which a funnel will go, to the other. Pour in water until full. Such a filter will be efficient in constant use four to six months.

FAMILY FILTER.

A Cheap and Durable Gate.

An article upon gates in May last has called out several descriptions of other patterns of gates; some of these have already been published, others are too complicated to be generally useful, while a few are, like the one here presented, simple and durable. The drawing is sent by Doct. A. Thornburgh, whose address we have mislaid. He says: "This gate will not—can not sag. The back part is made of a small sapling, with a fork or projecting limb, which answers for a brace. Through the top of this passes a piece of an old trace chain, tightened up by a bolt and nut, forming an arch over the gate; the remainder explains itself. I was pleased with the gates figured in the May No., but where one has no smith handy, he can, by using my gate, construct one equally as ornamental and much cheaper."

DR. THORNBURGH'S GATE.

FALL SOWN GARDEN CROPS.—The weather during the month of September, has been so dry in most parts of the country, that those things which were sown for late use or for keeping over winter, have had a poor chance. Spinach, sprouts, and other winter crops, in most cases, did not germinate until the rains came. If we have a favorable November they may still make good-sized plants before winter, and a chance should be given them as long as the growing season lasts, by cultivation and thinning.

Dwarf Pear-Trees.

There is no doubt that those who, like our friend Quinn, follow "Pear Culture for Profit," will find standard trees best suited to their purpose. But we must not forget that large class who have small gardens, with whom the pleasure of cultivation is the leading object, and to whom a dozen pears from their own tree is of more value than a bushel of fruit grown elsewhere. To such as these, the dwarf pear-trees are invaluable. They may be kept as dwarf as one pleases, be trained as small bushes or pyramids, or even grown to a single horizontal or vertical stem in what is called the cordon style. They are capitally suited to tenants; as when properly managed they may be taken up and moved to another garden as easily as a rose-bush. To do this the tree must be kept within certain limits. A dwarf pear-tree with the junction of the quince stock and the pear set several inches below the surface, soon ceases to be a proper dwarf, as roots are thrown out from the pear stem. In order that the tree shall remain truly a dwarf, the junction should be so near the surface that the pear roots cannot be found. If it be desired to keep the specimens very small, root-pruning should be resorted to, and the roots be thus kept within narrow limits. Rivers recommends transplanting the trees on alternate years, as an aid in dwarfing. It is generally considered injurious to a tree to allow it to bear fruit the season following its removal.

Outlet for an Ice-Pond.

Considering all the trouble we had last winter to get ice, and the prices we have had to pay, we may well look out beforehand, that our arrangements are made for securing a stock next winter. The coldest water lies at the bottom of a pond in summer; not so in winter. Then the warmest is at the bottom, for as soon as the temperature of the pond falls below 40°, the coldest water stays on the surface, and the water below remains at 40°, until it is either warmer, or becomes colder through the freezing of that above. Warren Leland, of Highland Farms, Westchester Co., N. Y., reports, that he doubled the thickness of his ice, by having the outlet made at the bottom instead of at the top, thus drawing off the warmest water instead of the coldest. He sent this statement to the papers last year, too late to be of any use to our readers. It is so in accordance with the facts above mentioned, that we do not doubt the correctness of Mr. L.'s conclusions.—He says: "The usual way of constructing outlets of ice-ponds is by an overflow at the surface. In this way the top of the water is always in motion; and after ice has formed, the warm water from springs, and underdrain flowing along its under surface, prevents its thickening. To obviate these difficulties, the outlet should be from the bottom of the pond—leaving the surface undisturbed to cool and freeze more readily, by the cold air and action of the ice. This style of outlet also tends to keep the water free from sediment. It is easily constructed, by having an iron or wooden tube leading directly from the bottom of the pond, or by a flume, arranged to draw from the bottom. Having in this way doubled the thickness of ice on my pond, I wish through your columns to give my brother farmers the benefit of my experience."

The Crested Turkey.

Among all domesticated poultry and other birds, so far as we know, except, perhaps, Guinea fowls, geese and swans, we find either natural crests, or a tendency of the feathers about the head to assume extraordinary shapes. Among doves we have frills, hoods, and crests; among fowls, turban-like crests, and mufflers about the throat; among canaries, frills and crests. Peafowls have peculiar crest feathers. Pheasants, beautiful and diverse crests. There is a crested breed of common ducks, and beautiful natural crests upon several wild species. Among the various genera of wild gallinaceous fowls are found beautiful crests of very different forms. The Curassow, for example, a bird of South America, nearly as large as the turkey, has a superb narrow crest, capable of being spread like a fan, with its edge to the front, and depressed, or folded, in like manner. So it does not seem an extraordinary freak of Nature that a turkey should occur with a crest, especially when we consider the varying circumstances under which our domestic turkeys are raised.

In the number of the London *Field* for July 17th, a gobbler was figured, having a crest much like the one represented in the engraving here given. The accompanying memoranda by Mr. W. B. Tegetmeier, gave a brief sketch of other reported crested turkeys. It seems certain that they have repeatedly occurred, and have received so much attention from poultry fanciers as to have become at different times, more or less established as a breed, but now they are, and for many years have been, altogether lost.

The specimen represented in the *Field*, and in possession of Mr. Tegetmeier, was said to have come from Zanzibar, and to have been sent to a Hamburgh dealer in Zoological specimens, by a collector in Africa. It is of the common species native to this country. How it should have occurred in Africa, is strange. Knowing, as we do, the very imperfect knowledge of natural history which zoological dealers usually possess, and the lack of accuracy which characterises their statements concerning their animals, even when truth would serve their purposes better than fiction, we respectfully don't believe a word of its having come from Africa.

Almost simultaneously with the appearance of this interesting bird in the yard of Mr. Tegetmeier, one is found in this country which closely resembles the other in many respects. We noticed it as shown at the exhibition of the Connecticut State Poultry Society, and again at that of the New York State Poultry Society, and as having been purchased by Mr. D. E. Gavit, in whose possession it still remains. The cock is a medium-sized one, weighing perhaps 18 or 20 pounds; of a blackish bronze color upon the body, fading into gray below, and into brilliant light chestnut bronze on the tail and wing feathers, these being edged with broad bands of black and white. The legs are dark, flesh color; the spurs indicating at least a 1½-year-old bird. The carunculations upon the neck, and the beard, are well developed. The crest is like the Hamburgh bird, "of a dull, uniform gray, the feathers composing it being soft in texture." It is a beautiful appendage, adding a peculiar grace to the bird. It is in a measure erectile at the will of the wearer, and gives the gay fellow quite the air of a Broadway belle, with her extraordinary *chignon*.

CRESTED TURKEY—"VICTOR EMANUEL."

We hope this notice may elicit some positive information from disinterested parties who know by whom and where this bird was bred. From all we can learn, he came from somewhere near Birmingham, Connecticut. The parties who sold him to Mr. Gavit are not disposed to be communicative in regard to his origin. If he lives he will be bred with care, and we hope and expect his progeny, some of them at least, will take after him in this beautiful peculiarity.

Sheep Require Water in Winter.

That sheep will eat snow and get along tolerably well without water, especially when the snow is light and clean, is a good argument in favor of the hardiness of these patient and useful animals; but it proves nothing in regard to the economy of compelling them to do without water. In our cold climate a large amount of food is required to keep up the animal heat, and all the water a sheep drinks has to be raised to the temperature of the body. The heat required for this purpose is derived from the food. Now, it requires twenty-one times as much heat to raise a pint of water when frozen, to the temperature of 40°, as it does a pint of ice-cold water, not frozen. Those who have ever undertaken to melt snow to wash with will readily believe this statement. What would be said of the farmer who should use wood or coal for such a purpose when it could be avoided, and yet wood and coal are far cheaper sources of heat than hay or corn. When we compel a sheep to eat snow, we, in effect, undertake to melt snow by burning straw, hay, and corn in the animal stove. The same remarks will apply to feeding animals on frozen roots. A sheep, weighing say 100 lbs., will eat, in addition to a liberal allowance of hay and grain, 10 lbs. of roots a day, containing nearly or quite 9 lbs. of water. Now, before the carbon in the turnips can produce fat, or furnish material to keep up the heat of the body, it must first raise this frozen water to blood heat. A flock of a hundred sheep will drink nearly two barrels of water a day, and there are few things more important to the economical and successful wintering of sheep than to see that they are constantly or at least frequently and regularly supplied with water. A running stream, brought by pipes into the barn-yard, is doubtless the best, but when this cannot be obtained, the water should be pumped up fresh for the sheep; and not allowed to stand until frozen, and then the surface broken and the ice left to float in the water that the sheep have to drink. We do not know that it would pay to heat water for sheep, but certainly it will not pay to make them eat snow, or drink water with melting ice in it.

Where early lambs are raised, it is important to provide the ewes with a constant supply of fresh water. If from accident or neglect such sheep have been kept without water for a day or two, nothing is more dangerous than to permit them to drink their fill of cold water. We once lost several sheep from this cause, and have since heard of many similar instances. Take the chill off the water and no danger need be apprehended. Suckling ewes require more water than other sheep, and when roots are not fed, it would undoubtedly contribute much to obtaining a good supply of milk for the lambs if the ewes were furnished with warm water, or, better still, warm, cooked food, such as oatmeal and bran mash, or boiled barley, etc.

Rats and Mice—Vermin-proof Walls.

Constant complaints are made through the press of the increase of rats and mice throughout the country. The only way to get rid of them *effectually* is to so construct our houses, barns, and other buildings, that they *cannot harbor* them. To insure this, cellar bottoms should be cemented, and the walls also, unless they are solidly laid up with brick or stone and mortar. Every basement of a house where the sleepers are close to the ground should be filled up between these to a level with their tops with cement or grouting, and the boards or plank then nailed on, leaving not a hair's breadth, if possible, between the grouting and the floor. Neither rats nor mice can then get under. We built a set of stables in this way fifteen years ago, and no vermin have ever found harbor under the floor nor have gnawed through it.

The Chaste-Tree.—(*Vitex Agnus-castus.*)

If our readers share our love for old plants, we present in the Chaste-tree one long enough known to satisfy the warmest lover of antiquity in plants. It has been in cultivation in England for two hundred years, and its use in the festivals of the ancient Greeks is mentioned by Pliny. Aside from its historical associations the shrub, —for it is not a tree—has in itself, much to commend it. The foliage is pleasing, and its flowers have the merit of appearing late in the season when there are but few shrubs in bloom. Our plant flowered this season late in September, but in less favorable seasons it does not bloom until October. The flowers are borne in interrupted, slender spikes at the ends of the shoots of the present season's growth. They are purplish-blue, and have a rather agreeable fragrance, while the odor of the leaves is not very pleasant. The engraving shows the flowers and leaves reduced about one-half in size. The fruit is a small four-celled nut, which is rather peppery, and to which, in former times, various medical virtues were attributed, but in common with many similar things they have long ago passed out of use. The Chaste-tree is a native of the shores of the Mediterranean, and is with us somewhat tender, and the shoots are partly winter-killed, but not enough to injure the vitality of the plant. Mr. Meehan informed us that his plants, in the more favorable climate of Philadelphia, were cut back in a similar manner. The shrub belongs to the Verbena family, of which it is the only woody representative that is hardy—or rather half-hardy, in our climate. It is propagated by cuttings and by layers. There are several other species of Vitex, but they are greenhouse or hot-house plants.

THE CHASTE-TREE.—(*Vitex Agnus-castus.*)

A Double-flowering Blackberry.

The common Blackberry, or Bramble of Europe, *Rubus fruticosus*, is a strong-growing, erect species, with hooked prickles, and bears a small fruit which is but little prized. It has given rise to several varieties, differing in their foliage, and the color of their fruit, and two, in which the flowers are double. The double white and pink varieties are very old, and we suppose that on this account they are seldom seen in gardens. Mr. Wm. Chorlton, of Staten Island, a lover of fine plants, whether new or old, last spring brought us specimens, which were so pleasing that we had the engraving made from them which we here present. The specimens were completely covered with blossoms which looked like miniature roses. The double brambles are valued in England, as they will thrive where more delicate, ornamental plants cannot be made to grow; and they make themselves at home at the roots of trees, and in rocky places, and are used to train against walls, etc. They have one disadvantage for some ornamental purposes; like our native species, the stem is biennial, and the old wood must be removed each year. Still, in many situations, this would be of little matter. We do not know if it is as much disposed to spread as our species. It is altogether a very pleasing shrub, and we are glad that Mr. Chorlton has brought it over and given us an opportunity to make it better known.

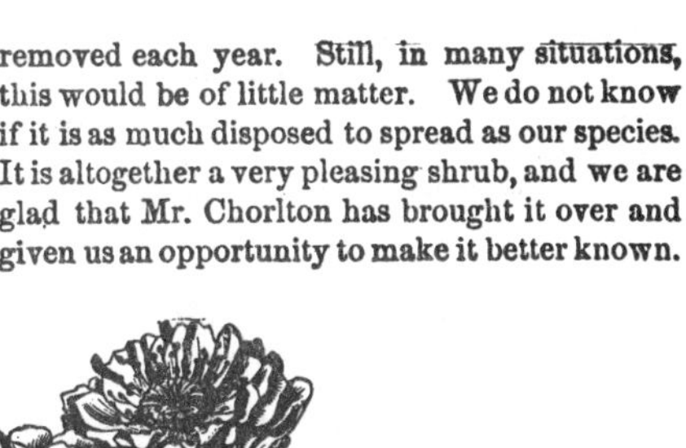

THE DOUBLE-FLOWERING BLACKBERRY.

A variety of this species, with divided leaves, is the *Rubus laciniatus* of the catalogues, and sometimes cultivated under the name of Cut-leaved and Parsley-leaved Blackberry. Another double variety of Blackberry is the well-known Bridal Rose, which is *Rubus rosæfolius*, with double flowers. It is a greenhouse species, and comes from Mauritius. Specimens of Wilson's Early Blackberry have sometimes shown a tendency to produce double flowers.

Keep the Small Potatoes until Spring.

—Farmers frequently feed their small potatoes to fattening pigs in the fall. It would be much better to keep them until spring, and then cook them, mix them with a little meal and feed them to suckling sows and young pigs. In the spring, before the clover is ready to turn into, we are generally short of succulent food, whereas in the autumn we have apples, pumpkins, cabbage leaves, and a variety of vegetables that will not keep until spring. The value of potatoes as food for stock does not lie so much in the mere nutriment they contain as in their giving tone to the stomach; and they will prove much more useful when fed out to young pigs and breeding sows in the spring, as is usual, than when fed to fattening pigs in the fall.

THE HOUSEHOLD.

☞ *(For other Household Items, see "Basket" pages.)*

Time for Reading.

BY FAITH ROCHESTER.

Let us *take* time for reading. It will never come if we wait to have every piece of work finished, and every speck of dirt removed from each article we use. We can always find something else to do, and conscientious house-keepers, with little taste for mental pursuits, are apt to make a great blunder. "The life is more than meat, and the body than raiment," which means—if I may be allowed to preach a wee bit of a sermon—that you yourself, with all your immortal faculties, are of vastly more importance than your house and furniture, and clothing and cookery; and these are utterly worthless if they serve as hindrances instead of helps to your individual human culture. No kind of labor is degrading if done from a worthy motive, and no motive can be nobler than the womanly desire to make a *pleasant home.* With this end in view—with love as a prompter—washing and darning and scrubbing are all elevated from drudgery to a nobler place. But our homes cannot be properly attractive and profitable to our families if we ourselves are dull and harrassed. Our brothers and fathers and husbands and sons need cheerful and intelligent companions at home, far more than they need nice dinners and spotless linen. It is necessary that good home-makers and keepers should read and reflect, and listen and converse.

What shall we read? Whatever really helps us along, whether it be history, science, philosophy, or morals. I can't read hard books when I am tired. Sometimes for several weeks the cares and labors of housekeeping and maternity take so much of my strength that mental *labor* is impossible, and then I take what I call easy reading—good stories and the lighter newspaper articles. But if we live on light reading entirely we cannot expect to gain in mental strength and growth.

Do you know what a joy it is to feel that, though your school-days are long past, your intellectual growth is still going on? The lessons of our own experience are most valuable, I know, but good books are great helps. From them we get the results of the experience and observation of others.

Fig. 1.—SIMPLE MATCH-SAFE.

Matches and their Safe-Keeping.

Friction matches allow us to kindle a fire with the least possible trouble. House-keepers of the present day know but little of the difficulty those of older times had in getting a fire, and of the care exercised in keeping it. Perhaps some can recollect how cautiously the coals were covered with ashes at night, that the embers might be ready to start the morning fire; and when with all the care the

Fig. 2.—DOUBLE MATCH-SAFE.

coals were found dead, how irksome it was to go of a cold morning to a distant neighbor's to "borrow some fire," and convey it home in a pan or

Fig. 3.—CARVED MATCH-SAFE.

shovel, in order that the household machinery might be once more set in motion. But the cheapness of matches has done away with all this, and now a scratch with a match brings the needed servant to do our bidding. The increased facility in obtaining fire has brought with it increased danger. The scratch that will bring the friendly fire will also bring the fire fiend to burn and destroy. Investigations by the proper officers in our large cities have shown that a considerable share of fires is to be attributed to carelessness with matches. There cannot be too much caution exercised in keeping matches, whether the stock for future use or those kept at hand to supply the daily needs. A match carelessly dropped may be ignited by the tread of the next passer, and give rise to a conflagration, or it may fall into the hands of a young child, and thus be the cause of disaster. Rats have been known to carry matches to their holes and thus add to their general mischievous ways by turning incendiaries. When a stock of matches is kept on hand they should be stored where none but the proper persons can have access to them, and in such a manner that there can be no possible danger of ignition by accident. We keep them in a tin box. Equal care should be taken with those kept at hand for frequent use. Leaving them about upon the mantle-shelf and in similar places should not be tolerated. There are match-safes of various designs sold at the stores which may be had at little expense, but any box or receptacle that can be affixed in a convenient place will do as well. The important thing is to have the match-safe in a fixed place, and have it understood that matches are to be found there and no where else. It is advisable that the receptacle have a rough surface for scratching the matches upon, as this will prevent much disfiguring of the woodwork. In rooms where there is no fire it is better to have a place to receive the ends of burnt matches. A half-consumed match is a small thing, but neat persons are often puzzled to know what to do with it, and all doubt upon the matter is done away with if a receptacle is provided. We give figures of match-safes in which some ornamental work is introduced. They will serve as suggestions to those who would like to make such articles. Their woods of various kinds, including cigar-box material, are easily worked by the use of a sharp knife and a fine saw. The cylindrical receptacles shown in fig. 2, may be made out of the round wooden match-boxes.

Training Children to Work.

BY FAITH ROCHESTER.

A little girl of twelve came to live with me a few weeks, doing household chores for her board. I thought I should teach her just how to wash the dishes, but some engagements prevented my doing so on the first occasion of her officiating at the dish-pan. I went into the kitchen after she had finished the task, and looked to see how the work was done. The sink was dry and clean. It actually *was* a plain kitchen table, but "sink" reads so much better! And every body knows that, theoretically, dishes should be washed in a good sink with racks and drains, and convenient supplies of hot and cold water. I felt of the dish-pan and dish-cloth. Both were guiltless of grease and properly cared for. The tumblers and spoons and forks looked bright and clear, and the plates and knives were free from spot. I did not ask the child how she had done her work. The result was satisfactory, and I let her go on as her mother had taught her.

Then I hired a "young lady" to do my work. She knew how—oh, yes! of course she did—to do everything; so I did not venture any suggestions at first. But such discomfort and uncertainty as we all felt about everything she undertook! And somehow, I had an impression all of the time, that she disliked work, and was unused to it, and might lame her back, or hurt herself in some way; so thought I had better do everything that was hard and disagreeable myself.

No doubt the difference in the working habits of these two girls was owing in great part to a difference of natural temperament; but it was easy to see that their home training had not been at all alike. There is every variety of talent among children. Some have much more executive force than others. Such children are "always getting into mischief," and it is hard to bear with their restlessness and its results. But this tendency to activity may be made to take a useful turn quite early in life, if properly directed. Whatever the child's temperament, it has a right to such an education of its faculties as will enable it to take care of itself and be useful to others in after life. A parent has no right to mark out a child's future career and educate it especially for that; but there are certain things that are necessary to be done anywhere, in order to secure the general and individual welfare.

It is not so necessary that a child be taught how to do each particular thing, as that it be trained *to do something very carefully and thoroughly*, and that it be taught perseverance. It is a great lesson when a child has once learned how much more enjoyment there is in doing a piece of work with accuracy, or with skill, than in doing it in a half-way, bungling fashion. Don't require too much at first. It may be a bit of over sewing. The first stint should be a small one, a single inch, perhaps, with the promise of some little treat when the stitches are all made evenly. Criticise kindly and sympathizingly, but never say "well done," until it is well done. Pick it out over and over, if need be, until the thing is right, and if the work gets soiled in the process, say nothing about it; children are so easily discouraged in their first efforts. Over-praise is just about as bad. Genuine *sympathy* is the right thing, joined with firmness in what we know to be for the child's good. We ought to use judgment in giving a task to a child, so that it may be able to carry the task through without getting perfectly sick of that kind of work. If we become convinced that the task set is really too great for the child's powers of endurance, either physically or mentally, we had better tell the little one that we have changed our mind and that it need not go on; for it is not well to simply cease insisting upon the performance of the task, leaving the child to a guilty consciousness of having failed in what was required of it. But, oh dear! how children will "beg off!" Katie knows it will take her all day to pick up the buttons she has just spilled, and Sammy is just as sure that it will make him sick to pick up a basket full of chips. Don't be impatient with them. Children will be childish—those, at least, who are destined to grow up. It is our business, patiently to teach them to be reasonable. Encourage them. Let them know of something very nice that is going to take place as soon as the buttons or chips are picked up. Under the influence of happy expectation, the task—over which they had been dawdling with tears—will be finished in a twinkling. When it is done, before the treat comes, call their attention to the smallness of the task when resolutely undertaken.

Children who have been trained to observe what is going on around them, to be careful in the details of their work, to plan their work before beginning it, and to persevere in what they have undertaken, will make good, faithful workers in almost any department. They should be encouraged to work with rapidity, but not at the expense of thoroughness. A great wrong is done to children when they are overworked by avaricious parents; but it is also wrong to allow children to grow up with untrained powers and shiftless habits.

To Catch, Dress, and Cook a Chicken.

It is not easy to catch a chicken in the day-time, unless you know how to do it. This is one good way: In the bottom of a pail put a few kernels of corn or a few bread crumbs. Tip over the pail on its side, and hold it so, standing behind the bottom of the pail. When Biddy walks into the trap to get the corn, reach over and seize her by the tail. If you wring the chicken's neck, do it so thoroughly, that the poor creature will not come to consciousness when half-picked. Hold it by the head, and swing it around and around, until it is quite dead. It can be picked more rapidly and with no trouble from vermin—if, unfortunately, the fowl is infested with them—after dipping it all over in scalding water. Those who save the feathers for their own use or for sale, prefer to pick them dry, and this is most easily done while the chicken is still warm. After all the feathers and pinfeathers are removed, pass the fowl through a blaze—a paper burned on the hearth or a lighted candle—to singe off the hairs from the skin.

Now there is a chance for skill and the pleasure that results from the use of skill, in separating the joints of the fowl. Take off the head, and cut away a small portion of the skin around the union of the intestinal canal with the outer skin. Make this opening large enough to introduce the hand, and you can gently draw out the entrails, crop and all, in one mass. Most people cook the liver and heart. In removing the legs and wings, you will soon see how neatly the muscles cleave apart so that there is comparatively little cutting to do. Cut through the flesh of these limbs at the joints, and then break the joints apart, cutting the ligaments. There is a small place in the backbone, where it may be easily broken. Try to separate the "wishbone" from the shoulders and breastbone, for the children's sake. Then it is not difficult to separate the breast and back, and to sever the neck from the back. The pieces should then be washed.

There are many good ways of cooking chickens. This is the last one I have learned, and very good. It is an easy way. Put a spoonful of butter in the bottom of a frying pan, lay in the pieces, cover it, and set it in the oven. Turn the pieces several times while cooking. When done, take out the chicken, turn off nearly all of the fat if there is much of it, and make a gravy of the remainder by adding a teacup of water; thicken, when boiling, with a thin paste of flour and water.

I doubt if this is quite as wholesome as the common *chicken-stew*. Put the pieces in a kettle with water enough to cover all. Let it come to a boil rather slowly and simmer gently until very tender, adding boiling water whenever any is needed to prevent burning. Before it is quite done, salt the whole. Take up the pieces, when very tender, and make a gravy of the broth by thickening it and adding a little sweet cream. This broth should always be skimmed when it first begins to boil, and all the fat floating upon the surface should be removed with a spoon before it is salted. Chicken fat is not very palatable, and when it is "tried," it makes a good oil for many domestic uses. R.

Cooking Egg-Plants.

Though the season for Egg-plant is about over, we print for future reference the following Creole style of cooking the vegetable, furnished by Mrs. A. L. Howard, Pike Co., Miss. We shall be glad to receive the recipes for other Creole dishes which Mrs. H. offers to send. She says: "Take six Egg-plants, cut lengthwise (through the stem as well), and soak them half an hour in salt water, boil until tender; it is a good plan to score them slightly around the edges before boiling. When tender dig out the center carefully so as not to injure the skins, which will then be left hollow with a handle like an oval frying-pan. Take about a pound of light bread (some prefer more and some less), soak in water until soft, and then squeeze out, mix with the contents of the Egg-fruit, kneading thoroughly together, season with salt, red pepper pods chopped, and one minced onion. Put into a frying-pan with a large spoonful of lard, and stir the mixture over the fire until the lard is well mixed in; then pack the mixture in the skins that have been emptied. Brown and roll bread crumbs, and sprinkle thickly over the top of each, put a few specks of lard over each, and put in the oven and brown nicely. Or you can pack the mixture in a shallow tin or pie-plate, and bake. Or you can keep it in the frying-pan, allowing it to brown into a cake."—

How to Roast a Pig.

To be in its prime, the pig to be roasted should be not less than a month old, and certainly not more than five weeks. The nearer to the length of a moon's journey, the better. It should be killed and dressed the day before it is required to be cooked and eaten. It will deteriorate in flavor if kept longer, and a shorter time would not allow of its getting cold and firm, which is equally important. Caution the butcher to be careful not to make too large an opening, as it is difficult to keep the pig in good shape if the bones of the hips and chest are divided. The head should not be removed, and the feet should also remain. The brains may be removed by opening the head from the under side, and returning the parts to shape with the aid of skewers and string. The feet should be doubled under the body, and the pig should sit on them while in the oven. When the force-meat is all in, stitch up the opening securely.

Put the pig into a hot oven, with no water in the pan, nor gravy of any kind. As soon as the surface of the skin is a little warm, rub it over with a little butter confined in a muslin bag; a tablespoonful will be enough. This process should be repeated every fifteen or twenty minutes until all the roasting is complete. From two to three hours will be required to cook it, but when of a fine brown—sides, feet and all, it is probably done.

If onions are relished by the family, the force-meat for stuffing, will be made as follows: Grate into crumbs a small loaf of bread, and add to them two good sized onions chopped very fine, and a tablespoonful of pulverized sage, a teaspoonful of pepper, and two teaspoonfuls of salt. Mix thoroughly, form into a large ball, and put inside the body of the pig. Onion sauce, apple sauce, and potatoes boiled whole, with the gravy from the pan after the fat is removed, are the usual accompaniments.

containing a great variety of Items, including many good Hints and Suggestions which we throw into smaller type and condensed form, for want of space elsewhere.

Moles—The Latest Cure.—A subscriber in Maryland, apparently in all seriousness, says: "Raise the ground in the center of $^1/_2$ to $^3/_4$ of an acre, one foot high and 5 feet in circumference, sloping, and put the head of a horse in the center of the mound, $^3/_4$ of the head sticking out, nose down. You will not be troubled with any more moles."—Mr. Fuller is expected to try this before he awards that $100. We can't say whether the head should have a horse attached to it or not: better try both ways.

Hubbard's Early Curled Leaf Tomato.—E. Hollister and D. L. Hull, in the report of the Alton (Ill.) Hort. Soc., speak in commendation of this variety. We have heard favorable reports from others.

Keeping Tuberoses, Caladiums, and Cannas.—"A. B. C. N." Tuberoses may be kept too hot and dry, if such an atmosphere is maintained as is common in some houses. To guard Tuberoses, Caladiums or Cannas, against shriveling, they may be wrapt up when dry, in cotton wadding, to the thickness of an inch, and placed in a drawer or upon a shelf. This will prevent the dry air from acting on them, and will keep them plump until spring.

Osiers on Sandy Soil.—We recently called upon Mr. Charles Clifton, a basket-maker, with horticultural tastes, who lives at Suffolk Station, on Long Island. The land at that place bears a natural growth of scrub-oaks and pines, and one would hardly select it as a suitable place for growing Osiers, which are generally thought to flourish only in a rich and moist soil. We were quite pleased to find a thrifty young plantation of willows which is already furnishing valuable basket material. A comparison of the rods with those from France and Belgium, confirmed Mr. Clifton's statement that those produced upon his own grounds were greatly superior to the imported ones.

The Agriculturist Strawberry.—For the benefit of the editor of The Horticulturist, we quote the following from the Report of the Committee on Small Fruits of the Alton (Ill.) Horticultural Society. "From M. W. Seaman, Shipman—very large and fine specimens of 'Agriculturist.' As a berry for the amateur we consider it one of the best; requires high cultivation and to be grown in hills."

Know your Enemies.—The Alton (Ill.) Hort. Soc., which is always doing some sensible thing, has ordered a cabinet of insects injurious to the fruit grower. Every Horticultural Society should have a "rogues gallery" of this kind.

Delaware Grapes.—The Delawares have been a great success this year. The crops have been fine, and the fruit of excellent quality. On Sept. 15, they were selling in the N. Y. market for 8c. per lb., at retail. So abundant and cheap were they that wine-makers have bought them for pressing. The finest specimens we have seen were from Mr. Capron, Walden, Orange Co., N. Y., and were not open to the fault usually found with the Delaware—that of being too small.

Grouse Cochins and Plymouth Rock Fowls.—Y. S. Sturgis, Boston, writes: "Will you please describe the plumage of the 'Grouse Cochin' fowl in the *Agriculturist*. I saw the other day some fowls with plumage like the 'Dominique,' and with feathered legs, and thought they might be 'Grouse Cochins.' If not, can you inform me what they were? I also saw a hen of the Brahma form, and with feathered legs, said to be a Dark Brahma, but I think she was not, as her plumage was not cloudy, but beautifully spangled like a Hamburg. Can you also tell me of what breed she is?"—The fowls were probably what are called in Massachusetts, Plymouth Rock Fowls. These, as we have seen them, are large Dominique-colored fowls, with single combs and yellow legs, more or less feathered. What the other fowls are we are at a loss to conjecture, unless, indeed, the tendency to spangled feathers often seen in Light Brahmas has been cultivated or has accidentally developed. Grouse, or Partridge, Cochins are described in the "Standard of Excellence," of the London Poultry Club: The cock as having a red face and deaf ear, a red head, red hackle and saddle with a black stripe down the middle of each feather, rich, dark-red back, and shoulder coverts, wings, rich, dark-red; the greater and lesser coverts, metalic, greenish-black, forming a wide bar; breast, under parts and thighs, black; tail, black; legs, dusky yellow.——The Grouse hen as having, face and deaf ear, red; head, brown; neck, reddish gold color, with a broad, black stripe down the middle of the feather; the rest of the plumage, rich brown, with distinct pencilings of darker brown following the outline of the feathers; legs, dusky yellow, with feathers of the same color as the body.——The Partridge hen is described as very similar to the Grouse hen, but having more brilliant contrasts of color, and having "the shafts of the feathers on the back, shoulder coverts, bow of the wing, and sides, creamy white. There is no difference between Partridge and Grouse cocks.

Smutty Corn.—G. Thompson, Leelenaw Co., Mich. The spores of smut are so very minute that we know of no way to prevent their distribution, and thus prevent your having smutty corn another year. The best you can do is to cut away the smutty ears as soon as discovered and burn them. This will prevent a scattering of spores (seed) from your own crop. Before planting next year thoroughly wet the seed in strong brine or solution of blue vitriol, and dry off in plaster or ashes, this will kill any smut that may be adhering to the seed. Of course you will not put corn next year upon the land where the crop was so badly affected this season.

Eggs—A good Average.—Mrs. I. J. B., of Angelica, N. Y., kept an account of the eggs laid by a lot of $^1/_4$-blood Dorkings in 1869. She begun with 14 hens, and ended with 9. The total number of eggs laid was 1,219. Had no hens been killed or sold, we calculate she would have had 1,517 eggs, which would make more than 108 eggs to each hen. They had only common care.

Thorn Seeds.—A. W. Comfort. The seeds do not germinate until the second year. Put them in a heap and cover with several inches of soil and let them remain a whole year; they may then be sown in the fall or in the spring, as may be most convenient.

The Freemason Peach.—Messrs. Kemp & Kerr, nurserymen, Denton, Md., send us specimens of a new variety called the Freemason, which originated with Mr. Nathan Todd, of Caroline Co., Md. It is a large, white free-stone, with a red cheek. It is of excellent quality, and very late, ripening after the Smock, and vastly superior to that variety. The tree is represented as hardy and an abundant bearer.

How to keep Eggs.—"Subscriber," of Plainfield, N. J., asks: "Can you inform me in what manner hens' eggs can be laid down for winter use?"—We have had tolerable success in keeping eggs when they were simply greased with sweet lard. Packed in crocks filled up with milk of lime and covered from the air, eggs will keep very well, but though sweet, are not like new-laid eggs. The nearest approach to perfect preservation of the eggs is accomplished by placing a few at a time in a wire basket (an ox muzzle will do), and plunging them into a kettle of actively boiling water for a few seconds, say while one can count 20 rapidly. It is well to raise the basket once or twice and lower it suddenly in the water so that the eggs shall float up and settle back again into a changed position. This surrounds the egg next the shell with a film of coagulated albumen, which is perfectly air tight.

Seedling Peach.—G. L. Osborn, Dobb's Ferry, N. Y., sends us a yellow-fleshed Clingstone, weighing 8¾ oz. It is from a tree which bore this year for the first time, and ripened 51 peaches nearly as large as the specimen sent. The fruit is beautiful in color, and of good quality. The variety, if it continues as it has begun, should be exhibited another year and brought to the notice of pomologists.

The Sylvester Apple.—Dr. E. Ware Sylvester, Lyons, N. Y., sends us specimens of his seedling apple, which we figured in January last. Though not as high colored as the one we figured, their eating qualities showed that we did not overestimate the variety.

Bone Manure.—"S. S. P.," writes: "I have the bones of 2,000 sheep near here, and I think it possible to convert them into manure. Please tell me how to do it through your valuable journal."——Sheep bones may be ground in an ordinary bark mill, at least they may be cracked up tolerably fine. They may be cracked up in a corn and cob mill, without much risk, if it is a strong one, especially if they are not fresh. They may be sledged upon a rock and so pounded quite fine. If they can be reduced to about the fineness of ground tanbark, they may be laid up in layers with horse manure, hen dung, and other heating substances, and so subjected to the action of fermentation, they will decay. The heap will require to be frequently wetted to keep the fermentation within bounds; and it will probably be best to make it over with fresh manure after it has ceased to heat up readily.

Board Roofs.—W. M. Carr asks: "Will a roof made of 1-inch pine boards, well pitched with pine-tar or pitch and well sanded, cost more than a pine shingle roof?"—Ans. No.—"Will it not last as long as any wooden roof?"—Ans. No. The sun will wring and warp it all to pieces, the pitch will drip from the eaves, and you will be sick enough of the job, no matter how well you rabbet or batten the edges. A well-made asphaltum or coal-tar roof, made with cheaper boards, covered with best quality roofing felt, would be cheaper than shingles and last with proper care nearly as long. A roof of good pine or spruce boards, planed on the under side, having a rather sharp pitch and covered with good, strong slate, will cost more than shingles, but will last a lifetime, and be a great safeguard against fire.

Hand-Power Machine for sawing wood, etc., is inquired for. Manufacturers can answer profitably by advertising it.

Mounting Maps.—B. Plumstead. We hope you will not find it necessary to mount any maps of the "seat of war" hereafter, but as the information may be useful for other maps, etc., we give it. The muslin, which should be an inch or two larger each way than the map, is to be tacked to a smooth board or table; then cover the back of the map with a good, smooth coat of boiled flour paste, made as stiff as it will work well with the brush, and place the map, paste side down, upon the muslin; if the map is large, it will require two persons to do it well; and it must be handled very carefully, as it will tear readily when wet with paste. When the map is properly laid down, smooth it with a soft cloth, rubbing gently from the center towards the edges, to remove all the air bubbles. Put down the edges securely, and let it dry. In drying it will wrinkle badly, but when perfectly dry, will be quite smooth. As the paper dries it shrinks, and brings a powerful strain upon the tacks, which should not be more than an inch apart. They need not be driven down to the heads, as they would then be difficult to remove. When thoroughly dry, take out the tacks and trim the cloth.

The Providence (R. I.) Journal is so well known as one of the best journals in the country, that it does not need our praise, but we must say a word in commendation of its column of "Rural Notes and Notions," which show not only an excellent knowledge of rural affairs, but exhibit a pleasing fancy and genial spirit not often met with in writings of this kind.

Old Plaster.—"What is the value of old plaster, taken from the walls of houses, used as a top-dressing?"—Old plaster contains lime and hair, both of which are useful upon the land; besides, a considerable portion of plaster of Paris is often present. To use it as a top-dressing on grass, it should be beaten small and run through a coarse sieve. In fact, this should be done at any rate, if half its value is to be gained.

Unfortunate.—Many complaints have reached this office concerning the dealings of Thomas B. Smith & Co., of Plantsville, Conn., for whom a single advertisement was inserted in the *American Agriculturist*, several months since. Their method of doing business appears to be such that we cannot advise parties to send orders to them.

Expensive Processes.—A correspondent in Michigan complains that some of the operations we describe are too expensive to be followed in a new country, and cites the account we gave of maple sugar making last spring, as one of these. He says: "The cost of sugar house and fixtures would be more than all the sugar we should make for years to come." And farther along he says; "Here, everything is new; we have to chop and log, and burn the timber, and then work among the stumps for years." Exactly so. You have started to make a farm in the wilderness, and are not ready for all the aids that machinery and inventive skill are ready to offer. The preliminary work must be done first, and in a few years you will be prepared to adopt improved processes which now appear, and really would be, expensive in your new surroundings. As far as the maple sugar article goes, it contains suggestions which will materially aid those who, from necessity, are obliged to make sugar in the primitive way. The same may be said of other articles.

AMERICAN AGRICULTURIST

FOR THE

Farm, Garden, and Household.

"AGRICULTURE IS THE MOST HEALTHFUL, MOST USEFUL, AND MOST NOBLE EMPLOYMENT OF MAN."—WASHINGTON.

VOLUME XXIX.—No. 12. NEW YORK, DECEMBER, 1870. NEW SERIES—No. 287.

HEAD OF PETTYPET.—FROM LIFE BY EDWIN FORBES.—*Drawn and Engraved for the American Agriculturist.*

Pettypet represents, in very high perfection, the very best points of the Channel Island cattle, produced by mingling of the Guernsey and high-bred Jersey blood. She was raised by Mr. James P. Swain, and is regarded by him as one of the best animals he ever bred. On the side of the dam she is nearly pure Guernsey, her dam, Pet, being out of Katie 2d, by a Guernsey bull, of the N. Biddle stock. Katie 2d's dam was Katie, and her sire Echo, imported from Guernsey by Thadeus Davids. Katie was out of Mr. Swain's old imported cow Guernsey, the mother of a line of the greatest milkers and butter makers we ever knew, by a bull called Colt Alderney, whose dam was the Alderney cow Curl-horn, imported by Mr. Swain, and his sire a bull bought of Roswell Colt, which came from the Island of Guernsey. So much for the Guernsey blood with one-thirty-second of Alderney, whatever that may be. The sire of Pettypet was Bashan, imported by R. W. Cameron, a nearly perfect type of the high-bred Jersey. This animal has impressed his characteristics upon his stock to the third and fourth generation with almost unerring certainty, and to-day we think a dash of Bashan blood worth more in a fancy animal than a cross of any other choice strain. His points were great style, beauty of form and carriage, superb head and horns, which were delicate, well set up, pointing forward, and black tipped, fine underpinning, (bony, but strong,) a deep carcass, well ribbed back, a very fine tail, with black switch, black mouth and tongue, very soft hide, with two distinct kinds of hair in his coat, changing his color more or less with the season, but being on the whole of a rather dark fawn, with hairs coming through tipped with gray, with very strong mealy ring about the muzzle.

THE AMERICAN BROILER.

(Pat. July 21, 1868, and October 19, 1869.)

THE FINEST CULINARY INVENTION OF THE AGE.

Embodying in a plain and cheap utensil

ALL THE PRINCIPLES INVOLVED IN THE PERFECT BROILING OF MEATS.

Operates upon the essential natural principles for broiling meats to perfection · prevents the escape of nutriment by evaporation, and retains all the rich juices and delicate flavor—which are mostly lost in all other Broilers, or by the process of FRYING.

Broils *in less than half the time* required by any other, and cooks the meat perfectly uniform, leaving no burned nor raw spots.

Does away with all smoke and smell of grease; requires no preparation of fire; and makes broiling, heretofore so vexatious, the QUICKEST and EASIEST, as it is the HEALTHIEST of all MODES OF COOKING MEATS.

Broils equally well over coal or wood; answers for all sized stove or range openings, and is equally good for BEEFSTEAK, CHICKEN, HAM, CHOP, FISH and OYSTERS.

It is not only

GUARANTEED

to fulfil each and every claim above set forth, but to any not satisfied after trial,

The Money will be Refunded, with Charges both ways.

Dealers throughout the city and country are invited to order on these terms, and families to test it for themselves, with the certainty of its costing them nothing, if not as represented.

Each Broiler will have the authorized label attached, with the Trade-Mark, "American Broiler," stamped thereon.

Retail Price only $2.

Liberal Discount to the Trade.

Now for sale in half-dozen and dozen packages, in New York City, by the houses below named, who will be responsible, to the full extent of the above guarantees, for all Broilers sold by them.

LALANCE & GROSJEAN MAN'F'G CO., Nos. 89 Beekman and 53 & 55 Cliff Sts.

E. KETCHAM & CO., No. 289 Pearl St.

RUSSELL & ERWIN MAN'F'G CO., Nos. 45 & 47 Chambers St.

N. E. JAMES & CO., No. 23 Cliff St.

For sale at retail by Stove, Tin, and House Furnishing Dealers generally.

TO PARTIES VISITING THE

AMERICAN INSTITUTE FAIR,

Don't fail to see

The Downer Mineral Sperm Lamp,

an entirely new article, to burn the Downer Mineral Sperm Illuminating oil—300 degrees fire test! Absolutely safe! Cannot explode, and the most brilliant light ever produced.

HALL'S PATENT

HUSKING GLOVES

ENABLE THE WEARER TO HUSK 50 PER CENT FASTER THAN WITHOUT THEM AND ABSOLUTELY PREVENT SORE HANDS. Made of the best leather with metallic claws attached. In ordering, state size; large, medium, or small. Send for sample or circular. Price, $1.50 per pair. A liberal discount to Dealers and Canvassing Agents. Address

The Hall Husking Glove Co.,
101 & 103 West Lake-st., Chicago, Ill.

COOPER'S PORTABLE ENGINES *with Steam Pumps* and *Lime Extracting Heaters* and *Saw Mills*, cutting 10 to 20 M. per day. STATIONARY ENGINES, *Boilers* and *Mill Machinery*. **$1,500** purchases a *complete Two Run Grist Mill*, with modern improvements. PRICES REDUCED. Circulars free. JOHN COOPER & Co., Mt. Vernon, O.

FARMERS, SAVE YOUR FEET,

AND WEAR

BALLARD'S CHAMPION SHOE.

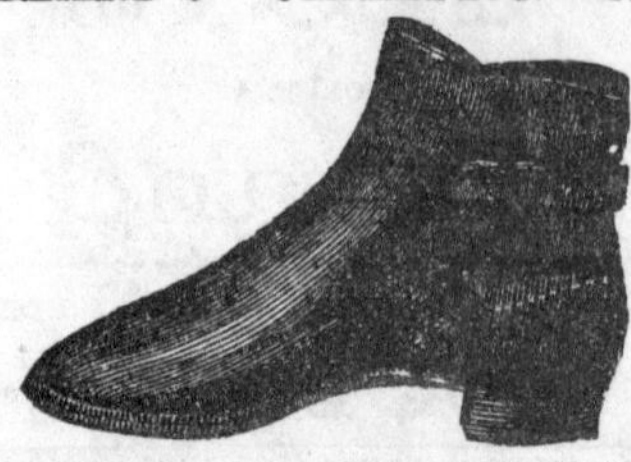

NEW YORK, May 10, 1870.

We have examined carefully "BALLARD'S CHAMPION SHOE," and, without practical use, should judge it to be a capital article. The peculiar cut gives it the set and bearings of a boot, with the ease and lightness of a shoe; and the strap brings it as closely as desired around the ankle without the trouble of strings. Those which we have seen are of good stock and well made. ORANGE JUDD & CO.

INDUCE YOUR SHOE DEALERS TO KEEP A SUPPLY FOR YOUR NEIGHBORHOOD.

A. BALLARD & SON,
32 and 34 Vesey Street,
P. O. Box 5,000. New York.

Gothic Furnace.

For warming Dwellings and Churches, is the best mode yet invented. Read what O. G. Howard, Esq., of Ithaca, N. Y., says: "The No. 12 Gothic Furnace bought of you last fall, answers *my expectations fully*. I selected it on my judgement from several varieties, after several years' experience with furnaces. Last winter I heated 7 rooms with it, each 11 ft. high, being together 1,352 square feet of floor, besides warming part of the time a Hall 9x24 feet, 21 feet high, and two upper bedrooms. I consider it superior in three respects. 1st. It can be rushed or hurried up easily and quickly. 2d. In warm, spring weather it will burn cinders more slowly and keep fire longer at a low heat than any Heater I know. 3d. It *does not and cannot leak gas in any form*. It is *simple, easily managed, economical, and does not burn the air*."

ALEX. M. LESLEY, Manufacturer, No. 605 Sixth Avenue, N. Y. Send for a Catalogue.

FARM MILLS!—Durable, efficient, and low-priced. Send for Circulars. Agents wanted.
LANE BROS., Manufacturers,
Washington, Dutchess Co., N. Y.

Holbrook's Patent Swivel Plows.

For Level Land and Side Hill.

These Plows leave no "dead furrows," "no ridges," and give an even surface for the Mowing Machine, Hay Tedder, and Rake. They turn sod ground flat 5 to 10 inches deep, disintegrate very thoroughly, and will not clog. Eight sizes, from a one-horse garden plow, to a 6-cattle plow. Changeable mould-boards for sod and stubble. *Send stamp for Circular*. Manufactured and sold by
F. F. HOLBROOK & CO., Boston, Mass.

Premium Farm Grist Mill.

Cheap, simple, and durable. Is adapted to all kinds of Horse-powers, and grinds all kinds of grain rapidly. *Send for Descriptive Circular*.
WM. L. BOYER & BRO.,
Philadelphia, Pa.

IVES' PATENT LAMPS.

"The next best thing to daylight yet discovered."—*New-York Tribune*.

"They are, without hesitation, the best in the world."—*American Institute*.

THE SILVER LAMP WICK—The best and cheapest ever introduced. Does not smoke. Requires no trimming. Lasts longer than any other wick. Does not incrustate, and therefore always burns with a clear, silvery flame.

The Folding Pocket Lanterns,

VERY LIGHT, STRONG and DURABLE. Can be folded and carried in the pocket or traveling bag.

Nearly 100,000 already sold.

THE IVES PATENT LAMP CO.,
Sole Agents for the United States,
37 Barclay-st., and 42 Park-place, New York.

Cane Mills and Sugar Evaporators.

The best and cheapest. Our improved Evaporator is licensed by the proprietors of Cook's, Cory's and Harris' patents, combined with our own improvements patented June 18th, 1869. The best Evaporator for Sugar Cane, Sorghum, and Maple Sugar. Send for Cane Circulars to Hartford, Ct.; for Maple Circulars, to Bellows Falls, Vt. Address
THE HARTFORD SORGHUM MACHINE CO.
State that you saw this in the *Agriculturist*.

Improved Foot Lathes,

With Slide, Rest, and Fittings. Elegant, durable, cheap and portable. Just the thing for the Artisan or Amateur Turner.

"Your $50 Lathes are worth $75." Good news for all! Delivered at your door. Send for descriptive circular.
N. H. BALDWIN, Laconia, N. H.

SELF-ACTING GATES.

Nicholson's Patent Self-Acting Carriage Gates and Self-Shutting Hand Gates are the latest improved and best in the world.

For illustrated Circulars and Price-list send to the

American Gate Company, Cleveland, O.

$250 a Month, with Stencil and Key-Check Dies. Don't fail to secure Circular and Samples, free. Address S. M. SPENCER, Brattleboro, Vt.

H. KILLAM & CO.,

Chestnut St., New Haven, Conn.

We manufacture the finest class of carriages for city use, consisting of Landaus, Landaulettes, Clarences, Coaches, Coupes, Coupelettes, Barouches, Bretts and Phætons, which we warrant equal in point of style, finish and durability to any built in this country.

Messrs. DEMAREST & WOODRUFF, 628 Broadway, are our Agents in New York City.

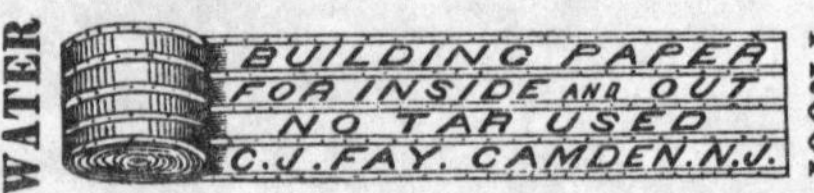

INVALIDS' TRAVELING CHAIRS, from $15 to $40, for in and out door use. Any one having use of the hands can propel and guide one. Having no use of the hands, any child of five years can push a grown person about. Invalids' Carriages to order. Pat. Carrying Chairs. State your case, and send stamp for circular.
STEPHEN W. SMITH,
No. 90 William Street,
New York.

Hawley's Air Pistol,

Pat. June 1, 1869.

No dirt, dust, or danger: uses compressed air. *No cost for ammunition*; from one charging, it shoots from five to ten shots, from thirty to fifty feet, accurate as any pistol. Price $3.50. Sent by mail on receipt of price and $1.35 for postage, or by express, C. O. D., and charges.

AGENTS WANTED. P. C. GODFREY,
119 Nassau St., New York.

India Rubber Gloves

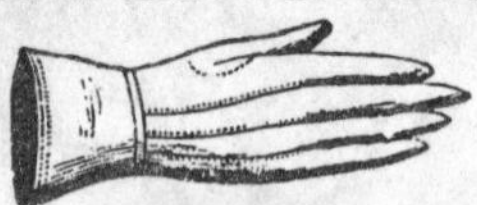

For Gardening, Housework, etc., etc. A perfect protection for the hands, making them soft, smooth, and snowy white. A certain cure for Salt-Rheum, Chapped Hands, etc. Ladies' short, $1.50; Gauntlets, $1.75 per pair. Gents' short, $1.75; Gauntlets, $2.00 per pair. Sent by mail on receipt of price, by GOODYEAR'S I. R. GLOVE M'F'G CO., No. 205 Broadway, New York, Manufacturers of all kinds of Rubber Goods.

TRINIDAD ASPHALTIC PAVEMENT.

For Streets, Carriage Ways, Walks, Floors, etc.
E. W. RANNEY, Patentee.

One continuous piece, not injured by heat or frost. Impervious to grass, water, and noxious vapors. Town, County, State, and Individual Rights for sale. Also the "*Patent Trinidad Asphaltic Cement*," with which every one can lay his own pavement. Send for Circular to E. W. RANNEY, General Agent, 440 West 23d St., New York City.

FRANK MILLER'S

Leather Preservative and Water-Proof Oil Blacking for *Boots and Shoes*. Frank Miller's prepared *Harness Oil* Blacking, for Harness, Carriage Tops, &c.; by the use of these articles one-half may be added to the durability of leather. For sale in nearly every city and town in the United States and Canadas.
FRANK MILLER & CO., 18 & 20 Cedar St., New York.

$30 to $50 per week sure.—CANvassers!

Look at this for your own benefit.

"THE UNIVERSAL FAMILY SCALE is worthy the special attention of all housekeepers. It is a new and desirable article for use in families; is the most complete thing of the kind we ever saw, and should be in every house in the nation. It has a face like a clock, and is so simple that a child can use it."—*New-York Independent*."

Canvassers wanted to whom liberal inducements will be given, and territory guaranteed by certificate. For pamphlet and terms, address G. W. LEONARD, Gen'l Ag't,
P. O. Box 2,833. 32 Cortlandt St., New York.

$732 IN 31 DAYS

Made by one Agent, selling SILVER'S PATENT BROOM. 75,000 in use. Recommended by *Horace Greeley* and *American Agriculturist*. One county reserved for each agent. C. A. CLEGG & CO., New York, or Chicago, Ill.

WANTED.—An experienced Agent to travel for an Agricultural Implement Manufacturer. Address JOHN CANTINE & CO., Schenectady, N. Y.

AGENTS LOOK!—$12 a day made, selling our SCISSORS SHARPENER and other wares. Sample 25 cts. Catalogue free. T. J. HASTINGS & CO., Worcester, Mass.

AMERICAN AGRICULTURIST.

NEW YORK, DECEMBER, 1870.

The short days are growing shorter, the cold nights colder. We draw our chairs nearer the fire by day, and an extra blanket over us by night. We must not forget that the furnaces that keep our animals warm are fed by fodder, and that warm quarters are as great a comfort to them as blankets are to us. The poor beasts, that shiver the long nights through behind the stacks of corn fodder, really cost the farmer far more to keep, than if housed and fed on the best of hay. This practice involves a loss in every way—fodder is wasted, vastly more is consumed, the manure is lost, the cattle are pinched, some perish, others come through lousy and "spring poor," and are not in good order again before June. We know there are few such farms, and they are happily growing fewer; but there are thousands of farmers who fodder on the ground; tens of thousands whose cattle never saw a stall; and hundreds of thousands who make but a mere fraction of the manure they might. When the comfort of the animals and the profit of the owner are so obviously coincident, it is a wonder that men do not care more for both. The season has been a peculiar one; the hay crop is short, and in many parts of the country hay is likely to be high. The mowing lands have not only been parched by the drouth, but farmers have been forced to feed them off close, and this involves, almost surely, a short crop next year. To remedy this, we have sowed corn for fall fodder, and rye for fall and spring pasturage, and more turnips than common. To remedy it still further, we must make twice the usual quantity of manure this winter and apply it upon land intended for corn, potatoes, and roots.

We are near the close of a year which will be long known as the one of the great drouth. In many parts of the country no such severe nor long continued period of dry weather has been known since meteorological records have been kept; and "the oldest inhabitant" has neither recollection nor tradition of a parallel summer to adduce. There was an anticipation of a short fruit crop on account of the severity of the winter in some sections, and of late frosts; but never were gloomy forebodings more thoroughly dispelled, for the crop of almost all of our more important fruits has been very large. The drouth arrested the growth of late potatoes, and reduced the yield of corn a little. Hay is decidedly a short crop, but a large supply of sound corn and roots has on the whole done much in New England and New York to supplement it. The usual abundance for man and beast is essentially curtailed on some eastern farms, which necessitates economy, and a reduction of stock. Still, there is no prospect of suffering; and if cows are cheap now, they will be dear again in the spring, which indeed is poor comfort for those who have to sell and expect to buy.

December, according to the pleasant usage of our ancestors, to the natural promptings of kindly natures, as well as according to the universal practice of Christendom, is the period of the year when men give most freely from basket and store to those not so bountifully supplied, or to those bound by ties of association, friendship, and consanguinity. It is a beautiful custom, giving great happiness if unselfishly carried out, and fast becoming characteristic of American life as it always has been of that of the rest of the Christian world. The strictest Puritan now no longer fears the keeping of Christmas as a papal encroachment, and the most fervent Catholic enjoys it none the less because Protestants keep it too. If all the readers of the *Agriculturist* greet the new year, having the experimental knowledge of the fact that "it is more blessed to give than to receive," how many happy people there will be in the world!

Hints about Work.

Manure making, as hinted above, is one of the most important labors of the season. For material to mingle with the dung of cattle large and small, we have straw, swamp grass, muck, forest leaves, dry earth, sawdust, and sand.

Buildings.—Barns, and especially stables, should be made snug and tight, so that snow will not blow in. Old barns, hen houses, etc., may easily be made warm by stuffing the sides between the inner and outer boarding with straw or litter.

Orchard and Nursery.

The care of young trees is of the greatest importance, as they are liable to injury, not only from wild and domestic animals, but from the human animal as well.

Fences and Gates.—It is a very common custom with many, as soon as snow covers the ground to disregard all established roads and paths, and drive or walk in the most direct line to their destination. We have known many young trees to be injured by this custom. See that fences and gates secure young orchards from human marauders; and if trees are exposed to injury by passers through the grounds around the house, drive down protecting stakes.

Young Trees should have mounds of earth, a foot high, drawn up around them, as it steadies them during heavy winds, and helps to ward off the mice.

Mice are, however, effectually guarded against by the use of tarred paper.

Rabbits are also kept at a distance by the use of tarred paper. Blood is much used at the West for the same purpose, and various protective shields, such as corn-stalks and laths, bound to the trunk with wire, are found to be serviceable.

Cions may be cut at any time when the trees are not frozen, and may be preserved in sawdust or in damp moss; or they can be buried in a box in a place where water will not stand. Take every precaution to have the varieties properly labelled.

Heeled-in Trees.—There is no risk in heeling-in trees for the winter if it is properly done. The earth must be thoroughly worked in among the roots, and no spaces be left for air or water. Bank up the earth well and smooth it off, and see that there is no litter near by to afford shelter to mice.

Water.—Open the needed surface drains, to prevent water from standing in the orchard.

Root-grafting is carried on in-doors when the weather will not allow of outside work being done. The greatest care should be taken to keep varieties distinct, and the grafts, when done, should be put in boxes with earth and placed in the cellar.

Seeds.—Peach, plum, and cherry stones, if they have been dried, must be exposed to the action of frost. When the quantity is small, it is sufficient to mix them with earth and expose them in a box in a place where they will be frozen and thawed.

Fruit Garden.

Grape-vines should have been pruned last month; but the operation may still be performed in mild weather. Young vines should have a mound of dry earth drawn up to them. It is better to lay down all vines, whether tender or hardy, as long as they are young and flexible. It preserves the buds and increases the general vigor of the vine. Bend them down and throw a few inches of earth over them. Wood from the prunings should be cut into suitable lengths and buried where it will not dry.

Strawberry Beds.—The covering should be done as soon as the ground is slightly frozen. The object is to prevent the injurious effects of freezing and thawing; and the earth, rather than the plants, is to be covered.

Raspberries.—Bend down the tender kinds, and throw a few inches of earth over the canes.

Kitchen Garden.

Roots.—If the hardy roots, such as salsify, parsnips, horseradish, etc., are not all dug, some litter may be put over them to keep the ground from freezing and to prolong the season of digging. Store a lot in earth in the cellar for use when the out-door stock is not accessible.

Spinach, Sprouts, Shallots and Leeks will need a covering where the snow does not afford one.

Covering of Pits in which roots are stored should proceed gradually. Do not cover too soon, but when really freezing weather comes, put on enough litter to keep out frost.

Cabbages.—If the crop is not already secured, store as directed last month.

Cold Frames.—Avoid too much heat. Air every day, and when the temperature is above freezing, remove the sash altogether.

Seeds.—Thrash and clean all that remain uncared for, and see that all are correctly labelled, not only with name but date. Store where the temperature is low and even, and mice cannot destroy them.

Flower Garden and Lawn.

Protection should not be overdone. Half-hardy plants do better if a few evergreen boughs are stuck around them than if they are strawed up in the old-fashioned way. Tender roses are best treated by laying them down and covering with sods. This is only practicable in places well drained.

Climbers, not entirely hardy, like the Wistaria, in northern localities, should be taken from the trellis, laid down and covered with earth.

Evergreens.—Young specimens often need a slight protection until they become well established. A little care for a few years will often save a specimen which will afterwards prove perfectly hardy. Cedar or other evergreen boughs tied in among the branches answer an excellent purpose.

Snow often injures evergreens if allowed to remain in their tops. Shake it out while still light.

Trellises, Seats, and all wooden garden ornaments should be put under cover. They will last much longer if they have a good coat of petroleum.

Pits, Frames, and Cellars.—Plants wintered in either of these need to be kept in a dormant state. Do not allow them to be too wet. It is safer to let them get too dry. Do not look for growth, but simply endeavor to sustain life.

Clear up whenever an opportunity occurs, and do not wait until spring to remove the debris of last season's operations.

Greenhouse and Window Plants.

Heat, water, and air are the things to be provided. Dust and insects are to be guarded against.

Heat.—Most plants will survive if the temperature gets below 40°; but no healthy growth nor bloom can be looked for at a lower average daily temperature than 60° or 65°. It is better that the night temperature should fall 15° less than this. One difficulty with plants in our dwellings is that they are as hot at night as during the day.

Water.—Give water only when the plants need it. A plant with its roots constantly in mud cannot thrive. It is better to wait until the flagging of a plant shows that it needs water than to keep it constantly soaked. Sprinkling or showering should be done as often as possible. Take the plants to a sink or a bath tub and give them a good dousing.

Air.—Do not be afraid of opening the windows whenever the outside air is not freezing. Give air every day when it is safe to do so. Not only will the plants be benefited, but the atmosphere will be the better for human lungs.

Dust is a great obstacle to the successful growing of plants in the house. The showering we have recommended will help remove it, but all smooth-leaved plants, such as Camellias, Ivies and the like, should be occasionally sponged to keep the foliage clean and healthy.

Insects.—Don't have any. If the plants are daily looked over, and the thumb and finger properly applied, they will be kept in check. If a plant is badly infested by the green fly, put it in a box or under a barrel and smoke it thoroughly. If the red spider appears, as it will be apt to do in hot and dry rooms, smoke will not help it. Remove the plant into hospital and shower it daily. Frequent wetting and a moist atmosphere is the best remedy.

The Prairie-dogs.

There never was a greater misnomer than that of calling the little animal, of which we present an engraving, a Prairie-dog. It has nothing canine about it, and its true relationships are among the rodents with the Squirrels, Marmots, and Woodchucks. We generally speak of *the* Prairie-dog, though there are two and possibly three species within our territory; but the most abundant is the one we have figured, which is found from Nebraska to Texas. Various names have been given to it by different authors, but the one adopted by our best naturalists is *Cenomys Ludovicianus.* It is so much like our common woodchuck in general appearance, that it is well enough described by calling it a woodchuck in miniature. Its length varies from ten to thirteen inches, and its weight from one to two and a half pounds. The ears are very short, and the tail about one-third as long as the body. The general color is reddish-brown; the shade varies with the season, and is lighter upon the lower part of the body than elsewhere. The tail for one-third or one-half from the tip is black, the rest being of the color of the body. The Prairie-dogs are social animals, and live in large communities known as "dog-towns." The animal burrows, and with the earth from the excavation forms a dome-shaped hillock which contains as much as two cart-loads of earth. These mounds are often seen upon the prairie as far as the eye can reach. The writer once, in Western Texas, was three days in passing through one of these dog-towns, and as the distance made by the train was estimated at twenty miles a day, the colony was at least sixty miles in length and extended on each side of the road as far as we could see. In each mound is an opening, sometimes two, extending downwards at an angle of 45 degrees. How deep the animals burrow we cannot state, but some assert that they go down until water is reached. The animals are evidently on visiting terms, as well-beaten tracks extend from one burrow to another. Where these animals abound in such numbers, the herbage is very scanty, and we have frequently seen "towns" where the surface in every direction was so barren that it was difficult to imagine how such a multitude could subsist. The animals are very fond of a species of purslane, which grows in great abundance in some localities. It is said by some travelers that the towns are extended as food becomes exhausted, and that a large share of the mounds are uninhabited, the animals having left for better pasturage. No one seems to have watched the Prairie-dogs sufficiently close to be able to give a very full account of their habits. If the traveler approaches a town cautiously, he will see the animals in constant motion and chirping to one another in the liveliest manner. As soon as he is seen, some animal gives the alarm, and away they all go, each for its own hole, where it stands with its head projecting and uttering its shrill bark. The Prairie-dogs are very difficult to shoot, as they dodge with great rapidity, or if shot, they tumble into their holes out of reach. Those who have eaten the flesh of the animal pronounce it excellent. The young are easily tamed, but make rather uninteresting pets, as they sleep a great portion of the time. In the dog-towns one meets with rabbits, numerous rattle-snakes, and a small burrowing owl. The prairie men assert that these live in common with the prairie-dogs, but it is more probable that they find it convenient to occupy the vacated dwellings of the prairie-dogs. The owl (*Athene hypogæa*) is a most comical bird, and may frequently be seen upon a mound, bowing and bobbing its head in an amusing manner. It is believed by the prairie travelers that it serves as a sentinel, and gives the prairie-dogs warning of the approach of danger. It is more probable, however, that the fondness of the owl for young prairie-dogs, and the convenience of finding ready-made burrows, are the reasons for its being so generally found in the dog-towns. Rattle-snakes, frequently mentioned as inhabiting dog-towns, are so abundant in parts of the far West, that it is difficult to say if they are more frequent there than elsewhere.

THE MISSOURI PRAIRIE-DOG.—(*Cenomys Ludovicianus.*)

Japanese Bantams.

The Japanese seem, even more than the Chinese, to have a taste for things *outré*, and for monstrosities, both in nature and in art. Probably the most remarkable contribution to our poultry yards, received from this interesting country, is the breed of bantam fowls, of which we give an engraving, taken from drawings made of specimens exhibited by J. H. Fry & Co., at the last exhibition of the New-York State Poultry Society. The whole form and style of these unique fowls are such, as to distinguish them from all others, although they vary greatly in color and markings among themselves. The group shown in the engraving were young fowls, and the sickle-feathers of the cock not fully developed. Their bodies are small and very compact; the legs short and smooth; combs single, erect, and very large in both sexes; heads carried well back, making the fowls remarkably "pigeon-breasted." The backs almost disappear between the necks and tails. The tails are carried more than erect, leaning forward, and in the case of male birds frequently extending further forward.

JAPANESE BANTAMS.—*Property of Messrs. J. H. Fry & Co.*

The Closed Gentian.

Among the many plants sent us for a name, there is no one we more commonly receive in autumn than the Closed Gentian. It is frequently met with late in the season, and attracts notice on account of its conspicuous, dark-blue flowers, and their persistent way of never opening. The engraving we give will show this peculiar appearance, which has caused some to call it the Bottle-Gentian. The most common species with this general aspect is *Gentiana Andrewsii*, and it sometimes occurs with perfectly white flowers. There are two or three other species similar in appearance, found in the Southern and Western States, and only to be distinguished upon close inspection. We are not aware of any attempts to cultivate either this or any other of our species of Gentian. Some of the Alpine forms of Europe have been introduced into cultivation, but they are difficult to grow with us. Probably our native ones, some of them very beautiful, if their requirements were properly studied and experimented upon, might be brought into cultivation.

THE CLOSED GENTIAN.

A Double-flowered Datura.

When one wishes to popularize a flower that is little known, how he is troubled to find a taking name. *Datura chlorantha flore pleno*, as they have it in the seedsmen's catalogues, is certainly formidable enough to repel any one; it might be called "Double-flowered Datura," but as there are other Daturas that have double flowers, this seems to be hardly specific enough. The Daturas we are sufficiently familiar with in our common *Datura Stramonium*, popularly known as Stink-weed, Apple of Peru, Devil's Apple, Jamestown and Jimson Weed, which is a common enough weed in all waste places. Even this is not without beauty, and did it require care to develop its fragrant funnel - shaped white flowers, we should probably prize it as an ornamental plant, despite the sickening odor of its leaves. There are several species of Datura cultivated for ornament; while their flowers are very showy, their leaves have all a most unpleasant odor. One should enjoy them without coming into too close contact. *Datura meteloides*, oftener incorrectly called *D. Wrightii*, is a very showy, single - flowered species, and a well grown plant of it makes a great display of enormous flowers. A friend of ours who makes the greatest possible show in a small garden, produced a fine effect last summer with a few plants of *Datura chlorantha flore pleno*—which we may translate as the "Double greenish-flowered Stink-weed." It grows four or five feet high, and makes a spreading, bushy plant, which is covered with a profusion of bloom. The individual flowers are six or eight inches long, and each curiously double. Each corolla has another within it, to the number of three or four, as shown in the engraving. The flowers are of a delicate, greenish-yellow color, and are fragrant, while the foliage, when bruised, is as offensive as that of our common species. For making a show, this is a desirable plant, and as it is one of those things which flourish well in hot weather, the best way is to start the seeds early in a hot-bed and grow the plants as rapidly as possible.

DOUBLE-FLOWERED DATURA.

Soils for Gardening and Farming.

BY PETER HENDERSON.

On many occasions I have referred to the great importance of selecting a proper quality of soil for all gardening and farming operations. But as year by year you are increasing the number of inexperienced readers, the fact cannot be too often nor too forcibly impressed that success hinges upon this alone more than on any thing else. Thousands are every year ruined by a bad selection of soil. I have scores come to me in the course of every season for advice in this matter of soils, but in most instances the advice is asked too late; the majority of the applicants having been unfortunate enough to buy or rent land that they had been led to believe was excellent, but only "run down." In my opinion this wide-spread notion of "exhausted lands" is, to a great extent, a fallacy, and that most of the lands said to be so exhausted never were good, and no power on earth short of spreading a good soil over them a foot thick, would ever make them good. In a visit to the suburbs of Richmond, Virginia, a few weeks ago, I encountered a man from the Eastern States that had gone down there four years ago. He had bought an "exhausted farm" of 80 acres, and with northern energy and northern capital, hoped to resuscitate it to what it had been (as the former owner had told him), a fertile farm. An expenditure of nearly $3,000, and the hard work of four years, had as yet failed to give him a crop of corn that paid for the labor. Not a stalk could I see that had been more than five feet high, and many of them not two feet. No wonder! his poor yellow soil in no place exceeded four inches in depth, and was underlain by a hard pan of clay. Ruin in all such cases is inevitable, the labor put upon such a soil can never pay, so long as there is any thing better within twenty miles of it. Our country contains millions of acres of lands, that are bought and sold annually, which are of but little more use for farming purposes than if the title deeds were given for the same area in mid-ocean.

"But," asks the reader, "how are we to select soils?" First, never buy a farm without personal examination—never take the seller's word about it; he may honestly believe that what he asserts is true, or he may know it to be false; but in either case if you are deceived you suffer. Let the examination be thorough; note the surroundings, if the district is settled and cropped; carefully examine the condition of crops on the farm and upon those adjoining it. If they are sickly looking and weak, if the corn-stalks instead of being seven or eight feet in hight are but two or three, you had better lose your time and expenses and get home again, than take the farm as a gift. But should there be no corn or other crops by which to test the quality of soil, an examination should be made by digging down at short distances all over the ground. The top, or "true soil," should not be less than nine inches in depth; the best color is a dark brown. The subsoil, or stratum lying immediately under the top soil, should be of a porous nature, and it is usually, in first-rate soils, of a yellowish, sandy loam. Occasionally we find a gravelly subsoil underlying soils of good quality, but this is not so common. Less frequent still is a subsoil of blue or yellow clay, such a clay as might be used for brick making. A subsoil of this kind when near the surface is a certain indication of a poor quality of soil for the purposes of either farming or gardening.

To illustrate the value of different soils for our market gardening purposes—we have in our immediate neighborhood men, who now pay $100 per acre annual rent, and who in the past ten or twelve years have made snug little fortunes upon eight or nine acres in cultivation. At a distance of not more than half a mile, there are others paying less than half that amount in rent, who have during the same time been struggling to make both ends meet. Though equally industrious and having an equally good knowledge of the business, their success has been quite different, and all simply for the reason of a difference in the normal condition of the soil. In the one case the land would be cheaper for the occupant at $200 per acre annual rent than the other would be if it could be had for nothing.

Our best lands for vegetable growing in this district (Jersey City, N. J.,) which is a mere suburb of New York City, are rapidly getting absorbed for building purposes, so that before many years the market gardens, for the supply of the great Metropolis, must be on Long Island. There the land is generally well fitted for the purpose. Immense tracts of level prairie-like lands are being devoted to ordinary farm crops in the vicinity of Flatbush and Flatlands, L. I., which in a few years will, without doubt, be occupied by garden vegetables and fruits.

Structures in Rustic Work.

The term "rustic work" is now used for many objects made of materials, the surface or the shape of which is left in the natural condition. The smallest flower-baskets, consisting of a bowl ornamented with cones and crooked sticks, and large, even elegant, edifices, such as are seen upon our parks, are classed under the rather comprehensive name of rustic work.

Probably no finer specimens of this style of architecture can be found anywhere than at New York Central Park; the shelters, summer houses, seats, arbors, boat-landings, and bridges, built in this manner, are numerous, and are tasteful in design and executed in a workmanlike manner. It is probable that the successful introduction of rustic work at the Park has done much towards popularizing it, for we now seldom visit a neighborhood where any attention is given to rural adornment that we do not see more or less ambitious attempts at this kind of decoration and frequently excellent examples.

Fig. 1.—BIRD-HOUSE.

Work of this kind should present the expression of durability and solidity. Its very rudeness of exterior demands that there should be nothing shaky about the structure. There is no wood so well suited to the purpose as the Red Cedar, not only on account of its great durability, but because the natural growth of its branches presents a great diversity of angles and curves, twists and knots, that in the hands of a skillful workman give most pleasing effects; besides these, its color is a harmonious one. No instruction can make one a clever builder of rustic work, he must have a natural ingenuity that will allow him to combine irregular shapes into something like symmetrical forms. A mere association of grotesque branches is not pleasing. There must be an architectural design, and the details of this worked out by the ingenious use of natural materials. We give a few illustrations of simple structures. In fig. 1, we have a bird-house and a support for climbers combined. The central pillar is made sufficiently strong to support the structure, and the vines are trained to the corners by means of wires.

Fig. 2.—RUSTIC BRIDGE.

Fig. 2 is a bridge upon the estate of Edwin A. Saxton, Esq., at Tenafly, N. J. Rustic work is often used with fine effect in small bridges, and though this is less regular in its design than some we have seen, the effect is very pleasing.

How far to Haul Peat for Manure.—When peat, or muck, is thrown out of the bed in summer and allowed to dry, it usually becomes quite hard and not easily broken up, even by the freezing of winter, unless again thoroughly water-soaked. Peat thrown out in the autumn or winter will be thoroughly crumbled by the action of the weather by spring, and may be drawn away as soon as it is dry enough, at any convenient time during the summer. It is very poor policy to haul water very far, especially if it is enclosed in peat, which you will have to dry wholly or in part before it can be used. In the condition called dry, it contains usually some 10 or 12 per cent of water; but this is little, compared with the 50 per cent or more, it contains, when freshly dug. The composition of peat, as regards the comparative quantities of organic and inorganic matters, varies greatly; but in those peats in which sand and earth are not obvious constituents, we may assume that there is about 60 per cent of vegetable matter, which contains nearly 2 per cent of ammonia. And it would be fair to assume an average of $1^1/_2$ per cent of this substance in the peaty deposits which are accessible to most farmers. Barn-yard manure as usually hauled does not contain half so much ammonia, for it is much wetter, but a good compost heap is frequently equally rich. Barn-yard manure is richer in potash and phosphoric acid, and is really better as a regular plant-food; but the variation in analyses and the general experience of farmers, we think, clearly lead to the decision that load for load, one is worth as much as the other; and hence the farmer can afford to haul one as far as the other, and to pay as much for it, provided it be dug and nearly dry. If it can be taken home as a return load when otherwise the teams would come back without one, it will pay well to haul it four miles. To go for it alone, would make it cost a good deal more, and few farmers would think they could go much more than half that distance, if so far.

Pumpkins as a Stolen Crop.

We have seen fewer of the yellow orbs than usual this season, on account of the drouth. We could wish that we had seen the last of them. It is about time that pumpkins were retired from service, and entered upon the fossil list. If any fossil farmers still wish to cultivate them, let them devote a piece of ground specially to the purpose, rather than cumber the cornfield and the potato patch with them. Even when they are planted at the first weeding in June, they soon spread over the intervals between the rows, and seriously interfere with cultivation. The profit of raising corn depends very much upon the thorough cultivation it gets in the month of July. Vines cannot help obstructing the hoe and cultivator. Then they make their growth at the same time as the corn, and must draw upon the same constituents in the soil that nourish the corn. The yield must be diminished. Turnips sown at the last cultivating in August only just get started when the corn is finished, and make nearly all their growth in the fall. Then pumpkins are of very little value when they are raised. For pies they are worthless beside the Hubbard or Marrow squash. The squash should have the ground on the principle of survival of the fittest. The pumpkin is used for making milk and beef. The corn that could be raised in its stead is worth more. Squashes are better. A stolen crop of turnips would be twice as valuable, and would be better for the land.

Tim Bunker on Good Neighborhood.

"I could a got along well enough with turkey shootin ef he had n't gone down to the store and brag'd on't;" said Seth Twiggs, rapping the ashes out of his third pipe, as he stood by my garden fence.

"You don't say that Jake Frink killed your turkeys, dew ye?" inquired Tucker.

Yes, he did, and brag'd on't, tew. Ye see I kept turkeys, and Jake also, and sometimes Jake's got into my garden and sometimes mine into Jake's field. Sometimes he'd bring in a bill for damaged corn, which I allers paid like a Christian man. Sometimes I druv his turkeys home and asked him to take care on 'em. But I never thought o' killin on 'em, more'n I wud one of Jake's sheep. And now the critter 's killed five of my young turkeys, and had n't the face to come and tell me on't like a man, but went down to the store and brag'd on't as ef he'd done suthin kind o' grand."

Home-made Horse-powers.

One of our readers at the far South, who lives off the lines of railway and water-carriage, wishes to avoid the purchase and costly transportation of a horse-power, and asks for a plan by which one may be made. Horse-powers for moving agricultural machines, such as thrashers, saws, feed-cutters, etc., require speed rather than great power, hence the gearing is accommodated to that object, and is very different from the slower and more powerful motion required upon a capstan or in the brick-yard. The application of the power of the horse is, however, the same, and it is not improbable that our readers who wish to set up a horse-power for thrashing, sawing wood, or grinding apples may find pieces of machines in their neighborhoods, now of little value, which they can turn to account.

We give engravings of two sweep horse-powers, from one of which the power is taken from above, leaving the ground clear, while from the other the power is taken from the ground level as nearly as possible. Both are intended for use under a building or shed. In both, also, the posts are of a light to accommodate the room—say 7 or 8 feet. They are held in position by strong pins of iron, with cross-arms let into the posts at each end, and revolving in oak bearings above and below. These posts are about one foot square at one end and rounded at the other. The sweep to which the horse is attached, shown in the engraving, fig. 1, should be made of a crooked stick of any hard wood, ash or oak would probably be best, and either would stand the strain if it were to be worked down to about 4½ to 5 inches square at the upper end, and 3 inches square at the other. It is much more convenient to use such a sweep, than one attached to the post so low down that a man cannot stand erect under it. The sweep should be let half its thickness into the post and secured in place by two strong bolts with nuts. The periphery of the bevelled wheels should either be of iron, cast in segments and bolted to the wooden wheel, or cast-iron segments with sockets into which teeth of oak are set. This makes not only a very durable gearing, but one which may be easily repaired, should a tooth become much worn or broken. It is very much to be preferred to the pin-gearing shown in figure 2. The castings may be obtained at any good foundry. The rod which conveys the power may be either of wood or of iron, the latter is preferable, and it should be set low, in order that the horses may easily step over. The track may properly be raised at the point where the rod crosses.

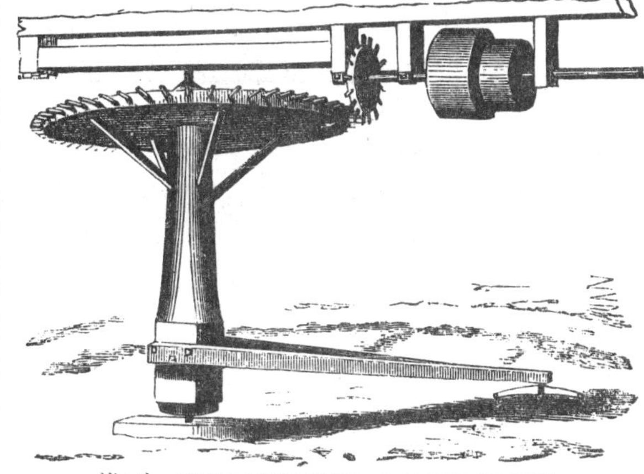

Fig. 2.—SWEEP HORSE-POWER WITH PIN-GEARING.

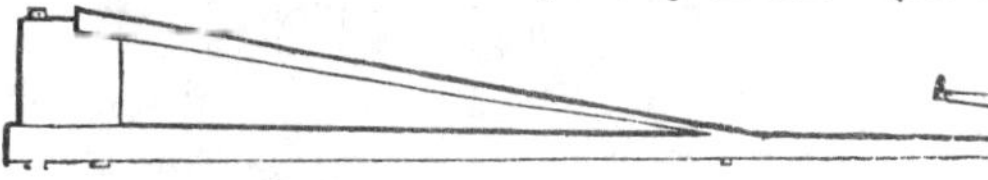

Fig. 3.—POLE AND BRACE FOR SWEEP.

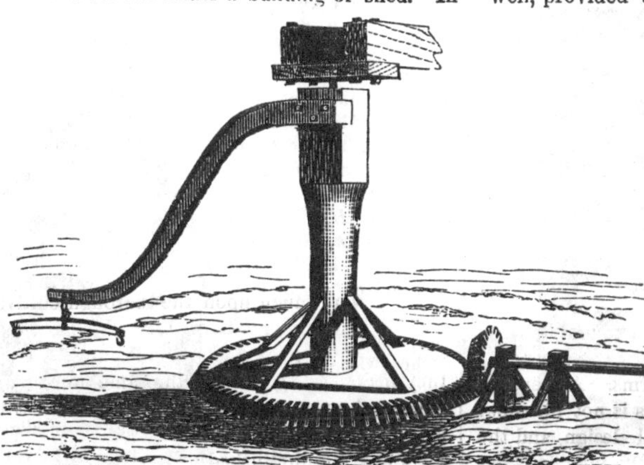

Fig. 1.—SWEEP HORSE-POWER WITH BEVEL-GEARING.

The pin-gearing shown in figure 2, works well, provided the strain upon it is even and not very great. The holes for the pins must be bored with great accuracy or the motion of the machinery will be unsteady. It is much better to have the gearing above than on the ground, if the power is to be employed upon the floor above. The form of sweep used in this, with the manner of bracing and bolting to the shaft, is shown in figure 3. The sweep may be of ash, hickory, cedar, or any moderately tough and somewhat elastic wood. The brace stiffens it greatly, and one of the bolts should pass through both the pole and the brace. The length of the sweeps is generally about 10 feet, though larger ones would in some cases be better, and much shorter ones are often used.

When sweep horse-powers are set up in the open air, a strong frame is made of four posts, connecting an upper frame with a lower and larger one, made of timbers embedded in the earth or in stone-work. The frame encloses a gearing like the one shown in figure 1.

Washing Roots.

The habit of half washing roots, which we feed to cattle, pigs and sheep, is too prevalent; we confess to many infractions of the rule of cleanliness and economy ourselves in feeding, but are entirely satisfied that it is a losing practice. Earth, soil, dirt, is very well in its place, and no doubt all of our domestic animals enjoy and are benefited when they "bite the dust," as they do now and then when they have an opportunity. It is, however, a very different thing to *boil* the dirt, or to season turnips with gravel. The discomfort which every one feels when he finds a particle of grit in his food ought to suggest the humanity of washing thoroughly all the roots fed to stock.

We have in the agricultural stores a root-washing apparatus, made of a cylinder of coarse, woven wire, which revolves by means of a crank in a trough of water. This is rather expensive, and we have found a very simple contrivance equally effective. A common coal or sand screen is laid flat upon two wooden horses, a bushel of roots at a time is thrown upon it, and spread out. Then, with a short hose and pipe attached to our submerged pump, (the same offered upon our premium list,) we direct a stream of water upon them, which thoroughly cleans off all adhering soil. If one end be lifted and jarred down a few times the roots will change places and expose new surfaces to the water. We shall have occasion soon to wash a cart-load at a time, and shall proceed thus:

The cart will be brought to the pump, and the roots thoroughly wetted and allowed to soak for an hour. The screen will be placed conveniently, and slung by a rope and chain to a limb over head. Then a sledge with a box upon it will be drawn to the spot. The roots will be shoveled from the cart into the screen, and as soon as washed, the screen will be swung around and they will be dumped into the sledge.

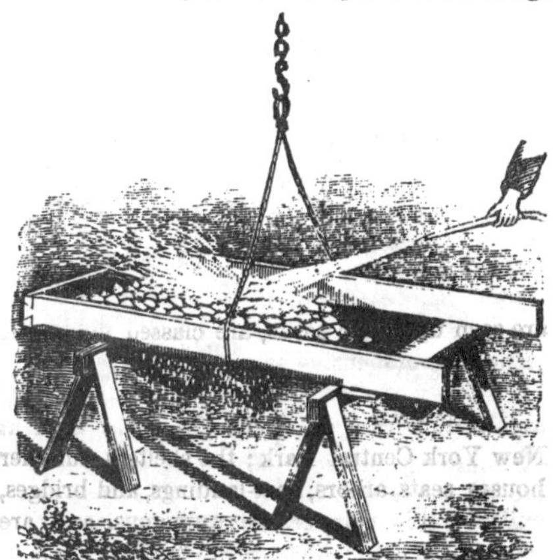

ROOT WASHER.

In drawing them over bare ground to the barn or cattle sheds, the horses will hardly draw as much as they will upon a cart; but it will be twice easier than the roots can be handled and moved in any other way.

☞ For other Household Items, see "Basket"

Christmas Presents.

BY MRS. A. B. MCK.

Christmas presents for four persons and only five dollars with which to buy them. O, dear! Could any one tell me how to purchase a foot rest for father, paper holder for James, lamp shade for Aunt Ann's poor eyes, and a bracket to hold the pretty statuette that cousin Mary gave Millie on our last visit to town, with only five dollars? Each of these articles had been by me priced and repriced, in the vain hope that five dollars could be stretched to cover a surface requiring fifteen; until at last I came to the conclusion that I must *make* them *myself*, if they were to be forthcoming at all. After that, for a whole week, whether sweeping, dusting, churning, or baking, the pros and cons of these coveted possessions, and the most plausible way of setting about their construction, constantly floated through this very feminine brain of mine; when, about the seventh day, a happy idea dawned, all became clear, and order was evolved from chaos.

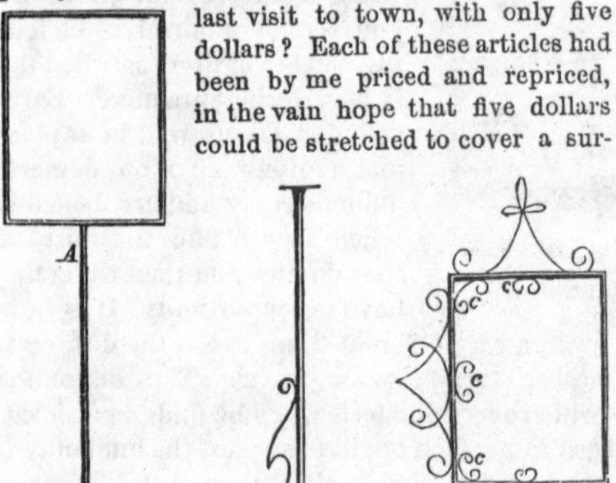

Fig. 1.—PARTS OF LAMP SHADE.

The first article attempted was Aunt Ann's lamp shade. Now the common round ones when used, left the whole family, as well as herself, in the dark. I wanted something to shield her eyes, and yet allow the rest of the room to be flooded with light For this purpose I had looked longingly upon a porcelain standard shade, displayed very temptingly in a certain shop window, but alas! it cost far more than the whole contents of my purse. However, I concluded to get some good out of my disappointment by using it as a model. Having procured a heavy wire frame from the tinman (*A*, figure 1), I moistened a little plaster of Paris, filled an old, tin blacking box with the creamy substance, and placed the lower end of the standard in the center, where I held it firmly with my left hand until the mixture hardened; meanwhile with my right hand I expeditiously arranged a row of coffee berries around the edge of the box, which, as well as the standard, were soon securely held in place by the hardened plaster. I next took four pieces of wire, which I bent and fastened to the standard and box as at *B* (figure 1). This I accomplished by means of very fine wire and small bits of putty. The edge of the upper part, or screen, I also ornamented with fancifully coiled wire, making the figures *c*, *c*, *c*, (at *C*, fig. 1,) double. Upon the edge of a round board, two inches larger in diameter than the blacking box, I puttied a row of coffee berries; fastened wire feet to the bottom, and glued this second and broader base to the first one, already at the bottom of the standard. (Had I screwed the box and wood together before using the plaster of Paris, it would have been much better.) Then I gave to the whole three coats of dark brown paint, and as many of varnish, letting each one dry *thoroughly* before applying the other. The last coat of varnish was mixed with a quantity of hair powder called "gold dust," before being applied, which gave to the frame the appearance of having been made from the golden sealing-wax, so much in vogue years ago. Having cut a transparency from bristol-board, I gave to each side a coat of white glue; afterward, two of varnish, and inserted it between the double wires at fig. 1 (*C*). When complete, it presented the appearance of fig. 2, and was pronounced by all, beautiful.

Fig. 2.—LAMP SHADE.

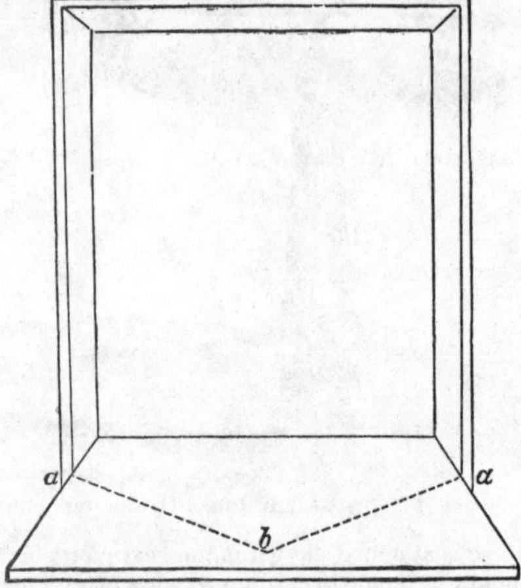

Fig. 3.—DIAGRAM OF FOOT REST.

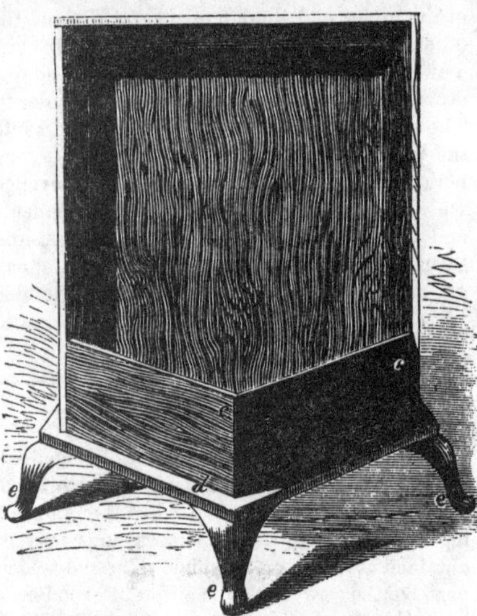

Fig. 4.—BACK OF FOOT REST.

Fig. 5.—FRONT OF FOOT REST.

A FOOT REST.—An old saleratus box 14½ by 17, by 17 inches, furnished the foundation for father's foot rest. These boxes are generally made stronger and smoother than ordinary dry goods boxes. I removed the top, cut the sides and one end down to 8½ inches, as in figure 3, and pointed the other end as at *a*, *b*, *a*, around which I nailed two small boards, *c*, *c*, (figure 4,) so as to form a box in which to keep the slippers. (A cover to this part would be an improvement, but I was not carpenter enough for that.) I nailed a second bottom or baseboard, 1½ inch projecting, to the rest when thus prepared, and screwed to the corners, feet 2½ inches in hight, *e*, *e*, *e*. These I purchased of a carpenter. They were stained to imitate black walnut. Then I carefully covered the whole with walnut figured wall paper, (except the front of figure, which had only a border of 1½ inches,) fastening the edges and corners securely, and being sure that no air bubbles were left under the paper. When dry, I gave it one thin coat of glue and three coats of varnish, after which it would have taken a skillful eye at a little distance to detect that it was not really walnut. In autumn I had put a new cover of reps upon the sitting-room lounge. A remnant was left from which I cut a piece 16½ by 14 inches. In the center of this I embroidered a medallion with initials, and tacked it over the front, as shown in the engraving (fig. 5), with upholsterer's gimp and white headed nails. Aunt Ann and father say it looks "just as boughten as can be," which is their highest term of praise. American men are noted for wanting their feet, while sitting, nearly as high as their head. This rest enables father to indulge in his favorite attitude without occupying an extra chair. He declares it a splendid affair for warming the feet. When not in use I keep it in the chimney corner, the pointed back fitting in so as to occupy but little room, and the front being very ornamental.

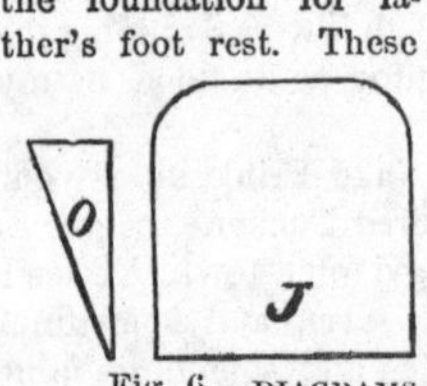
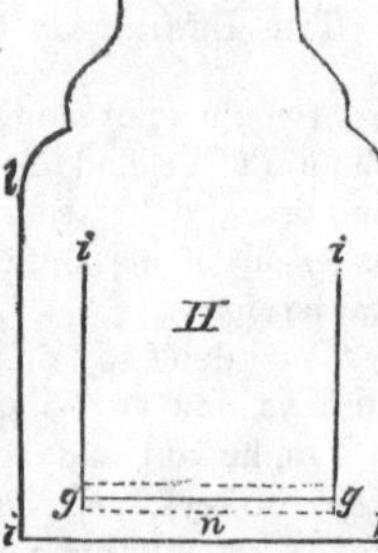

Fig. 6.—DIAGRAMS OF PAPER HOLDER.

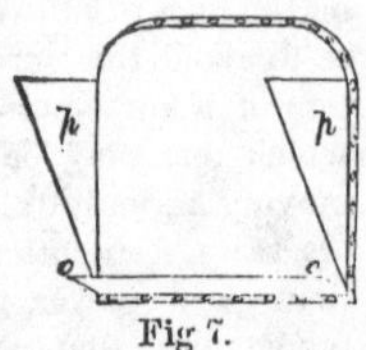

Fig 7.

PAPER HOLDER.—For the paper holder I procured two butternut boards the size of *H*, and *J*, figure 6. *H*, is 17 inches from *i*, to *k*, 16 inches from *i*, to *l*, 28 inches from *m*, to *n*. *J*, is 13 by 14 inches. *O*, is red kid, 13 inches long, and 4 inches wide at the top, where it is bound with narrow ribbon. There are two such pieces for the ends, and also a strip 13 inches long by 2 wide. After smoothing the boards with sand-paper I drew upon them the design, (figure 8). I stained the body of *H*, a rich brown, leaving the ornamental work of the original color; while on *J*, the corner figures and a wavy margin around the edge were stained, and the remainder of the board left in its natural state. To highten the effect of the designs I painted a

Fig. 8.—PAPER HOLDER.

narrow band of black around their inner edges. Then varnished *J*, and immediately pressed upon the center, face downward, an engraving previously soaked in water, from which I carefully rubbed off the white back. When it became thin enough to show the picture through, I allowed it to dry, after which it was again wet and still more of the paper back rubbed off, until only a very thin film of paper remained, which became so transparent by varnishing as to allow the grain of the wood to show through, and seemed to have been engraved upon the board. Both *H* and *J* received four coats of varnish, which gave them a very high polish. The three pieces of kid were then tacked upon the sides and bottom of *J*, with gimp and white headed nails, the gimp and nails being carried around the top. Held with the back toward me, it now presented the appearance of fig. 7. With carpet tacks

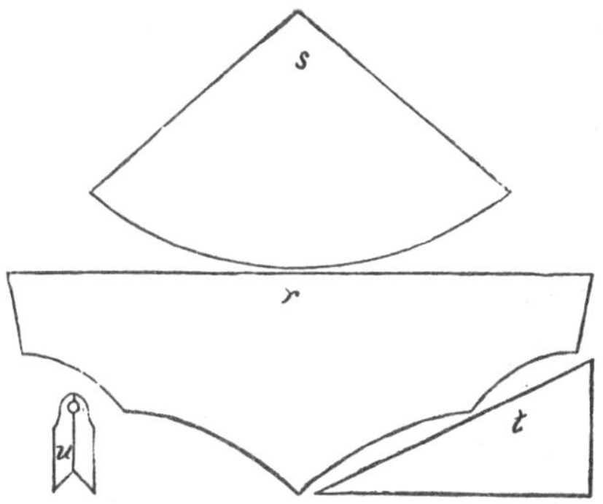

Fig. 9.—DIAGRAMS OF BRACKET.

I fastened *o*, *o*, upon the line *q*, *q*, (fig. 6,) in the manner shown by dots. When the pieces *p*, *p*, (fig. 7,) were brought up to *r*, *r*, (*H*, fig. 6,) this seam was left inside the holder. Gimp and white headed nails were used in fastening the kid to the lines *q*, *r*. This article was hung upon the wall by a large picture nail, and is for receiving newspapers.

A BRACKET.—Fig. 9 shows the parts of the bracket shelf, 13 by 17 inches, which I papered and varnished the same as foot rest; *t*, is one of the side pieces, 7 by 13 inches; *u*, a bit of tin, cut from an old tomato can, which I papered, varnished, and nailed over the junction of the side pieces, as a ring by which to hang the bracket, (fig. 10); *r*, (fig. 9) is embroidering canvas, 19 inches long, and 8 inches at the deepest point. There were also two other pieces of canvas, the same shape, but an inch larger each way than the side boards. In the center of each canvas, I embroidered a group of bright, autumn leaves, filling in the body with crystal beads, a bead in each stitch. Then tacked them over the side pieces, and around the front of the bracket, as in the engraving, hiding the tacks under a bead heading, and finishing off the lower edge of the curtain with a heavy fringe of the same.

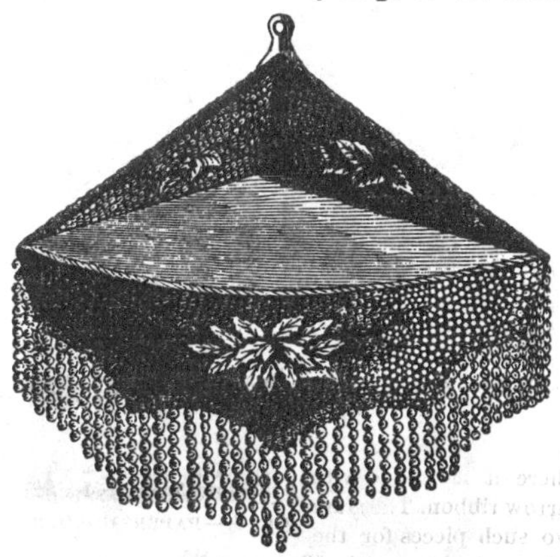

Fig. 10.—BRACKET.

When finished, this was far prettier than any of the brackets in the store, I had so coveted for sister. There is great satisfaction in having thus provided so many acceptable presents with so small an outlay of money. Perhaps these hints will help others to prepare gifts for friends at Christmas time.

Apple Jelly for Tarts.—By "W." Cut and core apples, without paring, cover them with water and let them cook slowly in an earthen dish until the apples look red; then pour into a bag and gently squeeze out all the liquid that will flow freely. Boil the liquor again about half an hour, then add half a pound of sugar to a pint of juice, and boil quickly for fifteen minutes. It will prove a firm, nice jelly, and requires but half the usual quantity of sugar.

Canaille Bread.—By Mrs. E. Lea.—Scald well two tablespoonfuls of Indian meal and add to it enough canaille to make a thin sponge with 2 quarts of milk, (or milk and water mixed). Add ½ a cup baker's yeast. Set this about 7 o'clock in the evening, and at bed time mix it stiff with canaille and two tablespoonfuls molasses. Bake in the morning.

Mrs. S. Buckley says: "I use canaille with buckwheat, or alone, mixed in the same way as that; I use it for common pastry, adding a little fine flour, and for molasses cakes. Minute puddings may be made by boiling any desired quantity of milk, and stirring in the canaille very gently to prevent lumps; let it boil a few minutes; eat with sauce, sugar and cream, or butter.

An Excellent Butter-worker.

The farmer of "Ogden Farm" described in one of the "Ogden Farm Papers," a new butter worker he had recently introduced into his dairy. There has been so much inquiry about

Fig. 1.—BUTTER-WORKER

it we have had engravings made to represent it which scarcely need any other explanation than to give the dimensions. The table is of white oak three feet long, and two feet wide, made very substantially. The side away from the dairy-woman, as shown in figure 1, is the lowest, and a groove runs around three sides of the table to conduct the butter-milk to a drip at one corner. The paddle or knife is shown at figure 2. It is a foot long and five inches wide, with handles six inches long, made from one piece of oak board, worked smooth and true to a blunt edge on each side, as shown in the figure.

The butter is formed and worked by this

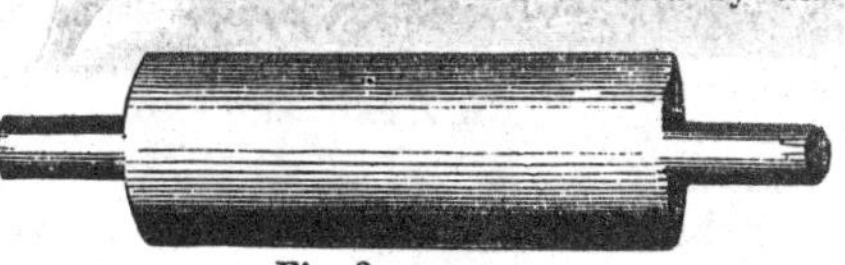

Fig. 2.—PADDLE.

knife, which is held in both hands. Any good butter-maker will quickly get the knack of using it, as it is much easier than working with the ladle or paddle commonly employed.

A Brick Smoke-House and Ash-House.

Safety from fire is or ought to be carefully considered in putting up any structure about the place, and in deciding where to put wood-ashes when removed from the stove or fireplace. Smoke-houses are peculiarly liable to take fire, if they are of wood, and such are a continual annoyance and care, during the season when they are most in use. At other times they are

Fig. 1.—BRICK SMOKE-HOUSE.

useless except to hang the hams and bacon in to keep the flies away. Smoke-houses ought to be built of brick, if we would not be made uncomfortable by them from fear of fire, and it is altogether best to have no wood about them. Fig. 1 represents a brick smoke-house, 7 feet high in the clear, 7 feet 8 inches high outside, and 7 by 8 feet on the ground. The walls and arch are one brick (8 inches) thick throughout, the whole standing upon a foundation bed of stone, not less than one foot deep, and extending one foot outside the walls on all sides. This is covered with gravel, and topped with clay, after the house is done, which secures a dry, hard floor. The roof is laid upon an arch of boards, in any good, strong mortar. When done, the arch of wood is removed, and strips of hoop-iron having tenter-hooks attached to them, are set in and fastened to the brick arch by key-bolts passing through and keyed on the inside. (See figure 2.) The roof is covered with a coat of mortar mixed with cement and "floated" down smooth. At the rear of the house a chimney should be placed, having an opening at the bottom, and also near the top. The lower opening is kept partly open all the time, but closed with wire-cloth, if mice find their way in. The upper one is opened only when the room is likely to get too hot. If stone slabs of two feet or more in width, and of convenient size are obtainable, they may be set up so as to give plenty of room upon the floor to stand when hanging meat and making the fire, while there will be room behind the slabs for two or more loads of ashes. Besides, this is very convenient, because smoked meats are better kept, when buried in ashes than in any other way.

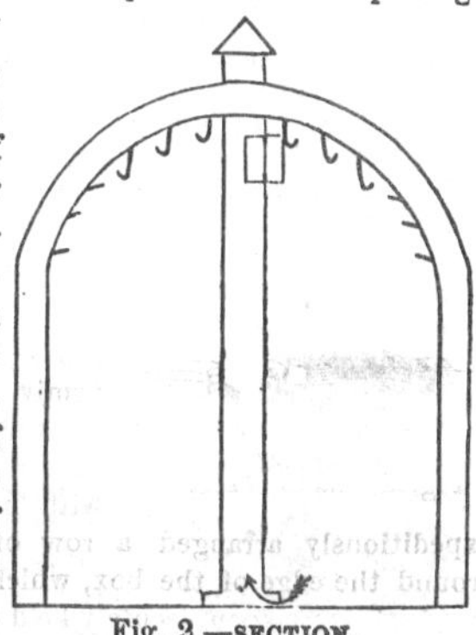

Fig. 2.—SECTION.

containing a great variety of Items, including many good Hints and Suggestions which we throw into smaller type and condensed form, for want of space elsewhere.

Postage 12 Cents a Year in Advance.—The postage on the *American Agriculturist* anywhere in the United States and Territories, *paid in advance*, is 3 cents a quarter, 12 cents a year. If not paid in advance, twice these rates may be charged.

Plum Trees.—M. C. Walton, Penobscot Co., Me., lost two plum trees. The stocks were 4 inches in diameter, and the graft 8 inches. The trouble was caused by grafting on a slow growing and unsuitable stock. With trees of this size we can suggest no remedy. Young trees, if grafted low, might have the junction set below the surface of the ground, and some longitudinal slits made through the bark at the point of union.

The Trophy Tomato.—Not many years ago a tomato was a tomato, and it is only recently that we have had named varieties. Leaving the matters of earliness and productiveness out of question, there is really a great difference in the quality of tomatoes—almost as much as there is in that of strawberries. By quality we include not only flavor but the texture and solidity of the flesh, which are characters of prime importance if the fruit is to be eaten raw, and of no little consequence if it is to be cooked. In the Trophy tomato we have a rare combination of excellent qualities. It is

as early as any, very productive, and for great weight and solidity and excellent flavor, it is unequalled by any variety we have tested. The Trophy grows to a large size, and in an engraving we can give only a reduced representation to show its regularity of form and solidity of flesh. In order to put this excellent variety within the reach of all who wish to try it, the publishers of the *Agriculturist* offer the seeds of the Trophy as a premium.

Swindling Nurserymen.—A New Jersey nurseryman writes that last spring he received an order for 500 trees from J. D. Wilson, of Fort Lee, who gave as reference Mr. Peter Henderson. Knowing Mr. Henderson, and supposing all right, he sent the trees, but can get no response to repeated letters asking pay for them. Mr. Henderson informs us that he gave no permission to refer to him, and moreover, that he never saw nor heard of any such Wilson, and that the above case is one of a score of similar cases. An operator in Plainfield, N. J., obtained over $5,000 worth of nursery stock from various nurserymen by a similar dodge. The swindlers shrewdly calculate that many nurserymen will not take the trouble to inquire directly of the parties referred to, and thus they pick up a good deal of stock which is often sold by them by another swindle; viz., by claiming, under other names, to be the agents of some leading nursery establishment. As their stock costs nothing, they can sell it cheap enough to secure a quick sale. The lesson taught is that nurserymen must learn the character of those ordering stock, either by direct application to the parties given as references, or by other means.

Quoting the Agriculturist.—We have, of late, seen notices to the effect that "so and so says in the *Agriculturist*," mentioning one of the editors. Such notices are of course kindly intended to be complimentary to the persons referred to, but *they* would prefer not to be individualized in this manner. The *Agriculturist* is an *institution*, and should be quoted as such. What credit is given it, belongs to all connected with it.